INSTRUCTOR'S MANUAL
PREPARED BY
Richard L. Mauger

ENVIRONMENTAL GEOLOGY
SEVENTH EDITION

EDWARD A. KELLER

PRENTICE HALL, UPPER SADDLE RIVER, NJ 07458

Executive Editor: *Robert McConnin*
Production Editor: *James Buckley*
Special Projects Manager: *Barbara A. Murray*
Supplement Cover Manager: *Paul Gourhan*
Manufacturing Buyer: *Ben D. Smith*

Copyright © 1997 by Prentice-Hall, Inc.
Simon & Schuster / A Viacom Company
Upper Saddle River, NJ 07458

All rights reserved. No part of this book may be
reproduced in any form or by any means,
without permission in writing from the publisher.

Printed in the United States of America

10 9 8 7 6 5 4 3 2 1

ISBN 0-02-363282-8

Prentice-Hall International (UK) Limited, *London*
Prentice-Hall of Australia Pty. Limited, *Sydney*
Prentice-Hall Canada Inc., *Toronto*
Prentice Hall Hispanoamericana, S.A., *Mexico*
Prentice-Hall of India Private Limited, *New Delhi*
Prentice-Hall of Japan, Inc., *Tokyo*
Simon & Schuster Asia Pte. Ltd., *Singapore*
Editoria Prentice-Hall do Brasil, *Ltda., Rio de Janero*

Table of Contents

This Manual; Its Contents and Suggested Uses	1
Environmental Notions/Ethics Inventory	10
Environmental Awareness Inventory: Foundations of Environmental Geology; Part 1	13

Chapter 1: Philosophy and Fundamental Concepts

Home Study Assignment	15	Commentary/Media Watch	24
Quiz	20	Supplemental References	28
Answers	23		

Chapter 2: Earth Materials and Processes

Home Study Assignment	30	Commentary/Media Watch	43
Quiz	38	Supplemental References	45
Answers	41		

Chapter 3: Soils and Environment

Home Study Assignment	46	Commentary/Media Watch	55
Quiz	51	Supplemental References	57
Answers	54		

Environmental Awareness Inventory: Hazardous Earth Processes; Part 2 — 58

Chapter 4: Natural Hazards: An Overview

Home Study Assignment	60	Commentary/Media Watch	66
Quiz	63	Supplemental References	68
Answers	64		

Chapter 5: Rivers and Flooding

Home Study Assignment	69	Commentary/Media Watch	79
Quiz	75	Supplemental References	90
Answers	78		

Chapter 6: Landslides and Related Phenomena

Home Study Assignment	91	Commentary/Media Watch	98
Quiz	95	Supplemental References	100
Answers	97		

Chapter 7: Earthquakes and Related Phenomena

Home Study Assignment	101	Commentary/Media Watch	110
Quiz	107	Supplemental References	111
Answers	110		

Chapter 8: Volcanic Activity

Home Study Assignment	113	Commentary/Media Watch	121
Quiz	118	Supplemental References	125
Answers	121		

Chapter 9: Coastal Hazards

Home Study Assignment	126	Commentary/Media Watch	136
Quiz	132	Supplemental References	141
Answers	135		

Environmental Awareness Inventory: Human Interaction with the Environment; Part 3 — 142

Chapter 10: Water: Process, Supply, and Use

Home Study Assignment	144	Commentary/Media Watch	158
Quiz	152	Supplemental References	161
Answers	156		

Chapter 11: Water Pollution and Treatment

Home Study Assignment	163	Commentary/Media Watch	173
Quiz	169	Supplemental References	181
Answers	173		

Chapter 12: Waste Management

Home Study Assignment	182	Commentary/Media Watch	189
Quiz	186	Supplemental References	193
Answers	189		

Chapter 13: The Geologic Aspects of Environmental Health

Home Study Assignment	195	Commentary/Media Watch	203
Quiz	200	Supplemental References	210
Answers	203		

Environmental Awareness Inventory: Minerals, Resources, and Environment; Part 4 — 211

Chapter 14: Mineral Resources and Environment

Home Study Assignment	213	Commentary/Media Watch	224
Quiz	219	Supplemental References	228
Answers	223		

Chapter 15: Energy and Environment

Home Study Assignment	229	Commentary/Media Watch	242
Quiz	236	Supplemental References	247
Answers	240		

Environmental Awareness Inventory: Global Change, Land Use, and   248
    Decision Making; Part 5

Chapter 16: Global Change and Earth System Science

Home Study Assignment	250	Commentary/Media Watch	261
Quiz	257	Supplemental References	263
Answers	261		

Chapter 17: Air Pollution

Home Study Assignment	265	Commentary/Media Watch	272
Quiz	269	Supplemental References	274
Answers	272		

Chapter 18: Landscape Evaluation and Land Use

Home Study Assignment	275	Commentary/Media Watch	285
Quiz	281	Supplemental References	288
Answers	285		

Reference Items

Sample World Outline Map	289	Sample U. S. A. Outline Map	290

This Manual; Its Contents and Suggested Uses

This manual includes the items shown in the following table. Clearly-focused learning objectives comprise the lead-in item for each textbook chapter; they can readily be adopted and/or modified to suit your particular needs and interests.

Each Text Chapter

Home Study Assignment   Quiz
Answers; Home Study/Quiz Questions   Commentary/Media Watch
Supplemental References

Each Grouping of Text Chapters (Parts)

Environmental Awareness Inventory (EAI)

Single Items

Environmental Notions/Ethics   Media Sources
Inventory (EN/EI)   Field Study Experiences

Home Study Assignments

The home study assignments are based directly on materials covered in the textbook. Some questions require attention to specific illustrations, photos, tables, and graphs. Observational, verbal/memory, and conceptual/integrative skills are exercised. In one way or another, information necessary to answer the questions is available in the text, although a few also require addition, logical, common-sense interpretations. Most questions are presented in multiple choice, matching, or true/false format. Every effort has been made to make correct responses "totally correct" and to make incorrect responses glaringly wrong and blatantly inaccurate. Some incorrect, multiple-choice responses are, frankly speaking, "convincingly-written misinformation". Your students deserve the opportunity to recognize nonsense when and where they see it! Questions involving subtle nuances and judgments among similar, nearly correct answers have been avoided. On their own, students should be able to answer a high percentage of the questions with a reasonable degree of confidence. Questions involving controversial and/or judgmental issues are noted as such, and students are asked to defend their positions and opinions on an informed, rational basis. With a little encouragement and nudging, these questions will generate class participation and discussion. Instructors can invite questions about the study assignments. Even if the typical student response is "What is the correct answer to number 10", the instructor can take some time to explain the appropriate verbal content and concepts before helping out with "the right answer".

This writer has routinely utilized similar assignments as homework in introductory geology and oceanography classes with positive results. Encourage students to work together in groups; learning is enhanced in a "social context". Questions that still stump your students after considerable individual and group effort deserve substantive classroom explanation and/or ought to be clarified in some way. Grade the assignments and give them some visible weight in the final semester grade; I assign from 10 to 20 %, depending on numbers of quizzes, exams, and other graded assignments. Most questions are formatted to allow for quick visual or electronic grading. We all know that grading is a tedious, time-consuming process and that grading time is not a hot-button positive in our annual reports and personal résumés. Instructors with multiple, large, class sections and no help from TAs or undergraduate students probably have no recourse but to use electronic grading or not grade at all. One solution to the latter problem is to organize in-class grading groups and have students grade each other's assignments. This will propel you, ready or not, into the forefront of the group-learning instructional movement and may reflect positively on your reputation for utilizing the latest, up-to-date, instructional practices.

Instructors ought to take full advantage of comprehensive, well-written and well-illustrated textbooks, such as Environmental Geology, seventh edition. The home study assignments will help do this. They encourage proactive student participation and discourage students from "settling in" for a purely passive, "free ride" in a course that may not, at first, seem to merge with their chosen or imagined academic majors and career goals. The home study assignments also function to "free up" precious class time from excessive attention to important, but highly detailed subject matter. On their own, students will pick up much of the basic course content, and instructors can set aside some class time for current events and other topics of interest such as the latest results concerning the spring, 1996, manmade flood in the Grand Canyon, the size of this year's hole in the ozone layer over Antarctica, and important legislation, regulations, and court decisions involving environmental and land use issues.

Each home study assignment includes an exercise based on geographic localities and features that should be part of an educated person's general knowledge or that are specifically noted in the textbook chapter. Students are asked to find these features/localities and plot them on an appropriate outline map (see last pages, this manual). Most standard, page-size, printed or computer-generated outline maps will do fine, and a copy of the map can be added as the last page of each assignment. The geographic questions can be formatted for quick visual or electronic grading by making each location/feature into a numbered question and placing different letters or numbers at appropriate locations on the map; students then have to match the letters/numbers on the map against the same features/locations listed in the assignment questions. Geographic knowledge and reasoning are basic to earth science studies; environmental geology is no exception. Some students may "know their way around the world" fairly well; others, especially at first, will have trouble finding South America, the Indian Ocean, and the Hawaiian Islands. Encourage your students to buy an up-to-date atlas and/or world map and include map questions in your quizzes and exams. By the end of the semester, even students who at first had trouble finding Japan, Iceland, and Iraq will have achieved a modest degree of literacy in World and North American geography. Today's "global marketplace" is not kind to nor tolerant of "geographic illiterates". And as an added benefit, the geographic content will be solid evidence of a strong "international component" in your course.

Quizzes

The chapter quizzes represent broad samplings of questions from the home study assignments. For the most part, however, the quiz questions are shorter and more succinct than parallel questions in the home study assignments and they are not predicated on visual access to specific items in the text. Weekly, substantive quizzes have the virtue of encouraging students to keep up; last minute cramming sessions may help students pass the course, but they do not foster long-term retention and insight. I compose my exams mainly from home study and quiz questions and include, if appropriate, a few questions of opportunity that have come up in class and are not necessarily addressed in the textbook, homework, or quizzes.

Learning is incremental and cumulative; effort is involve. Sequential attention to the substantive content of the home study assignments and quizzes should make your students competitive on more comprehensive exams and provide them with discernible evidence of a positive correlation between their study efforts and course performance. In my opinion, it is a mistake to turn *exams/tests* into a "carps shoot" wherein students' prior efforts or the lack thereof have little or no visible effect on performance. Remember, even world-class musicians and actors practice their notes and lines over and over again. Give your students some practice with your course content; their performances will also improve.

Answers; Home Study/Quiz Questions

Answers are provided to all home study and quiz questions. The majority of these are short answer questions formatted for quick checking and grading. Questions that involve subjective judgments, evaluations, or opinions are presented in an agree (A)/disagree (D) format. Written responses are invited in support of students' choices. Our own opinions and biases will obviously be at work here, so encourage your students to voice/defend their own ideas and positions on an informed, rational basis and place less emphasis on getting the right answer.

Commentary/Media Watch

In these sections, discussions of subject matter designed to supplement the textbook are interwoven with coverage of pertinent items for each chapter as reported in the media. For the most part, media coverage was limited to the time interval March to December, 1996. Environmental matters have gradually achieved legitimacy as "news" and as such, are widely covered on radio, television, and in the *print media*. Environmental issues cut across "party lines" and professional expertise; they affect us all. Poll after poll have shown strong, bipartisan support for environmentally-friendly legislative and regulatory policies. Environmental issues take many forms, from interdisciplinary in scope and global in scale to local, one-item issues such as groundwater contamination and the status of the municipal landfill. Media-reported environmental issues and events form a ready-made resource to be utilized in your classes and course-related activities. Keep an eye on the media; asked your students to do the same. Take a few minutes at the beginning of each class to comment on and discuss current media reports appropriate for an environmental geology class. For example, on Wednesday, November, 27, 1996, the EPA formally proposed new, tighter air-quality standards for ozone and small-size particulate matter (see textbook Table 17.2). This move is being aggressively resisted by auto manufacturers and other large, powerful industrial groups. The stakes are high; does the promise of a general improvement in public health, particularly among asthmatics and others suffering from various forms of respiratory impairment, justify investments of millions of corporate dollars to be in compliance with the new standards? Obviously, the EPA says "yes" and the impacted industries say "no" or at least "slow down"!

Much of the media-watch material was abstracted from local and regional newspapers. National and international news would be better covered in a respected, big city daily. Themes of the Times: The Changing Earth, distributed exclusively by Prentice Hall, is a collection of well-researched, recent articles from the New York Times focused on the earth sciences and on environmental aspects of the earth sciences. As a class project, assign readings from the "Themes" and have your students peruse library copies of major domestic or foreign, English language newspapers for items of interest to environmental geology. You could divide the class into regional or single-country groups depending on what sources are available. Devote some class time to brief explanations of the different events and issues reported by your students. These will include both supplemental reviews of previously-covered material and previews of material coming later in the course. Even the most recalcitrant academic administrators should be pleased with class study groups using the library to research national and international environmental issues!

Environmental issues are now deeply ingrained our culture and, as such, they are deserving of media attention. To illustrate this point, this writer found nine items with environmental content (shown in the following tabulation) in a randomly chosen example of USA TODAY's daily feature Across the USA: News from Every State. This section consists of one or two short news items from each of the 50 states and the District of Columbia. Remember, the environmental items are in competition with reports of murders, bombings, criminal trials, racial bias lawsuits, politics, health-related issues, narcotics arrests, and alien abductions.

Note that these issues are addressed directly or indirectly in Professor Keller's textbook. The item on dioxin contamination at Times Beach, MO, illustrates that media coverage continues long after the spotlight is focused on a major, breaking, environmental story. Many media reports are just latest installments in ongoing coverage of long-lasting, unresolved, environmental problems. For example, the settlement recently negotiated between Crown Butte Mining Company and the U. S. Government (Commentary/Media Watch, Chapter 14) is probably the final event in a process that began with eighteenth-century land explorations and mineral prospecting in and around today's Yellowstone National Park. This settlement addressed often competing, fundamental, legal and ethical issues such as private property rights, the public's right to an unspoiled, natural-lands recreational experience, the financial interests of investors and taxpayers, the profit motive, conservation of natural resources, and deep-rooted, civic concerns for preservation of natural lands and wildlife habitat. Media coverage will add historical and ethical content to your core subject matter, and it will broaden your course's appeal.

Environmental News Items
Across the USA: News from Every State, USA TODAY, November, 8, 1996

Massachusetts	A building in which state office workers were routinely sickened by bad interior air was sold to a private investment group (see text item 17.4, p. 496)

Mississippi	A tank car derailment threatened the Lake Cormorant area with a dangerous, toxic chemical spill (Chapter 12).

Missouri	Blood tests showed that dioxin levels had dropped in people living close to an incinerator in use since March to destroy dioxin in soils and other materials from the nearby, vacated Times Beach, MO, townsite (see text item 3.10, p. 77)

Montana	Wildlife researchers reported that for the limited span of time over which such data have been acquired, a record number of grizzly bear cubs, 70, were born in Yellowstone National Park this past year.

New Jersey	In a federal court settlement, Citgo Asphalt Refining Company agreed to reduce its sulfur dioxide emissions by ninety percent and to pay fines totaling $1.23 million for violations of the Clean Air statutes (see text, p. 498-502)

Oregon	On a state-wide basis, Oregonians recycled 34.7 % of their disposable, recyclable waste last year, an increase over the 32.5 % total of the previous year (see text item 12.2, p. 322)

Rhode Island	The EPA completed its cleanup of hazardous liquid wastes left behind and abandoned when Dytex Chemical went bellyup in March.

Tennessee	A single vial of mercury spilled in their home was blamed for sickening both parents and their three children. EPA paid for moving the family to a safe home and for cleaning up the site. The total cost to EPA's Superfund Program was quoted at $50,000 (Chapter 13).

Utah	Environmentalists are pressuring the U. S. Forest Service to determine how a proposed timber sale will impact threatened owl habitat and adjacent lands in the newly designated Escalante National Monument in southern Utah (Chapter 18).

West Virginia	The outgoing Governor signed an agreement to commit $2,000,000 toward the purchase and protection of 2,000 acres of woodlands adjoining Coopers Rock State Forest. The deal is designed to protect and preserve the natural setting, scenic beauty, and outdoor recreational values of state-owned forest lands (Chapter 18).

Supplemental References
A short list of useful supplemental references is provided for each chapter. The listed books and articles provide current, background information on many important environmental topics and issues and references to additional sources. Most have been published within the last few years, but some older citations, including a few "environmental classics", are also included. Some are intended for the professional earth scientist and earth science educator and others are aimed for the informed public. Many of the references are interdisciplinary and broadly informative, but are not written at such a highly technical level as to be unsuitable for environmental geology students.

Environmental Awareness Inventory (EAI)
This manual includes five "EAIs"; one for each major part of the textbook. These are presented in a quiz format for quick review and evaluation. Many of the questions are suitable for exams. Consider using the EAIs as ungraded, self-evaluated, in-class or homework assignments. The questions are forward looking when given as an early activity for each textbook part; students will not have had the opportunity to "be prepared". This process may, without any penalty, alert many of your students to an "awareness gap"; others can take some pride in their good showings. Try giving the same EAI when starting and finishing each part of the textbook. Your conscientious, hard-working students will recognize their progress. Questions focus on items of major significance to the appropriate chapters and on items of historical importance that may or may not be given significant coverage in the textbook. Environmental literacy should comprise an integral part of today's " public knowledge". Remember, your class may be the only focused, structured encounter that tomorrow's lawyers, writers, artists, health-care workers, teachers, accountants, and economists will ever have with earth science, environmental science, and a "land ethic". As teachers, it's our responsibility to make sure that this brief, one-term or one-semester encounter is a positive one!

Environmental Notions/Ethics Inventory (EN/EI)
Give this inventory as a first-day, in-class or take-home assignment. It is designed to assess your students' attitudes, opinions, and beliefs about public policy and environmental ethics. It is patterned after surveys of personal attitudes, opinions, and beliefs that are widely used in social science research. As such, the questions do not have right and wrong answers; each answer involves personal choices. Responses to questions are scaled in a consistent order so that total numerical scores place respondents in one of four groups (listed at the end of the inventory) ranging from passionate environmental activism to general indifference or overt hostility toward a "progressive" environmental agenda. An organization espousing similar ideas and notions and a fictitious INTERNET address are listed for each group. Ask your students to think about how their environmental attitudes and notions were influenced by specific individuals, events, and ideas. This exercise will generate some deep-reaching self-reflection and spirited class discussion. In this writer's experience, family is by far the most commonly-cited factor influencing students' attitudes toward Nature and the environment.

Media Sources

Suppliers of books other than textbooks, videos, films, video discs, and CD-ROMs on the earth and environmental sciences are assembled in the following list; it is not comprehensive. Today's advanced technological equipment and visual products can certainly add interest and zest to our courses, and publishers are competing head-to-head to provide the best, most extensive media packets to accompany their textbooks. Prentice Hall's "Geodisc" is available to support Environmental Geology, seventh edition, and other Prentice Hall earth science books.

For small classes, homework assignments centered on appropriate videos/CD ROMs etc. can be very effective. Instead of asking students to recall extensive lists of technical details and nuances, try having them each assume the role of a media critic; ask for brief comments on the educational value, substantive content, and an appropriate target audience for the product. Restrict their comments to a single page. Just like some popular, TV critics rate current movies, have your students rate effective, entertaining media products with a "thumbs up" and vice versa for products that lacked focus, were over dramatized or overtly biased, came across as trivial, or generally failed to impress. Instructors can also devise a short (no more than one page) exercise, based on the product, that includes questions and notations concerning items of specific interest and tie-ins with other classwork.

Instructors should give careful thought to 1) how these new products impact students' motivations and learning experiences, 2) what are the most effective delivery methods for these products, and 3) what percentage of class time and activities should be devoted to them. Keep in mind that we instructors are entrusted to educate our students and that good teaching involves much more than simple mastery over the latest technological wizardry.

Adult Learning Satellite Service, 1320 Braddock Place, Alexandria, VA 22314-1698; 1-800 257-2578; PBS Online, http://www.pbs.org/als

AGI (American Geological Institute) Publications Center, Box 205, Annapolis Junction, MD 20701; 301 953-1744; fax 301 206-9789

Annenberg/CPB Collection, Department SB3, Box 2345, South Burlington, VT 05407-2345; 1-800 532-7637, fax 802 864-9846; http://www.learner.org

Blue Sky Associates, 1208 Bridge Street, Grafton, WI 53924; 414 377-1398; fax 414 377-7750

Crystal Productions, Box 2159 Glenview, IL 60025; 1-800-255-8629

Educational Images, Box 3456, West Side Station, Elmira, NY 14905; 1 800-527-1090; 607 732-1090; fax 607 732-1183

EDUCORP Direct, 7434 Trade Street, San Diego, CA 92121-2410; 1-800-843-9497

EME Corporation, 41 Kenosia Avenue, Box 2805, Danbury, CT 06813-2805 1-800 848-2050; fax 203 798-9930; emecorp@aol.com

Films for the Humanities & Sciences, Box 2053, Princeton, NJ 08543-2053; 1-800-257-5126

Gould Media Inc., 44 Parkway West, Mount Vernon, NY 10552-1194; 914 664-3285; fax 914 664-0312

Insight Media, 2162 Broadway, New York, NY 10024-6642; 1-800 233-9910

Island Press, Box 7, Covelo, CA 95428; 1 800-828-1320; fax 707 983-6414; Island Press is the publishing house/sales outlet for the Center for Resource Economics, 1718 Connecticut Avenue N. W., suite 300, Washington, DC 20009

JLM Visuals, 1208 Bridge Street, Grafton, WI 53024-1964; 414-377-7775

Key Line Educational Materials, Box 166, Cedarburg, WI 53012-0166; 414 375-1999

Media Design Associates, Inc., Box 3189, Boulder, CO 80307-3189; 1 800 228-8854; fax 303 443-2882

MMI Corporation, 2950 Wyman Parkway, Box 19907, Baltimore, MD 21211; 410 366-1222; fax 410 366-6311; mmicorp@aol.com; http://members.aol.com/mmicorp

National Geographic Videos and Kids Videos, National Geographic Society, 1145 17th Street N. W., Washington, DC 20036-4688; 1-800-343-6610

National Geophysical Data Center, 325 Broadway E/GC4, Department 985, Boulder, CO 80303-3328; 303 497-6826, fax 303 497-6513; info@ngdc.noaa.gov

Society for Mining, Metallurgy, and Exploration, Inc., Product Sales; Box 625002, Littleton, CO 80162-5002; 1-800 763-3132, smeaime@aol.com

SPEM Society for Sedimentary Geology, 1731 East 71st Street, Tulsa, OK 74136-5180; http://dc.smu.edu/sepmsp/home.htlm

Stull, A, T. and Griffin, D., 1996, Life on the INTERNET; Geosciences, a student's guide: Prentice Hall, Upper Saddle River, NJ, 50 p. http://www.prenhall.com

Field Study Experiences

For many generations now, teachers and other enthusiasts of the earth sciences, ecology, and botany, etc. have been leading field trips to showcase important features, processes, and concepts in their "natural settings". Suddenly, educators have taken notice. Our old-time field trips, now redefined as out-of-doors, group-oriented, learning experiences, have leaped to the forefront of progressive educational methodology. Instead of trying to convince everyone that we've been doing this all along, take advantage of this "newly elevated status" and expand the role of field studies in your courses.

We are now predominantly an urban nation and have been so for at least a generation. Most Americans live in cities; as each year goes by, a smaller percentage live in rural areas and small towns. Formerly routine contacts with Nature such a starlight night sky, a favorite, pool-and-riffle trout stream, and an undisturbed, completely-forested, small-stream watershed are now rare and endangered. A student may have never seen a "natural stream" and thus will remain largely unaware of its rapid disappearance from the urban landscape.

Environmental geology presents many opportunities for field experiences, but in today's climate of tighter budgets and stricter accountability, think inexpensive and small-scale. Rural areas present the best opportunities to see less disturbed, more natural lands. Consider nearby national and state parks and forests; also many privately-owned, protected areas and municipal watersheds have been left 'undisturbed" for many years. For example, Duke University has its own Duke Forest, and the Wild Creek, Pennsylvania watershed, acquired by the city of Bethlehem in the late 1940s and managed for watershed protection since then, has suffered minimal land and ecological disturbances from farming, logging, and other human activities. If you have an accessible, "natural area nearby, utilize it; particularly in regard to surface water hydrology and flood control methods, natural areas constitute the "basic standard" to which highly modified, urban streams and drainage basins are compared.

In contrast, urbanized areas all exhibit storm-runoff retention structures, floodways, rerouted and/or channelized streams, water and sewage treatment plants, sanitary landfills and/or other waste disposal facilities, recycling centers, and problems with air quality. Develop a self-guided field experience suitable to your particular surroundings. Provide a list of localities with important and/or noteworthy features. Note erosion control methods and areas of excessive soil erosion associated with constructions activities and other forms of severe land disturbance. Point out modified stream channels, aquatic habitats, and flooding characteristics, floodplain obstructions and modifications, storm water runoff retention basins at shopping centers and apartment complexes, where rip-rap is being used to stabilize roadcut slopes and inhibit stream erosion, and borrow pits/quarries for sand, gravel, or crushed stone. Identify sites of permanent springs and transient, throughflow seepage following storms. Inspection of "dated" stone structures, such as curbing, cemetery monuments, and older building will give students some feel for rates of rock weathering; soil profiles may be nicely exposed in newly-dug trenches and other excavations. Feel compelled to utilize any nearby special geologic features such as fault scarps, glacial moraines, slope failures, sinkholes, and recent volcanic deposits etc. One afternoon and a three- to four-page trip guide should be adequate for starters. Don't feel inhibited in asking students to briefly comment on their personal feelings and emotions. Did they enjoy being out-of-doors for an afternoon? Did they notice things previously ignored? Did they enjoy the field experience? Was it a refreshing contrast to their other schoolwork? Have they ever considered volunteering for a summer position in a national park or national wildlife refuge, etc.? Encourage your students to recognize the broad public benefits and financial savings that accrue from earth-science based, rational, land use policies and zoning regulations. Contrast these benefits with the expensive losses, costly damages, and environmental deterioration associated with ignorance and indifference.

Environmental Notions/Ethics Inventory

Explanation

Please answer the following questions according to your personal preferences. Your answers are not necessarily right or wrong, but they should honestly reflect your personal opinions, judgments, and deeply-held philosophical beliefs.

1. The little remaining old-growth forest in the Pacific Northwest should be preserved as habitat for the spotted owl, for low-impact, public recreation, and as a national ecological treasure.
a. strongly agree   b. mildly agree   c. mildly disagree   d. strongly disagree

2. In the summer of 1996, the North Carolina Coastal Resources Commission, in a 6 to 5 vote, denied a variance to owners of Shell Island Resort, near Wilmington, NC, to construct a sand-tube coastal structure to stop an eroding inlet threatening the resort. The condominiums in question were built in 1985 after the current coastal management regulations had been adopted. The resort was valued at about $22 million and the condominium units had previously been sold to individual owners. What is your opinion of this decision?
a. strongly agree   b. mildly agree   c. mildly disagree   d. strongly disagree

3. Most creeks and streams I've seen are in a largely natural state and do not have upstream dams, manmade channel modifications, closeby roads or highways, or input pipes/channels for runoff from streets and other paved areas.
a. strongly disagree   b. mildly disagree   c. mildly agree   d. strongly agree

4. Companies operating coal strip mines are required by federal regulations to restore the land to its premining uses (agriculture, grazing, forests, etc.) and topographic configuration. These requirements obviously add to the cost of the coal. Which statement best fits with your assessment of these regulations?
a. I strongly support these regulations; individuals and corporate entities have to be forced to do the right thing and preserve land resources for future generations. Without forceful government intervention, the infantile mentality of the Wise Use Movement will prevail in this country.
b. I support these regulations; it's ethically correct and just good business to minimize land degradation associated with gainfully exploiting a natural resource.
c. I'm opposed to these regulations; they represent unnecessary governmental interference in the private enterprise system. If left alone, coal companies and other extractive industries would efficiently deal with land and water degradation arising from strip mining and produce lower-cost products in the process.
d. I'm strongly opposed to these regulations; they represent the worst kind of intrusive governmental meddling in our free enterprise system, cost jobs, and inflate production costs. They mainly reflect the self-indulgent whims of eastern liberals who haven't a clue concerning the coal industry.

5. Areas of barren soil, eroding, gullied slopes, and heavily silted streams make me think of a disturbed, stressed ecosystem and an unhealthy landscape.
a. always   b. sometimes   c. occasionally   d. never

6. The world's human population is projected to reach 6.5 billion by the year 2000; continued, annual growth of 2 % will have doubled the population to 13 billion by the year 2035. How do you feel about this glimpse of the future?

a. Sad and not very hopeful; Continued environmental pollution and habitat loss will result in more and more species extinctions; savage competition for natural resources and a higher standard of living will eventually cause collapse of Earth's terrestrial and aquatic ecosystems. Deadly new diseases will appear and once-controlled ones will reemerge with devastating results.

b. Guardedly pessimistic; International trade, democratic reforms, and emphasis on human rights might result in lower birth rates as living standards gradually improve in those "developing" countries that now have the highest birth rates.

c. Cautiously optimistic; Continued population growth will intensify competition for natural resources and land. Worldwide economic growth, local famines and epidemics, and manmade chemicals, herbicides, and pesticides in food supplies may eventually contribute to zero population growth.

d. Bubbling over with optimism; Economic growth, utilization of natural resources, international trade, and improvements in living standards will continue to accelerate, leading to better jobs, expanding markets for American products, and lower birth rates. Technological and genetic innovations will give us increasing mastery over our environment.

7. In the Black Hills of South Dakota, a cliff face of massive granite on the flank of Mount Rushmore has been sculpted into facial images of four U. S. Presidents. Which response best characterizes your feelings on this situation?

a. strongly object; The mountain should have been left undisturbed and returned to the Sioux people, its rightful owners.

b. mildly object; Now that we have the sculptures, incorporate them and the surrounding land into a National Monument with the mission of educating the public about the ecology and history of the Black Hills.

c. mildly in favor; However, Mount Rushmore should be expanded to include likenesses of Red Cloud, Crazy Horse, and Sitting Bull.

d. strongly in favor; Mount Rushmore stands as an eternal monument to individual initiative, tireless dedication, and deep commitment to America's proudest visions and most cherished dreams.

8. Which response most nearly characterizes your perception of property ownership and property rights?

a. All lands are held in a kind of communal public trust to be passed on unaltered and unchanged to future generations.

b. People are "temporarily" entrusted with land and its use; good stewardship demands that land resources and values be maintained or increased during the course of the owner's/user's lifetime.

c. Private property owners have wide latitude in how they use their lands; however, uses that infringe on the health and well being of others should be subject to governmental restrictions and regulations.

d. Private property owners have the right to use their land for whatever activities they see fit and to sell or deed that land "as is" to others.

9. Which response best describes your perceptions about waste and waste disposal?
a. Waste is an "out-of-place" natural resource; use both force and subsidies to accelerate recycling and conversion of all waste to safe, usable products.
b. Waste is an "out-of-place" natural resource; work hard to reduce overall waste volumes and stop production of especially dangerous/toxic materials; recycle as much waste as possible, and dispose of the rest in well-designed sanitary landfills.
c. Waste is an inevitable, if unwanted product of progress; collect and concentrate it; dispose of it in well-designed sanitary landfills and in the ocean.
d. Waste is a necessary by-product of production and economic growth; dispose of it as cheaply as possible by dispersion and dilution in the atmosphere, water, soil, and bedrock.

10. Which response best describes your opinions on domestic energy policy?
a. Enact a $3/gallon federal tax on gasoline, diesel fuel, and jet fuel; use the proceeds to subsidize energy conservation and alternative energy sources such as solar power and wind energy.
b. Enact a $0.55/gallon tax on gasoline, diesel fuel, and jet fuel to be shared equally between state and federal governments. Use the proceeds to encourage energy conservation and development of alternative fuels and energy sources.
c. Levy no additional fuel taxes; promote energy conservation and increased efficiencies in energy use; trust that political conditions in Saudi Arabia and the rest of the Middle East will allow for continued, ample, oil supplies in the international market place.
d. Repeal the 1992, 4.3 cent/gallon gasoline tax; subsidize expanded vehicle production by the "Big Three" and allow future energy demands and environmental concerns to work themselves out free of government imposed fuel taxes, environmental regulations, and international trade agreements.

Scoring; Evaluation

4 points for each a   3 points for each b   2 points for each c   1 point for each d
To determine your score, add up the total and divide by 10

Score	Environmental Attitude	Allies/Affiliates
1 to 1.5	Anarchistic	members, Wise Use Movement
1.5 to 2.5	Reserved	economists, U. S. Chamber of Commerce
2.5 to 3.5	Progressive	staff, Environmental Defense Fund
3.5 to 4	Strident	crew, the Greenpeace Warrior

Environmental Awareness Inventory
Part 1: Foundations of Environmental Geology

1. ( ) True (T) or False (F). By the year 2005, the world's human population will be near or will have surpassed six billion; between A. D. 250 and 1000, the world's population grew very slowly from 0.5 to 1 billion. 1 pt

2. "The Land Ethic", published in 1949, is an influential, widely read essay among naturalists and land management professionals. Who wrote the essay? 1 pt
a. Gifford Pinchot   b. Bruce Babbitt   c. George Perkins Marsh
d. Wallace Stegner   e. Aldo Leopold   f. Morris Udall

3. Which part of the world is often cited as a prime example of desertification, attributed to destruction of trees, overgrazing, and other unwise land use practices over the last 50 years or so? 1 pt
a. an area in central and eastern Siberia, Russia subjected to extensive timber harvesting, mainly by Japanese lumber and paper pulp companies
b. lands bordering on the Kalahari Desert, southern Africa
c. interior lands of central Australia surrounding Alice Springs
d. parts of Peru, Bolivia, and northern Chile west of the Andes Mountains
e. the Sahel; lands in North Africa between the Sahara Desert and the moist tropical parts of central Africa

4. Which **one** describes a major structural flaw in Earth's crust, California? 1 pt
a. the San Joaquin fault; a sliver of California on the Pacific plate west of the fault is slowing slipping toward the north northeast
b. the San Simon fault; a sliver of California on the North American plate east of the fault is slowing slipping toward the north northwest
c. the San Juan fault; a sliver of California on the North American plate east of the fault is slowing slipping toward the north northeast
d. the San Andreas fault; a sliver of California west of the fault is slowing slipping toward the north northwest

5. Which hypothesis specifically extols the dynamic nature and interconnectedness of ecological systems and planet Earth and has greatly stimulated interdisciplinary interest in environmental studies? 1 pt
a. Rama Hypothesis; named for the Hindu god of destruction and renewal
b. Ceres Hypothesis; named for the Roman goddess of the harvest
c. Gaia Hypothesis; named for the Greek Mother Earth goddess
d. Marduk Hypothesis; named for the Babylonian benefactor of man and god of life

6. Which surficial earth material is a sedimentary accumulation of windblown silt originally carried by glacial meltwater streams or derived by wind erosion of exposed desert soils? 1 pt   a. till   b. loess   c. gneiss   d. loam

7. Which areas produce the largest sediment yield per unit area, given equivalent topography, soil conditions, and duration and intensity of rainfall? 1 pt
a. heavily forested   b. planted in corn or soybeans
c. heavily urbanized   d. active construction/development site

8. What are ORVs and what are their negative environmental impacts? 1 pt
a. Overly Restless Vandals; wantonly destroy valuable land and biotic resources
b. Obnoxious Righteous Voices; preach incessantly for unwise, environmentally damaging land uses
c. Oversize Recreational Vans; convey so many people to camping areas and popular, outdoor sites that the ecological carrying capacity is threatened
d. Off-Road Vehicles; accelerate erosion and threaten survival of many native plants and animals

9. What happened at Times Beach, Missouri? 1 pt
a. soils were dangerously polluted by lead and arsenic from an abandoned smelter
b. widespread contamination by dioxin-laced oil, applied to reduce road dust, forced abandonment of the town
c. it was converted from an expensive, suburban, residential neighborhood to forest lands and pasture after repeated flooding by the Mississippi and Missouri Rivers
d. unusually high shrink-swell potential of local soils resulted in severe structural damage to numerous new homes in an upscale, suburban, residential development

10. Which statement concerning soils and soil profiles is blatantly erroneous? 1 pt
a. A weak soil profile or the absence of a profile indicate that the soil material has been exposed to weathering for at least a few thousand years, depending on local and past climatic conditions.
b. Red, yellow, and brown soil colors are imparted by finely divided iron oxides; gray and black colors usually indicate organic matter in the soil.
c. A residual soil is derived from the local bedrock or deep subsurface material; weathering and soil-forming processes produce great differences in appearance and composition between a soil and its fresh, unweathered parental material.
d. Cohesion, permeability, and plasticity are three commonly measured, engineering properties of soils.

Answers

| 1. F, False | 2. e | 3. e | 4. d | 5. c |
| 6. b | 7. d | 8. d | 9. b | 10. a |

Performance Evaluation; Scoring

Each correct answer is worth one point; add up your total. Now be honest, and subtract one point for each lucky guess!

Score	**Awareness Ranking; advice**	INTERNET Address
8-10	**nerd**; keep up the good work, don't let up	http://envnerd.com
5-8	**blue collar achiever**; earn an A with hard work	http://envblcol.com
2-5	**slowpoke**; don't look back, the clueless are gaining	http://envslpoke.com
0-2	**clueless**; extra hard work makes up for a slow start	http://envcluless.com

Philosophy and Fundamental Concepts • Chapter 1
Home Study Assignment

Refer to Figures 1.6 and 1.7; answer the following questions.

1. Which will have nearly half the World's people by the year 2000? **two answers**
a. India　　　　　b. United States　　　c. Brazil　　　　　d. China

2. Over how many years did the World's population double from 1 to 2 billion?
a. 100　　　　　　b. 1000　　　　　　　c. 10　　　　　　　d. 10,000

3. Assuming that the World's population reaches 6.2 billion by the year 2000, how many years were required for the population to double from 3.1 billion?
a. 38　　　　　　 b. 51　　　　　　　　c. 24　　　　　　　d. 77

4. In the year 2000, which area will represent the largest potential market for U. S. exports such as tobacco, agricultural products, vehicles, and technology?
a. Africa　　　　　b. South America　　c. Asia　　　　　　d. Europe

5. What effect did the millions of deaths during the Second World War, 1939 to 1945, have on world population?　a. big decrease　b. not much, if any

Questions 6-9; Matching: Colors On Infrared Images; Refer to Figure 1.8.
a. red　　　　　　b. white　　　　　　　c. black　　　　　　d. pale blue

6. ( ) water containing large amounts of suspended sediment
7. ( ) clear water devoid of suspended sediments or other particulate matter
8. ( ) forest vegetation　　　　　9. ( ) deforested lands or urbanized areas

10. What city is shown as the large "disturbed" area on the Rio Negro just upstream from its junction with the Amazon? Refer to Figure 1.8 and a good map atlas.
a. Mbandaka, Zaire　b. Iquitos, Peru　c. Khartoum, Sudan　d. Manaus, Brazil

11. Which two statements concerning the Aral Sea environmental situation are correct? See the *case history* Aral Sea, p. 7. **two answers**
a. The Aral Sea is situated in the republics of Uzbekistan and Kazakhstan, both parts of the former Soviet Union.
b. Water diverted for irrigation from the Ural and Volga Rivers has resulted in lowering of the lake level 20 meters since 1960.
c. Lands uncovered by the shrinking lake are fertile and well-suited for agriculture.
d. Cotton grown on the newly irrigated farmlands generates cash income; the now-defunct fishery in the much larger, former Aral Sea used to provide low-cost, good quality protein in local diets.

12. Which statement is true? Look over Figure 1.14, p. 19.
a. Animal and plant breeding programs at zoos, wildlife sanctuaries, and botanical gardens have dramatically reduced the number of species extinctions since the first decade of the twentieth century.
b. Since the mid-nineteenth century, rates of wildlife extinctions and world population growth have both accelerated; habitat alteration and excessive commercial exploitation are both responsible for the decline in species numbers.

13. Which statement concerning the Gaia Hypothesis is most reasonable?
a. In general, it has stimulated interest in and awareness of the complex interactions involving Earth's atmosphere, biosphere, hydrosphere, and surface environments.
b. In general, it has acted to slow growth of the modern global economy by forcing technologically advanced countries, such as the United States and Japan, to limit their use of petroleum and other vital natural resources imported from less developed countries.

14. Which **two** statements are correct and/or logically consistent with the Gaia Hypothesis? **two answers**
a. Oxygen in Earth's atmosphere is a waste product of photosynthetic living organisms; animal life based on oxygen respiration evolved only after excess oxygen began to accumulate in Earth's early atmosphere.
b. Gaia was the name of the Roman god of the harvest.
c. Lichens and mosses growing on fractured rock outcrops have little or no impact on rates of local rock weathering and soil formation.
d. Consciously or unconsciously, diversion of water from rivers entering the Aral Sea has led to local climatic changes and damaging dust storms.

Questions 15-21; True (T) or False (F): Refer to Tables 1.1 and 1.2, p. 14 and 15.

15. ( ) The uplift rate determined at Cajon Pass in southern California exceeds the average rate of erosional land surface lowering estimated for mountainous regions along the California coast and for lands in the drainage basin of the Colorado River.

16. ( ) Average annual sediment loads for the Mississippi River exceed those of the Colorado River because average erosion rates in the Mississippi River drainage basin are higher than those in the Colorado River drainage basin.

17. ( ) Aerosol particles injected into the lower stratosphere by powerful explosive volcanic eruptions can remain aloft for several years.

18. ( ) Atlantic Ocean deep water has a longer residence time than Atlantic Ocean surface waters, showing that circulation through the deep-water reservoir is faster than through the surface-water reservoir.

19. ( ) In general, highly soluble ions such as Na^{1+} and Cl^{1-} have longer residence times in the oceans than less soluble ions such as Ca^{2+} and Fe^{3+}.

20. ( ) The residence time of water vapor in the atmosphere greatly exceeds the residence time of liquid water in the oceans.

21. ( ) Long-term, continued growth in the mass of glacial ice on land "temporarily" removes water from the oceans and results in a lowering of sea level.

Read over the *special feature* Geology and Ecosystems, p. 22.

22. Which **two** statements are correct and/or probably true? The others are incorrect or logically or ethically flawed. **two answers**
a. Clear-cutting of old forest growth like that of the Prairie Creek drainage basin would provide a large infusion of cash to the owners of the timber rights; however, vastly increased sediment yields would seriously threaten local, breeding salmon populations.
b. Near-bank vegetation, soil, and bedrock have little influence on local variations in water depths in forest streams such as Prairie Creek.
c. Heavy forest canopy, such as is evident in the photo of Prairie Creek, acts to keep creek waters cool during low-flow conditions and warm days of the late summer and early fall.
d. Most Americans would like to see the redwood trees of the Prairie Creek drainage basin harvested to provide sustainable growth in forest-products and long-term, sustainable jobs in the local timber industry.

Read over the *case history* Soil Erosion in the Ancient World, p. 20.

23. Which **two** statements are essentially correct? The others are incorrect and/or inaccurate? **two answers**
a. Despite terraces and other conservation measures, conversion of forested slope lands to agriculture will accelerate soil erosion, especially from the steeper parts of the slopes.
b. Trees and brushy vegetation are gradually becoming re-established in lands surrounding present-day cities such as Khartoum, Sudan and Mogadishu, Somalia, because petroleum is displacing wood as the domestic fuel of choice.
c. Accelerated soil erosion and land degradation in areas converted from forested slopes to farmlands for export crops, such as the "mountain grown" coffee in Colombia and Costa Rica, are unique to nineteenth and twentieth century capitalism.
d. The disappeared, cedar forests of Lebanon, utilized for timber and fuel by the ancient Phoenicians, would be very difficult to restore because the soil has been eroded away in many areas, leaving bedrock exposed at the surface.

Questions 24-27; Matching: Specialized Earth Science/Geology Disciplines

a. petrology   b. tectonics   c. geomorphology   d. hydrogeology

24. ( ) Focus is on precipitation, surface waters, and groundwaters comprising the hydrosphere.
25. ( ) Focus is on minerals, bedrock, and unconsolidated surficial materials comprising the lithosphere.
26. ( ) Focus is on movements and deformations of rock comprising the lithosphere and deeper parts of the Earth.
27. ( ) Focus is on the evolutionary development of Earth's land surface.

Questions 28-31; Matching: Systems Behavior and Terminology

a. steady state b. feedback c. disturbance d. complex response

28. ( ) In a rainforest soil, basic nutrients released by decaying vegetation are soon taken up by living vegetation.
29. ( ) Manmade chlorofluorocarbon compounds in the atmosphere degrade Earth's protective layer of ozone.
30. ( ) Calcium carbonate deposition rates in the oceanic portion of the biosphere accelerate as the abundance of carbon dioxide increases in the atmosphere.
31. ( ) Fire or wind destroy most of the trees in a forested portion of a national park.

Questions 32-36; Which statements concerning *The Land Ethic and the American Experience*, p. 5, are essentially correct (C)? The others are seriously flawed (F) and/or lack scholarly support.

32. ( ) The Pilgrims who landed in New England, 1620, were well equipped and well prepared for survival in their new, relatively undisturbed, natural ecosystem.
33. ( ) The myth of superabundance states that despite a skyrocketing population, Homo Sapiens will never lack the basic resources necessary for survival.
34. ( ) Concepts of land ownership and use among Native Americans and early European settlers were quite similar; thus the settlers often bartered for or purchased exclusive ownership of lands previously owned by Indians.
35. ( ) American slave owners of the pre-Civil War era lived by a well-defined code of honor and ethics; this code however, denied basic human rights to specific individuals.
36. ( ) *The Land Ethic*, an essay read at least once by most of today's forest, wildlife, land management, and environmental professionals, was written by Aldo Leopold and published in 1949 as part of his essay collection entitled *A Sand County Almanac*.

Questions 37-46; In your opinion, which of the following statements concerning environmental matters deserves a "thumbs up" (U) for being substantive and informative or a "thumbs down" (D) for being nonsensical and/or rooted in ignorance. U and D choices may vary from student to student!

37. ( ) A sustainable global economy depends mainly on international marketing strategies for U. S. exports and on lowering transportation and labor costs associated with raw materials and consumer-oriented goods imported from abroad.
38. ( ) America's national parks and other popular, natural-lands, scenic areas are destinations of choice for millions of tourists, domestic and foreign. However, facilities such as golf courses, luxury hotels, gambling casinos, and lavish stage shows would reduce the terminal boredom afflicting many national park visitors.
39. ( ) Soils constitute the substrate for the terrestrial biosphere. Soil evolution, fertility, and thickness are strongly dependent on Earth's four, basic, process spheres.

40. ( ) The U. S. Government should encourage population growth in this country and abroad. Such growth assures a growing labor force, expanding markets for manufactured products, higher corporate profits, and increasing land and real estate prices. Thus all but the lazy and indolent will be rewarded with good incomes and high-quality lives.

41. ( ) Over the last 100 years, dam construction, increased sediment yields, overfishing, and chemical pollution have greatly diminished yearly salmon harvests from streams and rivers in the Pacific Northwest.

42. ( ) If for the next 100 years, annual death rates are lower than birth rates, the world's population will continue to grow and the mean age of the population will slowly increase.

43. ( ) The U. S. can greatly increase its annual harvests of basic food crops such as wheat, corn, and soybeans. Thus poor, impoverished peoples in parts of Africa and elsewhere are at little risk of starvation in the face of future drought or political chaos.

44. ( ) A uniformitarian look at Earth's past biosphere communities suggests that even very abundant, diverse, long-adaptive organisms such as the dinosaurs, can eventually become fatally vulnerable to evolutionary and environmental changes.

45. ( ) Long-lived, highly-stable, steady-state systems are dominated by processes with strong negative feedback; these limit departures from stable, operating configurations and suppress potentially destabilizing changes associated with positive feedback processes.

46. ( ) The so called environmental crisis is mainly a shrill cry from modern, doomsday prophets devoted to scare tactics and a pagan deity, Mother Nature. The deplorable conditions envisioned for future generations will, as today, be mainly limited to small, impoverished, overpopulated regions of Africa. In the advanced countries, increased utilization of technology, longer life spans, and a return to strongly held, fundamentalist, religious beliefs will eventually free us entirely from any ethical responsibility for the future status of Earth's biosphere.

47. On the world map provided, accurately plot and label the following countries, features, or locations.

a. Mesopotamia, Iraq
b. Mediterranean Sea
c. Ethiopia
d. Aral Sea
e. Kazakhstan
f. Uzbekistan
g. Brazil
h. Syria
i. Grand Canyon, AZ
j. Lebanon; home of Phoenicians
k. Israel and the West Bank; Palestine
l. Ducktown/Copperhill, TN, USA
m. Greenland; outline of its ice sheet
n. Yellowstone National Park, WY, MT, and ID
o. confluence of the Amazon and Rio Negro Rivers, Brazil
p. coastal redwood forests, northern California

Environmental Geology Quiz • Chapter 1

1. Assuming that the World's population reaches 6.2 billion by the year 2000, how many years were required for the population to double from 3.1 billion?
a. about 38   b. about 55   c. about 12   d. about 105

2. In the year 2000, which area will represent the largest potential market for U. S. exports such as entertainment, agricultural products, vehicles, and technology?
a. Africa   b. South America   c. Asia   d. Europe

3. Which color on infrared images indicates clear water devoid of suspended sediments and other fine-size particulate matter?
a. red   b. white   c. black   d. pale blue

4. Which **one** statement concerning the Aral Sea environmental situation is true/correct? The others are incorrect/untrue.
a. Cotton grown on the newly irrigated farmlands generates cash income; the now-defunct fishery in the much larger, former Aral Sea used to provide low-cost, good quality protein in local diets.
b. Lands uncovered by the shrinking lake are fertile and well-suited for agriculture.
c. Water diverted for irrigation from the Ural and Volga Rivers has resulted in lowering of the lake level 20 meters since 1960.

5. Which is the true/correct statement?
a. Since the mid-nineteenth century, rates of wildlife extinctions and world population growth have both accelerated; habitat alteration and excessive commercial exploitation are both responsible for the decline in species numbers.
b. Animal and plant breeding programs at zoos, wildlife sanctuaries, and botanical gardens have dramatically reduced the number of species extinctions since the first decade of the twentieth century.

6. Which **one** statement is correct and/or logically consistent with the Gaia Hypothesis?
a. Oxygen in Earth's atmosphere is a waste product of photosynthetic living organisms; animal life based on oxygen respiration evolved only after excess oxygen began to accumulate in Earth's early atmosphere.
b. Gaia was the name of the Roman god of the harvest.
c. In general, Gaia has acted to slow growth of the modern global economy by forcing technologically advanced countries, such as the United States and Japan, to limit their use of petroleum and other vital natural resources imported from less developed countries.

7. Which statement is essentially correct? The other is incorrect and/or inaccurate?
a. Trees and brushy vegetation are gradually becoming re-established in lands surrounding present-day cities such as Khartoum, Sudan and Mogadishu, Somalia, because petroleum is displacing wood as the domestic fuel of choice.
b. The disappeared, cedar forests of Lebanon, utilized for timber and fuel by the ancient Phoenicians, would be very difficult to restore because the soil has been eroded away in many areas, leaving bedrock exposed at the surface.

Questions 8-11; True (T) or False (F)

8. ( ) Average annual sediment loads for the Mississippi River exceed those of the Colorado River because average erosion rates in the Mississippi River drainage basin are higher than those in the Colorado River drainage basin.

9. ( ) In general, highly soluble ions such as Na^{1+} and Cl^{1-} have longer residence times in the oceans than less soluble ions such as Ca^{2+} and Fe^{3+}.

10. ( ) Long-term, continued growth in the mass of glacial ice on land "temporarily" removes water from the oceans and results in a lowering of sea level.

11. ( ) *The Land Ethic*, an essay widely read by today's forest, wildlife, land management, and environmental professionals, was written by Aldo Leopold and published in 1949 as part of his essay collection *A Sand County Almanac*.

12. Which **one** statement is incorrect and/or logically or ethically flawed?
a. Clear-cutting of old-growth forests like those of the Prairie Creek, CA, drainage basin can provide large cash infusions to owners of the timber rights; however, increased sediment yields could seriously threaten local, salmon populations.
b. Heavy canopy foliage, such as that in old-growth forests of the Pacific Northwest, acts to keep creek waters cool during the warm days and low-flow conditions of the late summer and early fall.
c. Most Americans would like to see all privately-owned redwood trees in California harvested to provide sustainable growth in forest-products and long-term, sustainable jobs in the local timber industries.

13. Which subdiscipline of the earth sciences is devoted to studying the landforms and evolutionary development of Earth's surface?
a. geomorphology   b. hydrogeology   c. tectonics   d. petrology

14. Assume that calcium carbonate deposition rates in the oceanic portion of the biosphere accelerate as the abundance of carbon dioxide increases in the atmosphere. What kind of system behavior does this process demonstrate?
a. a steady state   b. feedback   c. a disturbance   d. complex response

Questions 15-19; Do you agree (A) or disagree (D) with the following statements? Briefly cite scholarly evidence and ideas in support of your position.

15. ( ) The Pilgrims who landed in New England, 1620, were well equipped and well prepared for survival in their new, relatively undisturbed, natural ecosystem.

16. ( ) American slave owners of the pre-Civil War era lived by a well-defined code of honor and ethics; this code denied basic human rights to specific individuals.

17. ( ) If for the next 100 years, annual death rates are lower than birth rates, the world's population will continue to grow and the mean age of the population will slowly increase.

18. ( ) The U. S. can greatly increase its annual harvests of basic food crops such as wheat, corn, and soybeans. Thus poor, impoverished people in parts of Africa and elsewhere are at little risk of starvation in the face of future drought or political chaos.

19. ( ) The so called environmental crisis is mainly a shrill cry from modern, doomsday prophets devoted to scare tactics and a pagan deity, Mother Nature. The deplorable conditions envisioned for future generations will be mainly limited to small, impoverished, overpopulated regions of Africa and Asia.

20. On the world map provided, accurately plot and label the following countries, features, or locations.
a. Mesopotamia, Iraq   b. Mediterranean Sea   c. Aral Sea
d. Brazil   e. Syria   f. Greenland
g. Ducktown/Copperhill, TN, USA   h. coastal redwood forests, CA, USA

Answers; Home Study Assignment
Philosophy and Fundamental Concepts • Chapter 1

1. a, d	2. a	3. a	4. c	5. b
6. d	7. c	8. a	9. b	10. d
11. a, d	12. b	13. a	14. a, d	15. T
16. F	17. T	18. F	19. F	20. F
21. T	22. a, c	23. a, d	24. d	25. a
26. b	27. c	28. a	29. d	30. b
31. c	32. F	33. C	34. F	35. C
36. C	37. U or D*	38. U or D*	39. U	40. U or D*
41. U	42. U	43. D*	44. U	45. U
46. U or D*				

*

37. This may be a popular viewpoint among some marketers, economists, and free-trade proponents but the statement does not address any fundamental issues related to sustainable economies or other environmental matters, The statement essentially assumes perpetual economic growth without any limiting conditions.

38. Most national park visitors go for the natural setting and out door recreational opportunities. Thus staged entertainment is out-of-place and unnecessary. Most national park visitors are enthusiastic and excited about their park experience; crowding and too much noise may be important management issues, but terminal boredom is not a problem.

40. This statement may appeal to a mean-spirited subgroup of the ones who agreed with number 37. The reader should recognize that the crux of this notion does have wide political appeal.

43. The U. S. can certainly increase its harvests of basic food stuffs, especially in good growing years, but harvests will also drop in years with poor growing conditions. Even if American crop production can be maintained at high levels, that is no protection against famine and starvation in parts of Africa, North Korea, or elsewhere.

46. I hope that most students give this one a thumbs down (D). The whole thrust of the statement is to brush aside environmental considerations as unessential and insignificant.

Answers: Environmental Geology Quiz • Chapter 1

1. a	2. c	3. c	4. a	5. a
6. a	7. b	8. F	9. F	10. T
11. T	12. c	13. a	14. b	15. D*
16. A or D*	17. A or D*	18. D*	19. A or D*	

*

15. History tells us otherwise. Techniques and knowledge concerning agriculture, subsistence living, and winter survival, willingly imparted by Native

Americans, were quickly assimilated by the early colonists. European tools and technology improved their chances for survival in this strange, new land. Relatively speaking, the North American ecosystem was natural and undisturbed, although human activities such as farming, hunting, fishing, gathering, burning, etc., had been widely practiced by Native Americans for centuries prior to arrival of the first European settlers.

16. This statement could go either way, depending on one's preferences and prejudices. I agree with the statement. A very strict, well-known code of honor and ethics was in place in the pre-Civil War South. However, the notion that slavery was a denial of basic human rights was not part of that code. In Aldo Leopold's essay *The Land Ethic,* he retells the story of Odysseus, who upon returning home from the Trojan Wars, had servant girls put to death for various minor indiscretions committed during his time away from home. This action was not seen as cruel and inhumane, but as his absolute right and duty as a respected ruler. In both cases, basic human rights in a code of ethics had not yet been extended to include "slaves". Leopold compared the land to slaves. Basic ethical considerations had to be extended to "the land" before destructive land-use practices would be seen as "wrong" and stopped.

17. The statement is indeterminate but is probably true. In technologically advanced societies such as the United States, most European countries, and Japan, among others, falling birth rates and death rates have produced stable or relatively slowly growing populations with increasing mean ages. In less developed countries, Mexico and the Philippine, for example, traditionally high birth rates have accompanied falling death rates, producing rapidly growing populations with the highest percentages of people at the young base of the population pyramid. In this case the mean age will stay the same or actually decrease until birth rates decline and the population growth slows to levels typical of "advanced" countries.

18. The U. S. can certainly increase its harvests of basic food stuffs, especially in good growing years, but harvests will also drop in years with poor growing conditions. Even if American crop production can be maintained at high levels, that is no protection against famine and starvation in parts of Africa, North Korea, or elsewhere, especially if political and social chaos are part of the famine problem.

19. I hope that no one agrees with this statement; admittedly, some environmental communications may tend toward the apocalyptic, but most are well-reasoned and have strong social and scientific backing. The threat of "famine and pestilence" today may seem restricted to small, impoverished parts of the globe, but unforeseen worldwide environmental and climatic changes could quickly materialize and threaten the well-being of all peoples, rich and poor alike.

Commentary/Media Watch

Environmental Attitudes and Ethics

With Chapter 1, the instructor is presented with the formidable task of introducing the notions of environmental awareness and environmental ethics, and of looking into the future of Earth's biosphere and its living inhabitants. Based on Man's past and present-day activities, it's easy to slip into a deeply pessimistic, apocalyptic view with plenty of environmental degradation, famine, and pestilence in the future. In my experience, students don't react well to such negativism; after all, if things are essentially hopeless and continuing to deteriorate, what good can possibly arise from additional concern or worry? A better approach is to stress that an educated public, powerful new technologies, and a rational, common-sense approach to environmental issues offer hope for the future.

Environmental progress faces two major obstacles: 1) undreamed-of complexity; environmental issues are immensely complex in a scientific sense and in addition, they impact on virtually every segment of society and the economy; and 2) public ignorance and indifference; these are both widespread, even among otherwise, well-educated people. The later condition is particularly true regarding basic familiarity with the earth sciences. Biology and ecological science are widely taught and heavily enrolled in schools and colleges. Earth science courses are less widely available and are typically ignored by college students with chemistry and/or biology required as cognate sciences for their chosen majors. Thus, for example, many scientists and professionals in the health-related fields have had minimal academic exposure to geology or earth science.

Early in your course, give the Environmental Notions/Ethics Inventory (EN/EI) and the first Environmental Awareness Inventory (EAI) quiz as in-class or homework assignments. These should encourage students to think about current environmental issues in terms of their own knowledge, awareness, and attitudes; the self-grading should dispel some of their normal pre-quiz and pre-test anxiety. In addition, students will "be exposed" to basic facts and ideas that ought to be familiar stuff to any "modern, educated person". The Gaia Hypothesis focuses directly on the inherent complexities of environmental issues; and notions such as the Land Ethic, that state our obligations to the future, should begin to chip away at intellectual indifference and lethargy.

Curt Meine, biographer to the author of *The Land Ethic*, Aldo Leopold (Meine, 1988), spent considerable effort in tracking down long-since graduated students who had taken Professor Leopold's introductory game management course at the University of Wisconsin. This was a pre-professional class open to all. Although many of the details were long forgotten, Leopold's students vividly recalled leaving his classes thinking about newly-important issues and notions they had never, ever, previously considered as having any relevance or significance. Let's strive to achieve the same results in our environmental geology courses.

Remember, your course may just be the only formal exposure that future lawyers, developers, business executives, writers, bankers, bond traders, actors, congressmen/women, public educators, and musicians, etc. have to environmental geology and environmental issues. These are the people who will determine future policies toward the environment and in all probability, will be involved in future decisions and initiatives concerning environmental issues. So, make their class experience substantive and memorable. Extensive polling associated with the 1996 election campaign shows that environmental protection enjoys strong support among voters of both major parties. As teachers of environmental geology, let's do our part to keep it that way!

Environmentalism in the Media

An Associated Press article in North Carolina newspapers titled "The Big Green Machine", 7/19/96, described the $4 billion in annual industrial activity generated in North Carolina by environmental activities such as recycling, consulting, and engineering. Environmental activities were broadly defined. This breadth may have somewhat inflated the numbers, but the reader could not help but to be impressed by the scope, diversity, and dollar value of environmentally-based activities in the state's economy.

Television shows and print media featuring news and gossip about entertainment and entertainers often contain references or notations concerning celebrities' environmental activities. Such celebrity involvement may encourage your students to recognize that environmental awareness and basic knowledge of earth science are not solely reserved for learned scientists and professionals and that people of all ages, lifestyles, and professions can contribute to a cleaner, healthier environment. For example, Alexandra Paul, featured on the popular television show "Baywatch" was one of many environmental activists (7/19/96) gathered in an Oregon old-growth forest area protesting efforts to log the magnificent ancient trees. Television and movie actor Ted Danson, star of the popular, long-running television show "Cheers" and recently married to actress Mary Steenburgen, is well-known for his financial contributions, educational efforts, and public leadership in support of environmental causes, particularly on the West Coast. Danson is a co-founder the environmental organization American Oceans Campaign. See the article "Stars of the Sea", by Beth Livermore (Sea Frontiers, v. 40, p. 50-55, 1994) for more on celebrity involvement in environmental issues.

Robert Redford was featured "on stage" with President Clinton at Grand Canyon, AZ, in mid September, 1996, when the President proclaimed creation of a new national monument in southeastern Utah. Also, Mr. Redford's Sundance Center near Provo, Utah, supports scholarly studies, artistic endeavors, and educational activities dealing with conservation and environmental awareness. Well-known actors Meryl Streep and Roy Scheider are featured as host and narrator, respectively, in a video cassette series called "Race to Save the Planet", produced in conjunction with WGBH, Boston, and marketed by The Annenberg/CPB Collection. Entertainment mogul and Atlanta Braves owner Ted Turner has his substantial ranch holdings in southwestern Montana progressively managed toward the twin goals of restoring and/or enhancing the land's long-term ecological productivity and beauty and generating an acceptable, sustainable, long-term return on investment.

In a well-researched article for the New York Times Magazine (7/21/96; Tim Egan), "A new force for Nature", reprinted in the Raleigh (NC) News & Observer, Mr. Egan documents a growing sense of common purpose and cooperation concerning environmental issues between here-to-for antagonistic liberal environmental activists and traditionally conservative western outdoorsmen. Both groups are appalled by the extreme, anti-environmental, pro-development policies of the "Wise Use Movement" that were incorporated into budget cuts and deregulation efforts pushed in the House of Representatives by 1994 newcomers and a few, veteran members known for their outspoken opposition to progressive environmental policies. Their doctrinaire, uncompromising, anti-Nature and anti-outdoors positions have placed some western-states House members way out of step with constituents who otherwise support politically conservative candidates and lessened federal restrictions on commercial and recreational activities in federally-managed lands.

Grass-Roots Environmental Concerns

Communications (letters, phone calls, e-mail, etc.) to state and federal "officials", elected representatives, and corporate offices concerning environmental issues and pending policy decisions or proposed projects are usually answered, often promptly, with varying degrees of substance and insight. A class could take on a specific environmental issue, do some letter writing/e-mailing to corporate and public officials involved, and evaluate/discuss the replies. Some replies are deliberate obfuscations that assume the reader is easily manipulated and basically ignorant. Others address extraneous and/or related issues, and some focus directly on the issues at hand. One way or the other, however, these replies give fairly clear impressions of the environmental awareness and attitudes of the official or organization.

The obfuscation category is illustrated by a reply this writer's wife received from North Carolina's junior U. S. senator to a letter expressing concern about the serious water quality problems posed by the state's growing and largely unregulated hog industry. The situation was framed as "us", private enterprise providing jobs and income for all, versus "them", extremist environmentalists trying to shutdown the hog industry, destroy the "family farm", and spread poverty and joblessness. No mention was ever made of the threat to public health presented by polluted rivers and estuaries, of the serious declines in the fishing and shell-fishing industries, and of losses of tourist and recreation dollars caused by microbe-infested, overfertilized coastal waters.

On the other hand, this writer and his wife received an informative, well-documented reply, bearing the signature of Motohiko Numaguchi, President of Mitsubishi International Corporation, to a letter expressing concerns that his corporation's proposed large-scale salt evaporation/production facility at San Ignacio, Baja California Sur, Mexico, might negatively impact the prime breeding/calving grounds of the gray whale. Mr. Numaguchi's reply detailed the steps that his corporation, the parent corporation in Tokyo, and Exportadora de Sal S. A. de C. V., a subsidiary company jointly owned by the Mexican Government and Mitsubishi International, were taking to assess environmental problems posed by the project and to insure the public that construction would proceed only if a committee composed of prestigious international whale experts and marine scientists gave the project a green light. The substantive content of the letter, its up-to-date details, and the signature of the corporate president himself all convey a strong impression that Mitsubishi International is well aware of the situation and is committed to protecting the gray whale.

In many parts of the country, environmental groups and civic organizations provide help with issues of local, environmental concerns. Sportsmen and civic groups band together to improve fish and wildlife habitats, pick up trash, and promote recycling. Environmentalists join together to improve water quality in entire river basins and to protect ecologically sensitive habitats. Officers and members of these groups will almost always consent to meet with a class and inform students and faculty of their environmentally-friendly activities.

One such volunteer, community-based activity was described in an Associated Press article entitled "Sea turtle buffs rush to the rescue", carried in the Raleigh News & Observer (7/23/96). In mid-July, Hurricane Bertha struck an important nesting site for sea turtles in the North Topsail Beach area near Wilmington, NC. At the time the storm hit, sea turtle eggs had been laid and buried, but were not yet hatched. Vehicles and heavy equipment brought to the beach to remove debris and clean up roads presented a serious threat to egg-filled nests that survived the storm. To save the soon-to-be baby turtles, members of the Topsail Turtle Project dug up the surviving eggs and reburied them in safer, less-traveled areas along the beach. This work was particularly important because many turtle eggs had not survived beach erosion and prolonged submergence during the hurricane. Thus the only baby turtles to be born along this part of the beach this year would hatch from the relocated eggs.

Supplemental References

Babbitt. B., 1987, The Land Ethic: A guide for the world: *in* Tanner, T., editor, Aldo Leopold; The man and his legacy, Soil Conservation Society of America, Ankeny, IA, p. 137-142

Callicott, J. B., editor, 1987, Companion to A Sand Country Almanac: Interpretive and critical essays: The University of Wisconsin Press, Madison, WI, 308 p.

Dunn, S., 1996, Taking a green leap forward; In China, an environmental group moves into the mainstream: International Notebook section, The Amicus Journal, v. 18, no. 4, p. 12-14

Ehrlich, P. R. and Ehrlich, A. H., 1996, The betrayal of science and reason: How environmental anti-science threatens our future: Island Press, Covelo, CA 95428, 334 p.

Ehrlich, P. R. and Ehrlich, A. H., 1996, Biodiversity and the brownlash: Defenders; Conservation Magazine of Defenders of Wildlife, v. 71, no. 4, p. 6-17

Graham, F. Jr., 1995, Earth Day: 25 years old: National Geographic Magazine, v. 187, no. 4, April, p. 123-138

Harden, B., 1996, A river lost: The life and death of the Columbia: W. W. Norton & Company, 271 p.

Johnson, T., 1993, Sacred lands; Native intelligence; Environmentalists and Native Americans team up to protect the Earth: The Amicus Journal, v. 14, no. 4, Winter, p. 11-13

Leopold, A, 1949, A Sand County Almanac with essays on conservation from Round River: Ballantine Books, New York, NY; originally published by Oxford University Press Inc., 1949, 296 p.

Marsh, G. P., 1864, Man and Nature; or physical geography as modified by human action: Schribner, New York, NY: reprinted in 1965 by the Harvard University Press, Cambridge, MA

Meine, C. 1988, Aldo Leopold: His life and work: The University of Wisconsin Press, Madison, WI, 638 p.

Naar, J., 1993, The Green Cathedral: In this crusading congregation, ecology is God's work: The Amicus Journal, v. 14, no. 4, p. 22-28

Nash, R., 1982, Wilderness and the American mind, third edition: Yale University Press, New Haven, CT, 425 p.

Nash, R., 1987, Aldo Leopold and the limits of American liberalism: *in* Tanner, T., editor, Aldo Leopold; The man and his legacy, Soil Conservation Society of America, Ankeny, IA, p. 53-85

Nash, R., 1989, The rights of Nature; A history of environmental ethics: The University of Wisconsin Press, Madison, WI, 290 p.

Oliver, M., 1996, Low tide: What the sea gives to the human soul: The Amicus Journal, v. 18, p. 32-34

Owen, O. S. and Chiras, D. D., 1995, Natural resources conservation: Management for a sustainable future, sixth edition: Prentice Hall, Upper Saddle River, NJ, 608 p.

Sax, J. L., 1980, Mountains without handrails; Reflections on the national parks: University of Michigan Press, Ann Arbor, MI, 152 p.

Stegner, W., 1954, Beyond the hundredth meridian: Houghton Mifflin Company, Boston, MA; reprinted (1992) with the subtitle "John Wesley Powell and the second opening of the West" by Penguin Books USA, New York, NY, 438 p.

Taylor, P. W., 1986, Respect for Nature; A theory of environmental ethics: Princeton University Press, Princeton, NJ, 329 p.

Weiskel, T., 1993, New world, new values: Religion, belief, and survival on a small planet (essay): The Amicus Journal, v. 14, no. 4, p. 29

Earth Materials and Processes • Chapter 2; Home Study Assignment

Questions 1-4; Matching: Plate Tectonics and the Geologic Cycle

a. subduction zone   b. transform fault   c. spreading ridge   d. triple junction

1. (　) a convergent boundary between two lithosphere plates along which the denser, oceanic plate sinks into the asthenosphere and mantle
2. (　) a deep, vertical fracture zone that forms the boundary between two lithosphere plates moving horizontally in opposite directions
3. (　) a point on Earth's surface at which three, different, lithosphere plates are in mutual contact
4. (　) a divergent boundary between two lithosphere plates along which seafloor spreading takes place and new igneous rock is added to the edges of the plates

Questions 5-12; Which statements are true (T)? Which are false (F)? Correctly rewrite, edit, or rephrase the false statements. Study Figures 2.1 through 2.8.

5. (　) Upward convective flow and melting of mantle rock occur beneath a spreading ridge located more or less in the center of the modern-day Atlantic Ocean basin.

6. (　) The Nazca Plate forms part of the Atlantic Ocean basin south of eastern Brazil; the plate is moving west and sinking into a subduction zone beneath the Andes Mountain range at the western edge of the South American Plate.

7. (　) Earthquakes are preferentially localized along all plate boundaries; volcanoes are strongly concentrated along plate edges above subduction zones, and submarine volcanism is common along oceanic spreading ridges. Some volcanoes, such as those in Hawaii, are not localized along any plate boundary.

8. (　) Present-day, continental collisions are in progress where the Arabian and Indian subcontinents are pushing into the southern edge of the Eurasian Plate.

9. (　) On the basis of seafloor spreading rates, each year the distance between New York City and Madrid, Spain decreases by about 5 meters.

10. (　) The Juan de Fuca Plate is the small, oceanic plate currently moving to the northeast and sinking beneath the subduction zone along the southwest margin of the Caribbean Plate.

11. (　) The Dolomite Alps in northern Italy, are comprised of sedimentary rocks squeezed and folded when most of the Tethys Seaway was closed and obliterated during a continental collision between Africa and Eurasia.

12. ( ) Wegener's supercontinent of Pangaea began to break apart about 200 million years ago. By 65 million years ago, the South Atlantic Ocean basin had opened between Africa and South America and northernmost India was just crossing from the Southern into the Northern Hemisphere.

Questions 13-15; Matching: Rock-Forming Processes and the Rock Cycle. Study Figures 2.9 through 2.12.

a. metamorphic rocks   b. sedimentary rocks   c. igneous rocks

13. ( ) comprised of tightly interlocking mineral grains; formed at elevated temperatures and/or pressures from other, pre-existing rocks
14. ( ) comprised of tightly-interlocking mineral grains crystallized from molten silicate rock material
15. ( ) include mineral precipitates and cemented or compacted deposits of sand, silt, and clay particles derived by weathering and breakdown of other rocks

16. Which **one** is incorrect and false? The others are correct. **one answer**
a. Carbon dioxide liberated by metamorphism and melting of subducted marine sediments would eventually be released to the atmosphere through volcanism.
b. In most years, fossil fuel combustion now contributes more carbon dioxide to the atmosphere than volcanism.
c. Carbon in the atmosphere as CO_2 exceeds the quantity of carbon residing in shallow ocean waters as bicarbonate ion.
d. Buried marine sedimentary deposits contain more carbon than any other reservoir in or above the lithosphere.

Questions 17-28; Matching: Characteristics of Rocks and Earth Materials

a. bentonite	b. tuff	c. shale	d. conglomerate
e. limestone	f. schist	g. quartzite	h. basalt
i. granite	j. breccia	k. detrital	l. gneiss

17. ( ) sedimentary rock comprised of silt- and clay-sized particles; varies considerably in toughness depending on the extent of compaction and cementation
18. ( ) relatively soft shale rich in swelling clays derived by breakdown of original volcanic ash; a very unstable foundation material, especially when alternately wetted and dried
19. ( ) tough lithified rock to poorly consolidated materials comprised mainly of volcanic ash and other tephra
20. ( ) rock comprised mainly of coarse size, angular fragments broken from other pre-existing rocks or pyroclastic debris
21. ( ) very tough rock when well cemented; represents lithified deposits of rounded, gravel-size particles
22. ( ) chemical sedimentary rock comprised of tightly-interlocking calcite crystals; readily dissolves in acidic soil waters and groundwater

23. ( ) very hard metamorphic rock comprised mainly of quartz and formed by metamorphism of quartz-rich sandstone
24. ( ) intrusive, coarse-grained igneous rock composed of quartz, feldspars, and minor ferromagnesian minerals; typically very safe, stable foundation material unless weakened by fault ruptures and other fractures
25. ( ) type of sediment and sedimentary rock comprised dominantly of weathered, transported rock and mineral particles
26. ( ) mica-rich, crystalline metamorphic rock; generally a tough, stable foundation bedrock if unweathered; can readily slip along weakened foliation planes, especially if the mica minerals are deeply weathered and partly decomposed
27. ( ) stable, tough, crystalline metamorphic bedrock; weakened by fractures and may slip along local, mica-rich seams
28. ( ) fine-grained igneous rock typically erupted as lava flows; the bedrock is tough, but may be considerably weakened by weathering, brecciation, columnar jointing, and closely-spaced fractures; openings such as vesicles and lava tubes can result in local, highly porous zones

Questions 29-40; Matching: Minerals, Compositions, and Properties. Look over the *special feature* Minerals, p. 38.

a. pyrite	b. quartz	c. sulfides	d. carbonates
e. feldspars	f. oxides	g. silicates	h. clays
i. biotite	j. native element	k. hematite	l. bauxite

29. ( ) complex oxide ore of aluminum; forms by intense tropical weathering of mainly aluminum-rich igneous rocks
30. ( ) sodium, calcium, and potassium aluminum silicate minerals; very common minerals of igneous and metamorphic rocks
31. ( ) SiO2; tough, chemically resistant, highly insoluble, common mineral in igneous, metamorphic, and detrital sedimentary rocks and the major component of most sandstones and unconsolidated sands
32. ( ) dark-colored, iron-magnesium mica; common mineral in granite, gneiss, and schists
33. ( ) FeS2, an iron sulfide mineral; reaction with dissolved oxygen forms acidic waters containing high concentrations of sulfate ion and dissolved iron; common in coals, dark-colored, organic-rich shales, and veins and ore deposits containing lead, zinc, and copper sulfide minerals
34. ( ) sulfur crystals around a volcanic vent, gold nuggets and dust in a stream gravel deposit, and diamonds are examples
35. ( ) minerals comprised of one or more metallic elements and oxygen; the common iron and aluminum ore minerals are examples
36. ( ) minerals comprised dominantly of silicon and oxygen in combination with other elements such as aluminum, sodium, magnesium, calcium, potassium, and iron; comprise most of the rock-forming minerals

37. ( ) Fe2O3; the most widely mined ore mineral of iron; very tiny, finely-divided, particles of this mineral provide the red colors seen in many sandstones, shales, and soils
38. ( ) very fine size, platelike, mineral grains produced by weathering and chemical alteration of ferromagnesian minerals and feldspars; major component of most shales, muds, and soils
39. ( ) most common ore minerals of metals such as copper, zinc, lead, cadmium, antimony, mercury, and cobalt
40. ( ) compounds of elements such as calcium, magnesium, iron, and other elements with the [CO3]2- ion; calcite, the major mineral in limestone, is an example; slowly dissolves in water so caves and underground openings are common in areas with limestone bedrock

Questions 41-43; Matching: The Tectonic Cycle, Stress and Strain. Look over the *special feature* Stress and Strain, p. 40.

a. shear stress/strain   b. tensile stress/strain   c. compressive stress/strain

41. ( ) intensely deformed rocks along a transform fault, such as the San Andreas fault in California, that separates the Pacific and North American Plates
42. ( ) horizontal stresses of this type result in horizontal stretching and vertical thinning; rocks usually fail by fracturing and faulting; most common stresses/strains generated at divergent plate boundaries
43. ( ) horizontal stresses of this type result in horizontal shortening and vertical thickening; continental collisions generate these in enormous volumes of rock along the convergent margins of the two plates

Questions 44-45; Matching: Brittle and Ductile Behavior of Rocks Under Stress

a. ductile strain response to stress   b. brittle strain response to stress

44. ( ) strain is without failure (rupture) and proportional to stress for stresses up to the proportional elastic limit; strain increases rapidly for only small stress increases above the proportional elastic limit
45. ( ) strain is proportional to stress until failure (rupture) occurs

Questions 46-47; Matching: Deformations of Solid Materials

a. plastic   b. elastic

46. ( ) Strains are very small and the material reverts to its original unstrained configuration when the stress is removed; includes bedrock affected by most seismic vibrations.
47. ( ) Strains are finite and the material retains its deformed configuration after the stress is removed; is the common deformation in moist, partly weathered shales beneath the surface soils of a steep slope.

Questions 48-51; Do you agree (A) or disagree (D) with the following statements? If you disagree, very briefly note the reasons for your disagreement. Look over the photos, Figures 2.18-2.22.

48. ( ) The granite shown in Yosemite Valley, CA is probably highly fractured; otherwise, erosion would not have fashioned such steep cliffs from the bedrock.

49. ( ) Deposition of successive, sheetlike layers of sand in large, now-lithified, ancient sand dunes accounts for the observed, slabby nature of the sandstone in Zion National Park; the sandstone is stratified, not fractured. Most of the sand grains in the rock are quartz; thin hematite coatings on the individual sand grains account for the red color in parts of the sandstone.

50. ( ) The gravel-size sedimentary particles in the Death Valley, CA conglomerate are very well rounded, suggesting that they were strongly abraded and transported long distances in ancient streams.

51. ( ) The jagged, pinnacled rock surface shown from Yunnan Province, China probably formed by weathering and dissolution of limestone beneath a deep soil cover. Subsequently the soil was eroded away, exposing the formerly buried surface of the bedrock.

Questions 52-53; Do you agree (A) or disagree (D) with the statements? If you disagree, briefly note reasons for you disagreement. See Figures 2.23 and 2.24.

52. ( ) In some metamorphic rocks, the foliation planes or surfaces, especially in weathered rocks, can function as tiny channels for water seepage and as slip surfaces for potential, downhill block movements.

53. ( ) The left roadcut bank in Figure 2.23a and the right roadcut bank in Figure 2.23b are both susceptible to block slippage and weakening from soil water drainage.

Questions 54-55; Do you agree (A) or disagree (D) with the following statements? If you disagree, briefly note the reasons. Instead of representing foliation planes, imagine that the parallel-line pattern in Figures 2.23 and 2.24 represented bedding or stratification surfaces separating inclined, alternating beds of shale and sandstone.

54. ( ) The potential problems with subsurface water seepage and downhill block slippage would be eliminated if the strata were alternating sandstone and shale.

55. ( ) The potential for slope failure in road cuts through subhorizontal, interbedded sandstone and shale strata would be lower than for cuts through similar strata that are inclined as shown in Figure 2.23; however, seepage around one or both abutments could still potentially compromise stability of the dam (Figure 2.24) even if the strata were subhorizontal.

Questions 56-62; Do you agree (A) or disagree (D) with the following statements? If you disagree, very briefly note the reasons.

56. ( ) Glacial outwash refers to uncemented, surficial, sand and gravel strata deposited by meltwater streams flowing away from the edge of a glacier.

57. ( ) Tills are sediments dropped directly from melting ice of a glacier; till deposits are recognized by their well-defined, interlayered silt, sand, and gravel beds.

58. ( ) In areas covered by Pleistocene glacial deposits, tills are better sources for commercial sand and gravel than are outwash deposits.

59. ( ) The active layer of continuous permafrost thaws out and dries each summer as the liquid water infiltrates deeper into the frozen soil, evaporates, or is transpired by vegetation.

60. ( ) The Pleistocene ice ages ended for certain 10,000 to 15,000 years ago when the last of the continental ice sheets melted away. In North America the southern edge of the expanded ice sheets closely followed the present-day southern limits of areas with discontinuous permafrost.

61. ( ) The passive method of permafrost engineering basically depends on keeping the ground permanently frozen to prevent lateral and vertical movements in unconsolidated foundation materials subjected to summertime thawing and repeated freezing and thawing in the spring and fall.

62. ( ) See Figure 2.29, p. 52. The V-shaped cross-canyon profile indicates that Yellowstone Canyon continues to be deepened and widened by stream channel erosion and rapid erosion of the canyon slopes. Only an expert geomorphologist could recognize the erosional, glacier-cut origin of the Sierra Nevada canyon (Figure 2.29b) with its broad U-shaped, cross-valley profile and numerous lakes.

Questions 63-64; Do you agree (A) or disagree (D) with the statements? If you disagree, briefly note the reasons. Study the *case history* Hubbard Glacier, p. 50-51.

63. ( ) Continued surging of the Hubbard Glacier, Alaska, could seriously threaten environmentally sensitive stream and marine habitats in the area.

64. ( ) As recently as a thousand years ago, this area was essentially ice free.

Questions 65-68; Do you agree (A) or disagree (D) with the following statements? If you disagree, briefly note the reasons. Look over the *case histories* Baldwin Hills Dam, p. 42-43 and St. Francis Dam, p. 48-49.

65. ( ) The Baldwin Hills reservoir was designed and constructed to prevent infiltration and seepage. However, the dam failed when cracking and small, aseismic movements along a pre-existing fault allowed water to seep through the sealed bottom of the reservoir and under the dam. The cracks and the fault displacement were probably linked to man-induced subsidence centered over a nearby oil field.

66. ( ) From the surface downward, a deep, vertical borehole at the proposed west abutment of the St. Francis damsite would have penetrated the following:
    3) relatively thick zone of tough, resistant sedimentary strata; good foundation material when dry but weakened by moisture; minimal problems with infiltration and seepage
    2) thin zone of breccia, pulverized rock altered to clays, and gypsum; very high potential for slippage, infiltration, and seepage
    1) zone of schist with foliation surfaces inclined toward the west; poor foundation material; high potential for landsliding, infiltration, and seepage

67. ( ) At the St. Francis damsite, pre-construction, geologic site studies would have clearly shown that the boundary between the schist and sedimentary beds came to the surface in the canyon well below the proposed water level in the reservoir. Given the geologic setting of the Los Angeles area, it would be reasonable to assume that the contact was a fault, and that infiltration and seepage along the fault might endanger the dam after the reservoir was filled.

68. ( ) In retrospect, both dams were probably destined to fail, but the St. Francis failure was much more of a surprise and disappointment to engineers and geologists than failure of the Baldwin Hills reservoir.

Questions 69-74; Do you agree (A) or disagree (D) with the following statements? If you disagree, briefly note the reasons.

69. ( ) Sand dunes migrate in the direction of the slip face, the steeper, downwind slope of the dune.

70. ( ) Seif dunes have their long dimensions aligned at right angles to the prevailing wind direction. Individual dunes can be 90 meters in height and 60 miles long.

71. ( ) Deposits of windblown silt are known as loess. In the U. S., silt particles in most loess deposits were produced by abrasion in glaciers, transported and deposited by streams and rivers, then picked up and redeposited by the wind.

72. ( ) In the Lower Mississippi River Valley, loess deposits are thicker and more extensive east of the river than west of the river.

73. ( ) Loess deposits in which the silt grains are weakly cemented by thin films of clay and/or calcite are subject to hydroconsolidation if wetted.

74. ( ) The large, sand-covered area in western Nebraska, shown in Figure 2.32, has its long dimension roughly east-west. Its perimeter configuration is consistent with sand deposition from west-to-east blowing prevailing winds.

75. On the world map provided, accurately plot and label the following countries, features, or locations.
a. Yakutat, AK   b. Nazca Plate   c. Sahara Desert   d. India
e. Brazil   f. Andes Mountains   g. Madrid, Spain   h. Dolomite Alps, Italy
i. Los Angeles County, CA   j. New York City   k. Arabian subcontinent
l. Yunnan Province, China   m. Loess deposits of the Lower Mississippi Valley
n. Sand Hills area, western Nebraska   o. Yellowstone National Park, WY
p. approximate southern limit of the Pleistocene ice sheets in North America
    q. southern limit of discontinuous permafrost zone in Mongolia

Environmental Geology Quiz • Chapter 2

1. Which **one** is a deep, vertical fracture zone that forms the boundary between two lithosphere plates moving horizontally in opposite directions?
a. spreading ridge   b. double junction   c. transform fault   d. subduction zone

Questions 2-4; Which statements are true (T)? Which are false (F)?

2. ( ) The Nazca Plate forms part of the Atlantic Ocean basin southeast of Brazil; the plate is moving to the west and sinking into a subduction zone beneath the Andes Mountain range at the eastern edge of the South American Plate.

3. ( ) The Juan de Fuca Plate is the small, oceanic plate currently moving to the northeast and sinking beneath the subduction zone along the southwest margin of the Caribbean Plate.

4. ( ) Earthquakes are preferentially localized along all plate boundaries; volcanoes are strongly concentrated along plate edges above subduction zones, and submarine volcanism is common along oceanic spreading ridges. A few volcanoes in the Pacific Ocean basin are not localized along plate boundaries.

Questions 5-7; Matching

a. metamorphic rocks   b. sedimentary rocks   c. igneous rocks

5. ( ) comprised of tightly interlocking mineral grains; formed at elevated temperatures and/or pressures from other, pre-existing rocks

6. ( ) comprised of tightly-interlocking mineral grains crystallized from molten silicate rock material

7. ( ) include mineral precipitates and cemented or compacted deposits of sand, silt, and clay particles derived by weathering and breakdown of other rocks

8. Which statement is correct?
a. Buried marine sedimentary deposits contain more carbon than any other reservoir in or above the lithosphere.
b. Carbon in the atmosphere as CO2 exceeds the quantity of carbon residing in shallow ocean waters as bicarbonate ion.

Questions 9-13; Matching

a. tuff   b. granite   c. quartzite   d. schist
e. limestone   f. conglomerate   g. shale   h. gneiss

9. ( ) sedimentary rock comprised of silt- and clay-sized particles; varies considerably in toughness depending on compaction and cementation

10. ( ) very tough rock when well cemented; represents lithified deposits of rounded, gravel-size particles

11. ( ) chemical sedimentary rock comprised of tightly-interlocking calcite crystals; readily dissolves in acidic soil waters and groundwater

12. ( ) intrusive, coarse-grained igneous rock composed of quartz, feldspars, and minor ferromagnesian minerals; typically very safe, stable foundation material unless weakened by fault ruptures and other fractures

13. ( ) mica-rich, crystalline metamorphic rock; generally a tough, stable foundation bedrock if unweathered; can readily slip along weakened foliation planes, especially if the mica minerals are deeply weathered and partly decomposed

Questions 14-19; Matching

a. bauxite   b. hematite   c. quartz   d. clays   e. biotite   f. pyrite

14. ( ) Fe_2O_3; a widely mined ore mineral of iron; very tiny, finely-divided, particles of this mineral provide the red color seen in many sandstones, shales, and soils

15. ( ) complex oxide ore of aluminum; forms by intense tropical weathering of mainly aluminum-rich igneous rocks

16. ( ) SiO_2; tough, chemically resistant, highly insoluble, common mineral in igneous, metamorphic, and detrital sedimentary rocks and the major component of most sandstones and unconsolidated sands

17. ( ) FeS_2; reaction with dissolved oxygen forms acidic waters containing high concentrations of dissolved sulfate ion and iron; commonly occurs in coals, dark-colored, organic-rich shales, and metallic sulfide mineral veins and ore deposits

18. ( ) dark-colored, iron-magnesium mica; common in granites, gneiss, and schists

19. ( ) very fine size, platelike, mineral grains; major component of most shales, muds, and soils

Questions 20-22; Matching

a. shear stress/strain   b. tensile stress/strain   c. compressive stress/strain

20. ( ) convergent plate boundaries   21. ( ) divergent plate boundaries

22. ( ) transform fault plate boundaries

23. Which **one** characterizes ductile deformation in rock or other solid materials?
a. strain is without failure (rupture) and proportional to stress for stresses up to the proportional elastic limit; strain increases rapidly for only small stress increases above the proportional elastic limit
b. strain is proportional to stress until failure (rupture) occurs

24. Which **one** is an elastic deformation?
a. slow, downhill creep of weathered shale   b. earthquake-wave vibrations in rock

Questions 25-27; Matching

a. Yosemite Valley, CA   b. Zion National Park, UT   c. Yunnan Province, China

25. ( ) scenic, steep canyon walls, cliffs, and outcrops of white to reddish sandstone; the sand originally accumulated as ancient, wind-deposited sand dunes

26. ( ) deep canyon system with many very steep to near-vertical walls carved into relatively unfractured granite bedrock by glaciers

27. ( ) scenic area of jagged, rock pinnacles and outcrops formed by subsurface dissolution of limestone and exposed at the surface by extensive soil erosion

Questions 28-33; True (T) or False (F)

28. ( ) Tills are sediments dropped directly from melting ice of a glacier; till deposits are recognized by their well-defined, interlayered silt, sand, and gravel beds.

True (T) or False (F); continued

29. ( ) In areas covered by Pleistocene glacial deposits, tills are better sources for commercial sand and gravel than are outwash deposits.
30. ( ) The active layer of continuous permafrost dries out each summer as liquid water infiltrates into the frozen soil, evaporates, or is transpired by vegetation.
31. ( ) Seif dunes have their long dimensions aligned at right angles to the prevailing wind direction. Individual dunes can be 90 meters in height and 60 miles long.
32. ( ) Continued surging of the Hubbard Glacier, Alaska, could seriously threaten environmentally sensitive stream and marine habitats.
33. ( ) The Baldwin Hills, CA, reservoir and dam failed when cracks due to small, aseismic displacements along a pre-existing fault allowed water to seep through the formerly-sealed bottom of the reservoir and under the dam.

Questions 34-36; Do you agree (A) or disagree (D) with these statements? Briefly note reasons for any disagreements.

34. ( ) The southernmost edges of the Pleistocene ice sheets in North America closely followed the present-day southern limits of areas with discontinuous permafrost.

35. ( ) Deposits of windblown silt are known as loess. In the U. S., silt particles in most loess deposits were produced by abrasion in glaciers, transported and deposited by streams and rivers, then picked up and redeposited by the wind.

36. ( ) Loess deposits in which the silt grains are weakly cemented by thin films of clay and/or calcite are subject to hydroconsolidation if wetted.

37. On the world map provided, accurately plot and label the following countries, features, or locations.
a. Yunnan Province, China   b. Loess deposits, Lower Mississippi Valley
c. Sand Hills area, western Nebraska   d. Yellowstone National Park, WY
e. approximate southern limit of the Pleistocene ice sheets in North America

Answers; Home Study Assignment
Earth Materials and Processes • Chapter 2

1. a	2. b	3. d	4. c	5. T
6. F*	7. T	8. T	9. F*	10. F*
11. T	12. T	13. a	14. b	15. c
16. c	17. c	18. a	19. b	20. j
21. d	22. e	23. g	24. i	25. k
26. f	27. l	28. h	29. l	30. e
31. b	32. i	33. a	34. j	35. f
36. g	37. k	38. h	39. c	40. d
41. a	42. b	43. c	44. a	45. b
46. b	47. a	48. D*	49. A	50. D*
51. A	52. A	53. A	54. D*	55. A
56. A	57. D*	58. D*	59. D*	60. D*
61. A	62. D*	63. A	64. D*	65. A
66. A	67. A	68. D*	69. A	70. D*
71. A	72. A	73. A	74. A	

*

6. The Nazca Plate forms part of the eastern Pacific Basin, not the Atlantic, and lies off the west coast of Chile and Peru, not off the Brazilian coast. The plate is moving east, not west; however, the rest of the statement is correct.

9. The New York to Madrid distance is increasing, not decreasing, slightly each year due to seafloor spreading on the Mid-Atlantic Ridge. The 5 meters per year rate is way too large; 5 centimeters per year would be about right.

10. The first part of the statement is accurate. The Juan de Fuca Plate, however, is sinking beneath the western margin of North America along the Pacific coastlines of Washington, Oregon, and northernmost California.

48. The steep to vertical rock walls in Yosemite show that the granite bedrock has few fractures; highly fractured bedrock would crumble and not support such steep slope angles.

50. Gravel fragments in the photo are fairly angular and show little rounding; thus the gravel particles were deposited fairly close to their site of derivation.

54. The problems with potential seepage and block slippage would still be present with interbedded sandstone and shale; they may be more severe if the sandstone is fairly permeable and the shales are strongly pyritic and/or rich in bentonite.

57. The first part of the statement concerning tills is correct; tills have weak to absent stratification and rarely show much in the way of sediment-size sorting. The last part of the statement describes the bedding characteristics of outwash sand and gravel, not of till.

58. Outwash deposits are much better sources of sand and gravel than tills; many tills contain substantial silt- and clay-sized particles, may contain large boulder-

size blocks that require basting, and show little sorting among the sand- and gravel-size materials.

59. The active layer does thaw out each summer. However, it does not dry out, but remains moist or saturated because downward infiltration is prevented by the permafrost; cool days and sparse trees mean that evaporation and transpiration rates are low.

60. First of all, the North American continental ice sheets did melt away between 10,000 and 15,000 years ago, but the Pleistocene ice ages may or may not be over for good. The southernmost edges of the ice sheets were well south of the modern-day, discontinuous, permafrost limit. East to west, the limits of the ice sheet extended from Long Island through central Pennsylvania, the Ohio River valley, the Missouri River valley, the northern Rockies in Montana and Idaho, to central Washington.

62. The only thing wrong with this statement is the notion that only a well-trained, expert geomorphologist could recognize the glacier-cut origin of Yosemite Valley and other glacial valleys. The U-shaped, cross-valley profiles and other glacier-cut feature are very distinctive and anyone showing a little effort and practice can recognize these valleys as being cut by glacial erosion.

64. The whole area was probably ice-covered, not ice-free, 1000 years ago. The village of Yakutat is built on 1000-year-old glacial deposits.

68. In retrospect, both dams were probably destined to fail, but Baldwin Hills must have been a stunning surprise and disappointment because obvious shortcomings brought to light in the St. Francis failure had been corrected.

70. Seif dunes have their long dimensions parallel with the prevailing wind direction. The stated dimensions are correct.

Answers: Environmental Geology Quiz • Chapter 2

1. c	2. F	3. F	4. T	5. a
6. c	7. b	8. a	9. g	10. f
11. e	12. b	13. h	14. b	15. a
16. c	17. f	18. e	19. d	20. c
21. b	22. a	23. a	24. b	25. b
26. a	27. c	28. F	29. F	30. F
31. F	32. T	33. T	34. D*	35. A
36. A				

*

34. The southernmost edges of the ice sheets were well south of the modern-day, discontinuous, permafrost limit. East to west, the limits of the ice sheet extended from Long Island through central Pennsylvania, the Ohio River valley, the Missouri River valley, the northern Rockies in Montana and Idaho, to central Washington. The southern limit of the discontinuous permafrost zone extends from southern Labrador through southernmost James Bay, northernmost Lake Winnipeg, northwestward to the Pacific Coast north of the Alaskan Panhandle.

Commentary/Media Watch

Ideas and Notions

Chapter 2 presents us with the problems of providing more than "lip service" to topic such as plate tectonics, mineralogy, and petrology while still retaining the emphasis on environmental issues. Students in any earth science course should be familiar with the ideas and concepts of plate tectonics, with the three, different kinds of tectonic plate boundaries, with plate motions, and with boundary configurations associated with each. Worldwide distributions of seismic activity and volcanism are readily portrayed in the plate tectonic framework. Obvious exceptions, such as volcanism in Hawaii and Yellowstone National Park are conspicuous in their linkage to mantle hot spots; the New Madrid earthquake zone can be used to bring along the idea of geological history and the present-day significance of relict, geologic features left over from earlier times in Earth's geologic history.

The rock cycle and hydrologic cycle are of basic importance in understanding subject matter in following chapters. Again, rather than dwelling on each in great detail (admittedly, they are very deserving topics in their own right) use ideas and concepts to set up the major environmental topics to follow such as soils, site studies, foundation characteristics of earth materials, and traditional geomophological topics such as mass wasting, fluvial systems, and coastal processes.

For example, rising sea level exacerbates threats to coastal development; melting of the Greenland and Antarctic ice sheets is providing additional water to the oceans. In terms of the hydrologic cycle, water now existing as ice in glaciers and ice sheets is being "temporarily stored" on land. If global warming is a future reality, as many knowledgeable scientist believe to be the case, more ice will melt and sea level will continue to rise. Effects of Man's activities, such as adding CO_2 and other greenhouse gases to the atmosphere, hint at the "nonintuitive complexity" of global systems and illustrate such basic concepts as feedback mechanisms.

Mineral grains in crystalline, silicate, igneous and metamorphic rocks were originally formed at depth under elevated temperatures and pressures; the mineral grains are very strongly bound together. These rocks are mechanically hard and tough and, except in the context of extended weathering or hydrothermal alteration, they are insoluble. Mechanical strength is determined mainly by the density and orientations of potential slip surfaces such as fractures and foliation surfaces. Sedimentary rocks and deposits are far more variable in their mechanical characteristics. Stratification surfaces present additional avenues for slip. Grain-to-grain binding is highly dependent on the degrees of compaction and cementation. Well-cemented quartz sandstone and unconsolidated quartz sand are completely different in their mechanical characteristics; yet they are virtually identical mineralogically and chemically.

Shales and mudstones have strong tendencies to deteriorate and weaken upon weathering (hydration of clays, oxidation of pyrite, dissolution of carbonates, etc.). As Professor Keller succinctly notes, "Shale causes many environmental problems, and its presence is a red flag to the applied-earth scientist" (p. 42).

Calcite and dolomite are about the only common, rock-forming minerals that are eroded at significant rates by dissolution. Dissolution rates increase drastically in acidic solutions. Thus in contact with the abnormally acidic rains and fogs of today, limestones and dolomites deteriorate at noticeably faster rates than do dominantly silicate-mineral rocks such as granite, slate, and basalt. Again, as a way of highlighting concerns about land subsidence and groundwater conditions in coming chapters, remind students that areas of limestone bedrock are to planners, developers, and geologists what minefields are to the military, fraught with danger and deserving of very careful study and consideration.

Certain minerals and rocks deserve special mention because of their widespread importance in many facets of environmental science. Pyrite, FeS_2, is one of these. It is the only really common metal sulfide mineral to occur in rocks and sediments; it is an abundant mineral constituent of many metallic ores, and also is a common, minor, mineral constituent of coals. Pyrite oxidizes readily in oxygenated, aqueous or moist media, generating dissolved iron (ferrous, Fe^{2+} and ferric, Fe^{3+}) ions, soluble sulfate, and acidic water. Acidic mine drainage is a serious water quality problem in many metal- and coal-mining areas. Smelting of sulfide ores and combustion of coals can introduce huge quantities of SO_2, a major precursor to acid precipitation, into the atmosphere. Limestone and other calcite-bearing rocks are in the front lines defending against acidic precipitation and mine drainage. As calcite dissolves, it neutralizes acidity in water and promotes precipitation of excessive dissolved metal contents. On contact with SO_2 in industrial stack emissions, calcite decomposes to CO_2 and CaO; the CaO, a powerful chemical base, reacts with SO_2, forming calcium sulfate and preventing the SO_2 from entering the atmosphere. Lime, a widely used soil conditioner, is manufactured directly from limestone and/or marble.

Other rocks and minerals deserve special mention because of their unusual properties or important commercial uses. Limestone is the basic raw material for Portland cement; and, gypsum and anhydrite are major components of plasters and plasterboard. Barite, barium sulfate, is used as a component of commercial drilling muds because of its high density and chemical stability. Garnet, because of its high density, crystal shape, and chemical stability, is often used in water filtration systems. Its granularity and hardness make garnet a widely-used mineral constituent of sandpaper and other abrasives.

Bentonite, like barite, also a common component of drilling muds, is an expanding clay. In a dry state, bentonitic soils and shale strata seem perfectly stable and innocuous; when wet, they swell, lose their bearing strength, and turn a docile landscape into a seething quagmire. Moving vehicles slip and slide uncontrollably and stopped vehicles sink to their axles; even walking is nearly impossible. With each step, glob after glob of gooey mud sticks to your boots; after a few steps, your feet are so heavy that you might as well be nailed to the ground!

Glaciated areas, bentonitic terrains, loess-covered terrains, permafrost regions, and areas of active or stabilized, windblown sand dunes each present their own, unique set of environmental conditions concerning surficial materials, ease of excavation, foundation stability, hydrologic conditions, and site engineering. Local geologists and contractors will be familiar with these; newcomers and out-of-state interests may not. In these localities, ignorance of "local conditions" can be shamefully embarrassing and very expensive!

Television Commercial for Fairfax, VA

Chapter 2 topics are clearly not "media darlings". Mineral and rock names are seldom mentioned; neither are slip surfaces, foliation, loess, and bentonite. However, plate tectonics and continental drift do have their media proponents. A television commercial shown on CNN, fall, 1996 depicts the western hemisphere continents "first splitting and drifting apart" as poor communications and isolationist thinking pushed the world apart; then the viewer sees the drift motion reverse and the continents move closer together to reunite in a pre-Atlantic Ocean, Pangaean configuration. This latter scene is then tied to Fairfax, Virginia, touting its preeminence as a communications and business center for bringing the world together and for launching your business and its products into the new global market place.

The Vietnam War Memorial

Cracking and deterioration of the Vietnam Memorial, Washington, DC, have attracted media coverage. The relatively thin slabs of "back granite", probably a diorite, that comprise the monument are subjected to stresses resulting from architectural design and out-of-doors weathering processes. Opening and lengthening of hairline cracks in the facing slabs are of serious concern to National Park Service personnel and the public at large. Acidic rainfall and fog, differential thermal expansion and contraction, freezing and thawing, air pollution, and visitor touching all contribute to the problem. The monument is well known to most students and would provide a good focus for presenting material on rock properties, rock strength, and weathering processes. Pyrite was not mentioned in media reports concerning the monument, but a mafic rock like the "black granite" probably contains a small percentage of metal sulfides. Oxidization of these would certainly accelerate the weathering process.

Supplemental References

Manning, D. A. C., 1995, Introduction to industrial minerals: Chapman & Hall, USA, One Penn Plaza, New York, NY 10119

Prentice, J. E., 1996, Geology of construction materials: Van Nostrand Reinhold, New York, NY, 216 p., Box 668, Florence, KY 41022-9979 for U. S. orders

Robinson, D. A. and Williams, R. B. J., editors, 1994 , Rock weathering and landform evolution: John Wiley & Sons, New York, NY 10158-0012, 519 p.

Twenty-eight chapters on rock weathering and deterioration under different climatic conditions

Schultz, C. B. and Frye, J. C., editors, 1968, Loess and related eolian deposits of the world: v. 12, Proceedings VIIth INQUA Congress, Univ. of Nebraska Press, Lincoln, NE, 369 p.

A collection of 34 papers dealing with loess deposits mainly in the United States and Europe

Soils and Environment • Chapter 3; Home Study Assignment

Questions 1 and 2; Engineering professionals and natural scientists have their own, different definitions of soil. Which is which?

a. engineering definition of soil   b. natural science definition of soil

1. ( ) Soil is an earth material, unconsolidated or to some extent consolidated, that can be excavated, recontoured, or removed without recourse to drilling and explosives.
2. ( ) Soil is a surface accumulation of physically and chemically altered rock and mineral particles and organic matter; biochemical activity is generated by soil organisms and plant roots.

Questions 3-9; Matching: Soil Horizons; See Figure 3.1.

a. A   b. B   c. C   d. R   e. E   f. O

3. ( ) fresh, unweathered, local parental earth material or bedrock
4. ( ) weathered and partly decomposed, local parental earth material or bedrock
5. ( ) dark to black surface accumulation of decomposing vegetable matter; humus
6. ( ) horizon of residual, tough, chemically resistant mineral particles and organic matter
7. ( ) light colored zone enriched in clays, iron oxides, and carbonates derived from overlying soil horizons; the zone of accumulation
8. ( ) light colored horizon composed of tough, chemically resistant mineral grains; generally devoid of clays and organic matter
9. ( ) ( ) these two comprise the zone of leaching

Questions 10 and 11; Matching: Distinctive Soil Horizons and Soil Types

a. caliche   b. hardpan

10. ( ) unusually well-compacted and or well-cemented clay-rich soils; may be impermeable and will prevent or greatly slow the downward percolation of water
11. ( ) B horizon soil strongly impregnated with calcium carbonate; forms under semiarid to arid climatic conditions; may be cemented strongly enough to require blasting

Questions 12-14; Matching: Dominant Soil-Coloring Agents

a. black, dark gray   b. red, yellow, brown   c. white

12. ( ) calcium carbonate; typical of caliche and well-developed K horizons
13. ( ) iron oxides and hydrated iron oxides   14. ( ) organic matter

Questions 15 and 16; Matching: See Figures 3.2 and 3.3.

a. soil texture   b. soil structure

15. ( ) description and classification are based on sizes of the individual soil particles
16. ( ) description and classification are based on sizes and shapes of soil particle aggregates called peds

17. A soil consists of 70 % sand, 15 % clay, and 15 % silt. What is its U. S. Department of Agriculture textural class? See Figure 3.2.
a. silt b. sandy clay c. clay loam d. sandy loam

Questions 18 and 19; Matching: Soil Profile Development; See p. 62.
a. weakly developed b. well developed

18. ( ) soil structure apparent; clays abundant in the B horizon and sparse to absent in the leached zones; B zone is reddened with hematite dustings on the soil particles; requires fifty thousand to over a hundred thousand years to form, depending on local conditions

19. ( ) B horizon is thin or absent; A zone rests directly on C and parental rock of the R zone may be visible in roadcuts, gullies, and streams; requires a few hundred to a few thousand years, depending on local conditions

20. Which **one** is not a major soil nutrient compound?
a. nitrogen as nitrate b. phosphorous as phosphate c. potassium d. aluminum

21. A soil chronosequence study would be based on which soil classification system?
a. the soil taxonomy system used by soil scientists including such terms as entisols and mollisols (Table 3.1)
b. the unified classification system used by engineers including categories such as clean gravel and high plastic clay (Table 3.2)

22. Which soil engineering property denotes how easily water will flow through the saturated, soil material?
a. erodibility b. cohesion c. permeability d. plasticity

23. Which engineering soil material would have the highest permeability?
a. silty clay b. clean gravel c. plastic clays d. clayey sand

24. Which engineering material would probably have the largest compressibility?
a. silty clay b. clean gravel c. plastic clay d. clayey sand

25. Which engineering soil material would have the highest cohesion?
a. dry silty clay b. dry clean sand

26. Which engineering property denotes the tendency for a soil material to undergo liquefaction when subjected to shaking or vibrations?
a. erodibility b. cohesion c. permeability d. sensitivity

27. Which are the two main factors determining soil strength? **one answer**
a. permeability, erodibility b. sensitivity, corrosion potential
c. shrink-swell potential, compressibility d. cohesion, friction

28. Which soil would be expected to show the higher corrosion potential?
a. a clean sand only occasionally saturated with water
b. a normally moist forest soil, such as a spodosol, with ongoing, extensive biochemical activity

29. Which soil engineering factor bears directly on the process of soil compaction and settlement of the land surface?
a. compressibility   b. cohesion   c. permeability   d. plasticity

30. Which engineering soil would have the highest plasticity?
a. dry silty gravel   b. moist silty clay   c. dry silty clay   d. moist silty gravel

31. Which description correctly characterizes the pore spaces in nonsaturated soil above the water table?
a. filled with liquid water and soil gases   b. filled with liquid water

Questions 32-37; True (T) or False (F): What is wrong with the false statements?

32. ( ) Easily erodible soils have low cohesion.

33. ( ) In some soils, water saturation may lower soil sensitivity and decrease the potential for liquefaction.

34. ( ) In general, moist silt and clay-rich soils will be less compressible than clean sands and gravels.

35. ( ) In general, percolation of water would be faster through a clean sand than through an organic-rich clay.

36. ( ) Peat, muck, and histosols are soils composed dominantly of organic matter.

37. ( ) As they lose moisture, expansive soils swell and increase in unit volume.

38. What is montmorillonite? What effects does it have on the engineering properties of soils?
a. a calcium carbonate mineral; raises strength and cohesion of B-horizon soils
b. a clay mineral; when wetted, can swell to many times its dry volume and dramatically weaken a soil

39. Assume a building is sited close to a large oak or maple tree. When would the shrinking and swelling associated with expansive soils be most apparent?
a. summer, when tree is growing   b. winter, when tree is dormant

40. As a generalization, which soils are most suitable for bearing loads associated with heavy structures?
a. poorly drained, clay-rich soils   b. well-drained, sandy or gravely soils

41. Cohesive forces between soil particles and soil moisture are strongest under which condition? **one answer**
a. pores are water saturated   b. fairly dry soil; no liquid water in pores
c. liquid water and soil gases occupy the pore spaces; the soil is unsaturated

42. Which **one** is not a component term of The Universal Soil Loss Equation?
a. soil erodibility index
b. erosion-control practice factor
c. rainfall/runoff quotient factor
d. hillslope/gradient factor

43. Which involves changes in evapotranspiration, runoff, and sediment yield similar to those produced when forested lands are clear-cut for timber?
a. conversion of forest to agricultural fields
b. conversion of agricultural fields to urbanized areas

44. Which **one** is not an expected consequence of clear-cutting sloping forest lands?
a. Runoff and sediment yields decrease.
b. Interception and evapotranspiration decline drastically.
c. Maximum, daily surface soil temperatures increase.
d. Soils dry out somewhat and compact; the potential for slides and other slope failures increases.

45. Which statement concerning urbanization of farmlands is reasonably correct?
a. Urbanization greatly increases runoff; sediment yields increase during active construction but decline thereafter.
b. Urbanization decreases runoff during active construction and greatly increases sediment yields after most of the construction is finished.

46. Which best reduces sediment yields from areas around construction sites?
a. protecting all disturbed areas with artificial coverings and grass, especially during the summer thunderstorm season
b. regrading disturbed areas after each storm to fill in erosional rills and gullies formed by the previous storm

Questions 47-52; Matching: Environmental Issues, Locations, and Technology
a. Mojave Desert Region, California
b. Times Beach, Missouri
c. San Joaquin Valley, California
d. Indio Hills, California
e. North Africa, lands bordering the Sahara Desert

47. ( ) severely contaminated with the cancer-causing chemical dioxin

48. ( ) documented severe environmental damage from excessive ORV use

49. ( ) beset with multiple environmental problems related to long-term overuse of groundwater, easily erodible soils, and extensive irrigation for agriculture

50. ( ) severe dust storm in 1977 that removed as much as 60 cm of soil from some areas; reduced visibility resulted in numerous highway vehicle accidents

51. ( ) severe land degradation resulting from excessive domestic animal populations, excessive groundwater withdrawals, and prolonged natural periods of drought; region in which the term desertification was first widely publicized

52. ( ) alluvial fan morphology and soil science studies allow the local slip rate on the San Andreas fault to be estimated for the past 20,000 years

Questions 53-58; True (T) or False (F): What is wrong with the false statements?

53. ( ) A detailed soil map is helpful in deciding what areas of a land tract would be suitable for construction and which areas should be left in their natural state.

54. ( ) Gradual buildup of salts in soils and of dissolved salts in irrigation effluent present a serious long-term threat to agricultural lands of the San Joaquin Valley, California.

55. ( ) Potential pollutants, such as heavy metals and pesticides, are held in place more tightly in clean sand soils than in clayey sand soils.

56. ( ) Following urbanization of a watershed, sediment yields increase significantly after remaining steady during the active construction phase of development.

57. ( ) The Universal Soil Loss Equation can be used to predict erosional soil losses from a proposed construction site and to suggest conservation practices that reduce sediment yields.

58. ( ) Because they are not motorized, mountain bikes have little environmental impact on back country mountain trails.

59. On the provided map of the U. S., accurately plot and label the following areas, locations, or features.
a. Times Beach, MO   b. Indio Hills near Indio, CA   c. Charlotte, NC
    d. Agua Fria River and Lake Pleasant, near Phoenix, AZ
    e. Lake Michigan shoreline, southern Michigan and northern Indiana
f. Yosemite National Park, g. Mojave Desert Region, h. San Joaquin Valley,
   California   California   California

Environmental Geology Quiz • Chapter 3

1. Which definition of soil is generally used by civil engineering professionals?
a. Soil is an earth material, unconsolidated or to some extent consolidated, that can be excavated, removed, and recontoured without drilling and blasting.
b. Soil is a surface accumulation of weathered rock and mineral particles and organic matter; biochemical activity is generated by soil organisms and plant roots.

Questions 2-5; Matching: Soil Horizons.
   a. A   b. B   c. C   d. R   e. E   f. O

2. ( ) ( ) these two comprise the zone of leaching
3. ( ) light colored zone enriched in clays, iron oxides, and carbonates derived from overlying soil horizons; the zone of accumulation
4. ( ) fresh, unweathered, local parental earth material or bedrock
5. ( ) dark to black surface accumulation of decomposing vegetable matter; humus

6. Which definition describes a hardpan soil?
a. unusually well-compacted and/or well-cemented clay-rich soil; is typically impermeable and prevents or greatly slows downward percolation of water
b. a B horizon soil strongly impregnated with calcium carbonate; forms under semiarid to arid climatic conditions; may require blasting for removal

Questions 7-9; Matching: Dominant Soil-Coloring Agents
a. black, dark gray   b. red, yellow, brown   c. white

7. ( ) iron oxides and hydrated iron oxides   8. ( ) organic matter
    9. ( ) calcium carbonate; typical of well-developed K horizons

10. What are peds?
a. aggregates of soil particles that determine soil structure
b. individual rock, mineral, and organic particles that determine soil texture

11. Which is a sandy loam, according to the U. S. Dept. of Agriculture textural classification of soils?   a. 70 % sand, 15 % silt, 15 % clay
b. 30 % sand, 15 % silt, 55 % clay   c. 45 % sand, 15 % silt, 40 % clay

12. Which description is characteristic of a weakly-developed soil profile
a. soil structure apparent; clays abundant in the B horizon and sparse to absent in the leached zones; B zone is reddened with hematite dustings on soil particles
b. B horizon is thin or absent; A zone rests directly on C and parental material of the R zone may be visible in roadcuts, gullies, and streams

13. Which one is not a major soil nutrient compound?
a. nitrogen, nitrate   b. aluminum   c. potassium   d. phosphorous, phosphate

14. A soil chronosequence study would be based on which soil classification system?
a. the unified classification system used by engineers including categories such as clean gravel and high plastic clay
b. the soil taxonomy system used by soil scientists including categories such as entisols and mollisols

15. Which soil engineering property denotes how easily water will flow through the saturated, soil material?
a. plasticity	b. permeability	c. cohesion	d. erodibility

16. Which engineering soil material would have the highest permeability and lowest compressibility?
a. clean gravel	b. silty clay	c. clayey sand	d. plastic clays

17. Which engineering property denotes the tendency for a soil material to undergo liquefaction when subjected to shaking or vibrations?
a. erodibility	b. permeability	c. sensitivity	d. cohesion

18. Which are the two main factors determining soil strength? **one answer**
a. cohesion, friction	b. permeability, erodibility
c. shrink-swell potential, compressibility	d. sensitivity, corrosion potential

19. Which soil engineering factor bears directly on the process of soil compaction and settlement?
a. plasticity	b. cohesion	c. permeability	d. compressibility

20. Which description correctly characterizes the pore spaces in nonsaturated soil above the water table?
a. filled with liquid water	b. filled with liquid water and soil gases

Questions 21-24; True (T) or False (F)

21. ( ) In general, moist silt and clay-rich soils will be less compressible than clean sands and gravels.

22. ( ) In general, percolation of water would be faster through a clean sand than through an organic-rich clay.

23. ( ) Peat, muck, and histosols are soils composed dominantly of organic matter.

24. ( ) As they lose moisture, expansive soils swell and increase in unit volume.

25. As a generalization, which soils are most suitable for bearing loads associated with heavy structures?
a. poorly drained, clay-rich soils	b. well-drained, sandy or gravely soils

26. Which **one** is not a component term of The Universal Soil Loss Equation?
a. hillslope/gradient factor	b. erosion-control practice factor
c. rainfall/runoff quotient factor	d. soil erodibility index

27. Which **one** is not an expected consequence of clear-cutting sloping forest lands?
a. Soils dry out and compact; potential for slides and other slope failures increases.
b. Interception and evapotranspiration decline drastically.
c. Maximum, daily surface soil temperatures increase.
d. Runoff and sediment yields decrease.

28. Which statement concerning urbanization is reasonably correct?
a. Urbanization decreases runoff during active construction and greatly increases sediment yields after most of the construction is finished.
b. Urbanization greatly increases runoff; sediment yields increase during active construction but decline thereafter.

29. Which best reduces sediment yields from areas around construction sites?
a. protecting all disturbed areas with artificial coverings and grass, especially during the summer thunderstorm season
b. regrading disturbed areas after each storm to fill in erosional rills and gullies formed by the previous storm

Questions 30-32; Matching: Environmental Issues and Locations

a. Times Beach, Missouri   b. San Joaquin Valley, California
c. North Africa, lands bordering the Sahara Desert

30. ( ) severe land degradation resulting from excessive domestic animal populations, excessive groundwater withdrawals, and prolonged natural periods of drought; region in which the term desertification was first widely publicized

31. ( ) severely contaminated with the cancer-causing chemical dioxin

32. ( ) beset with multiple environmental problems related to long-term overuse of groundwater, easily erodible soils, and irrigation

33. At which location have studies of soils and alluvial fan morphology led to estimation of the slip rate on a major fault over the past 20,000 years?
a. Indio Hills, CA; San Andreas fault zone
b. New Madrid, MO; New Madrid fault zone
c. Salt Lake City, UT; Wasatch fault zone   d. Charleston, SC; Charleston fault zone

Questions 34-36; True (T) or False (F): What is wrong with the false statements?

34. ( ) A detailed soil map is helpful in deciding what areas of a land tract would be suitable for construction and which areas should be left in their natural state.

35. ( ) Potential pollutants, such as heavy metals and pesticides, are held in place more tightly in clean sand soils than in clayey sand soils.

36. ( ) For a given area and proposed land use, the Universal Soil Loss Equation can be used to predict erosional soil losses and to suggest conservation practices that reduce sediment yields.

37. On the map provided, accurately plot and label the following areas or locations.
a. Times Beach, MO   b. Indio Hills near Indio, CA   c. Charlotte, NC
d. Yosemite N. P., CA   e. Mojave Desert area, CA   f. San Joaquin Valley, CA

Answers; Home Study Assignment
Soils and Environment • Chapter 3

1. a	2. b	3. d	4. c	5. f
6. a	7. b	8. e	9. a, e	10. b
11. a	12. c	13. b	14. a	15. a
16. b	17. d	18. b	19. a	20. d
21. a	22. c	23. b	24. c	25. a
26. d	27. d	28. b	29. a	30. b
31. a	32. T	33. F*	34. F*	35. T
36. T	37. F*	38. b	39. a	40. b
41. b	42. c	43. a	44. a	45. a
46. a	47. b	48. a	49. c	50. c
51. e	52. d	53. T	54. T	55. F*
56. F*	57. T	58. F*		

*

33. Saturation with water can greatly increase a soil's sensitivity and greatly raise its potential for liquefaction. Normally compacted, unsaturated soils usually exhibit low sensitivities.

34. In general, fine-grained soils rich in silt and clay particles are more compressible than those dominated by sand and gravel. As a soil compresses, pore space between the grains is closed. The rigid, sand and gravel particles resist deformation and pore spaces stay open; under loads, smaller clay grains deform to close up existing pore space and increase material compressibility.

37. Expansive soils swell when wet and shrink when dry.

55. As a general rule, fine-grained particles, such as clay and fine silt, adsorb and hold heavy metals much better than sand-size quartz and other mineral grains in sand. Thus a clayey sand would have a larger capacity for attracting and holding onto pollutants, such as metal ions and pesticides, than a clean sand.

56. Sediment yields skyrocket during the active construction phase, then decrease to low levels after development is completed.

58. Mountain bikes produce narrow, deep ruts which often are enlarged and deepened by runoff flowing along the trails, Thus in especially sensitive areas, mountain bike access and traffic need to be carefully managed and regulated to prevent damaging trail erosion.

Answers: Environmental Geology Quiz • Chapter 3

1. a	2. a, e or e, a	3. b	4. d	5. f
6. a	7. b	8. a	9. c	10. a
11. a	12. b	13. b	14. b	15. b
16. a	17. c	18. a	19. d	20. b
21. F	22. T	23. T	24. F	25. b
26. c	27. d	28. b	29. a	30. c
31. a	32. b	33. a	34. T	35. F*	36. T

35. As a general rule, fine-grained particles, such as clay and fine silt, adsorb and hold heavy metals much better than sand-size quartz and other mineral grains in sand. Thus a clayey sand has a larger capacity for attracting and holding pollutants, such as metal ions and pesticides, than a clean sand.

Commentary/Media Watch

Soils and "Soils"; Pedologic/Physical Characteristics

Make sure that you gets across the differences between soil as used in engineering and natural science senses. Engineering properties of soils (p. 65-69) basically convey information as to how the soil material will respond to excavation, compaction, and reconfiguration of the landscape and how it will behave as a foundation material or in some other load-bearing capacity. Soil evolution, textures, soil profiles, and agricultural or biologic processes require a natural science approach. Some soil properties, such as porosity and permeability, are equally important to both. Abundances of sand, silt, and clay-sized soil grains, porosity or void space, moisture content, and proportion of organic matter have major influences on soil engineering properties and natural soil processes. Moisture content directly controls shrinking/swelling of soils! Water saturation can dramatically "weaken" soils in the engineering sense and radically alter, at least temporarily, the normal biological and biochemical activities in soils. Indeed, wetlands determinations sometimes rest on evidence for regular, or at least occasional, extended periods of water saturation. Increasingly, soil biota are being recognized as having the capacity to break down pollutants such as petroleum products and manmade, toxic organic chemicals. Time and the soil microflora are strong environmental allies! Soils can also effectively immobilize metal ions, but, of course, not destroy them. Thus applications of metal-bearing sewage sludge and animal wastes, for example, must be carefully monitored to prevent overloading and possible metal poisoning of the soil (Commentary/Media Watch, Chapter 13).

The Universal Soil Loss Equation (p. 72) offers a good vehicle for educating students about soil erosion, its mitigation, and land management issues. Novel techniques and/or alternative practices can significantly reduce sediment yields and stream habitat damage from nearby land-disturbing activities associated with urban development, timber harvesting, and farming. On a practical basis, visual evidence for soil erosion is readily seen in newly-cut rills, gullies, and arroyos, and local soil erosion rates can be estimated from exposed tree roots and datable cultural features, such as old fence lines, stone fences, or family burial plots "perched" above the surrounding land surface.

No-Till Farming

Be sure to mention no-till farming in discussions of soils and land management. In many parts of the country, this may encourage some class participation and discussion. No-till is being touted as an everybody wins situation. Farmers save fuel costs and time by not plowing and not harrowing. Erosional losses from wind and running water are greatly reduced for two reasons; bare, disturbed soil is not exposed as it would be with plowing, and "understory cover crops" are commonly planted to protect the soil while the main crop is maturing and after it has been harvested. Soil porosity is maintained or increased by the root masses of both crops and by lessened impacts from heavy machinery. Reduced soil erosion evidently translates directly into reduced fertilizer requirements; nitrogen-fixing cover crops also act to recycle and retain vital nutrients. Increased populations of soil organisms, no longer damaged by plowing and other disturbances, further help to maintain soil porosity and to recycle nutrients.

Be on the lookout for environmental studies dealing with no-till and other, non-traditional, farming methods. No-till could produce significant improvements in farmland productivity and land use stewardship; and reduced soil erosion will eventually reward us with better aquatic habitats and water quality in rivers and streams. Television commercials for Archer Daniels Midland, the agribusiness giant that bills itself "Supermarket to the World" tout no-till farming as a way to reduce fertilizer requirements and "prevent erosion of our precious topsoil"; detailed GPS-based maps are also featured as the high-tech, modern way to manage fertilizer needs and other farming practices. Despite the company's legal problems with the Federal Government regarding its pricing practices, ADM's television commercials show merit for their environmental awareness, emphasis on soil as a valuable natural resource, and use of modern technology to reduce farming costs and improve long-term productivity.

Organic Farming

Many of today's students mirror the general public's interest in physical fitness, healthy lifestyles, and sound nutrition. This trend has stimulated rapid growth and interest in "organic farming". Fresh, visually-appealing, nutritious, organically-grown produce commands premium prices and is commonly included in televisions shows and program clips that feature healthy, upscale, "food preparation" and chefs from well-known hotels and restaurants. Organic farming methods encourage integrated, ecological methods of pest and weed control and thus lower environmental risks from pesticides and herbicides. Natural fertilization and soil conditioning methods reduce soil erosion, improve moisture retention, and reduce fertilizer runoff to streams. Thus organic farming practices contribute directly to improved surface water quality.

In addition to its steady growth in the United States, organic farming has, for better or worse, been forced on the entire Cuban agricultural community by the loss of inexpensive, subsidized chemical fertilizers and pesticides provided in the past by the former Soviet Union and other former, Soviet block countries. The successes, shortcomings, and environmental consequences of this nation-wide "organic-farming experiment" should be of great interest to agricultural experts and environmentalists around the world (see Simon, 1996).

Soil Surveys and Maps

Soil surveys and soil maps are standard fare for federal agencies such as the Soil Conservation Service. Despite utilizing the complicated U. S. Department of Agriculture classification system (see Table 3.1, p. 66 for a generalized summary), these maps are widely used in setting zoning policies and in addressing land use issues; typically, they are available free of cost to the public. These maps and other information on soils and agricultural practices are readily available through local

Soil Conservation Service and U. S. Department of Agriculture offices. Public, land-grant university campuses usually host extensive agricultural research and extension service organizations. Take advantage of them; keep up on the latest trends in crop science and agricultural land use practices.

Sediment Control

Sediment control activities and farming practices may merit coverage in local newspapers. Sedimentation and erosion control programs are in effect in most areas; administration is through state, county, or municipal governments. For example, in Greenville, NC, enforcement of sedimentation regulations concerning road building, housing developments, and drainage ditches is under local city control. Most such programs in North Carolina are under state administration. In this case, city control is favored by local developers and engineering firms. Statewide, major land-disturbing activities such as forestry, agriculture, and mining are still exempt from the state's sediment-control regulations, although farmers are encouraged to voluntarily comply with the state's "best management" practices.

Supplemental References

Bohn, H. L, McNeal, B. L., and O'Connor, G. A., 1985, Soil chemistry, second edition: John Wiley & Sons, New York, NY, 341 p.

Capper, P. L. and Cassie, W. F., 1976, The mechanics of engineering soils, sixth edition: Halsted Press; John Wiley & Sons, New York, NY, 376 p.

Colman, S. M. and Dethier, D. P., editors, 1986, Rates of chemical weathering of rocks and minerals: Academic Press, Inc., Orlando, FL 32887, 603 p.
A collection of 23 papers focused on chemical weathering and rock/mineral dissolution rates under different climatic and hydrologic conditions

Froth, H. D., 1990, Fundamentals of soil science, eighth edition: John Wiley & Sons, New York, NY 10158-0012, 360 p.

Klinkenborg. V., 1995, A farming revolution: National Geographic Magazine, v. 188, no. 6, December, p. 60-89

Martin, I. P. and Chesworth, W., editors, 1992, Weathering, soils, and paleosols: Elsevier, Amsterdam & New York, 617 p.

Nahon, D. B., 1991, Introduction to the petrology of soils and chemical weathering: John Wiley & Sons, Inc., New York, NY, 313 p.

Ollier, C., 1969, Weathering: American Elsevier Publishing Company, Inc., New York, NY, 304 p.

Simon, J., 1996, An organic coup in Cuba? The Amicus Journal, v. 18, no. 4, p. 35-40

White, R. E., 1987, Introduction to the principles and practice of soil science, second edition: Blackwell Scientific Pubs., Oxford, U. K and Boston, U. S. A., 244 p.

Environmental Awareness Inventory
Part 2: Hazardous Earth Processes

1. In the years 1982 and 1983, severe drought and raging brush fires hit eastern Australia, heavy rains and floods struck usually very dry parts of Peru, Bolivia, and Ecuador, and unusually severe winter storms affected the west coast of North America. These climatic events were associated with unusually warm surface waters in the equatorial Pacific Ocean from South America westward. What is the accepted name for these related climatic/marine events? 1 pt
a. a Lahore   b. La Pacifica   c. El Niño   d. a Rayleigh Monsoon

2. Which two were powerful nineteenth century volcanic eruptions in what is now the country of Indonesia? **one answer**, 1 pt
a. Unzen, Pinatubo   b. Vesuvius, Etna   c. Pelée, Trident   d. Tambora, Krakatoa

3. Which Cascade Range volcano has given rise to two catastrophic mudflow episodes in the last 6000 years? Equivalent events today would result in severe property damage and heavy casualties. 1 pt
a. Mt. Whitney   b. Mt. Rainier   c. Mt. Lamington   d. Mt. Shasta

4. In 1986, over 2000 people and numerous domestic animals were asphyxiated by gas suddenly released from a lake in the caldera of an extinct volcano in the central African country of Cameroon. What is the name of the lake and what is believed to have been the major gas emitted in the eruption? 1 pt
a. Nios; carbon dioxide   b. Victoria; hydrogen chloride   c. Nasser; sulfur dioxide
d. Chad; methane

5. Which **one**, in general, characterizes the common style of volcanic activity and eruptions of the Hawaiian volcanoes? 1 pt
a. non-explosive emissions of basaltic lava from vents and fissures; lavas move slowly enough to allow evacuation of areas threatened by the advancing flows
b. highly explosive eruptions of andesitic ash and pumice; fast-moving clouds of hot gases and volcanic particles can devastate an area within a few minutes

6. Which impoverished country on the Indian subcontinent is at high risk from devastating flooding associated with storm surges driven by tropical cyclonic storms in the Bay of Bengal and heavy monsoon rains to the north? 1 pt
a. Vietnam   b. Brunei   c. Oman   d. Bangladesh

7. Which equal length segments of the Atlantic and Gulf coasts of the U. S. have the highest probability of being hit by a hurricane in any one year? 1 pt
a. those from Cape Cod to Long Island   b. those on the east coast of South Florida
c. those along the northern Florida and Georgia coast
d. those from central New Jersey to the Virginia-North Carolina state line

8. Which scale and range is used to describe damage to buildings and other infrastructure due to earthquake shaking?
a. Saffir-Simpson; 1-5   b. Richter; 1-9   c. Mercalli; I-XII   d. Jordan-Rodman; 1-100

9. Which well-known, well-studied, well-publicized, prehistoric landslide complex began renewed movements in 1956? Extensive mitigation efforts apparently stabilized the mass, at least temporarily, since 1985. 1 pt
a. Portuguese Bend, southern California   b. Winter Park, Winter Park, Florida
c. Lake Peigneur, southern Louisiana   d. Muddy Creek, west-central Colorado

10. Which area of the United States experienced three powerful earthquakes and numerous, smaller aftershocks in the winter months of 1811 and 1812? Such events today would result in severe property damage and loss of life. 1 pt
a. San Francisco-Oakland-Silicon Valley area, northern California
b. New Madrid area, southeastern Missouri, western Tennessee, and northeastern Arkansas
c. Charleston area, coastal South Carolina
d. Wasatch Mountain front area from Ogden to Salt Lake City to Provo, Utah

11. What's the quiet, untold, story gradually unfolding in the lower Mississippi River Valley?
a. Finally, construction of numerous dams, floodways, and levees has insured flood protection for all but the most floodprone parts of the major urban areas such as Memphis, Baton Rouge, and New Orleans.
b. The slow, natural diversion of the Mississippi into the Atchafalaya-Red River system has been "stabilized" by the Corps of Engineers, but a severe, lower basin flood could threaten these stabilization efforts.

12. What is a tsunami?
a. explosive, fast-moving, destructive cloud of hot volcanic particles and gases
b. destructive, devastating, "unannounced" flood caused by sudden release of water melted by volcanic activity beneath a glacier or small ice sheet
c. long wavelength ocean surface wave generated by sudden displacement of large volumes of water along a coastal zone; seafloor earthquakes are the most common cause
d. huge submarine landslides known to have broken away from Hawaii and other active volcanic islands within the past 200,000 years

Answers

| 1. c | 2. d | 3. b | 4. a | 5. a | 6. d |
| 7. b | 8. c | 9. a | 10. b | 11. b | 12. c |

Performance Evaluation; Scoring

Each correct answer is worth one point; add up your total. Now be honest, and subtract one point for each lucky guess!

Score	**Awareness Ranking; advice**	INTERNET Address
9-12	**nerd**; keep up the good work, don't let up	http://envnerd.com
6-9	**blue collar achiever**; earn an A with hard work	http://envblcol.com
3-6	**slowpoke**; don't look back, the clueless are gaining	http://envslpoke.com
0-3	**clueless**; extra hard work makes up for a slow start	http://envcluless.com

Natural Hazards: An Overview • Chapter 4; Home Study Assignment

1. Which statement better characterizes the magnitude-frequency concept (p. 90) of natural hazards and resulting property damage?
a. In general, the more powerful the natural event, the lower its recurrence interval. Although impressive in its effects, the rare, very powerful magnitude natural hazard events seldom produce any significant, long-lasting consequences.
b. In general, lower-magnitude, less powerful, natural hazard events occur more often than higher-magnitude events. Thus in some parts of the United States, seasonal, short-term, low-energy processes such as expanding soils and frost heaving account for more property damage during a given year than earthquakes.

2. In the U. S., which result in costly damages each year without having much potential for turning catastrophic? See Table 4.1.
a. hurricanes   b. expansive soils   c. floods   d. coastal erosion

   Questions 3-5; Matching: Natural Hazards and Long-Term Consequences
a. boundary between North American and Pacific Plates, California
b. Nile delta lands, Egypt   c. Island of Hawaii

3. ( ) Prior to closing of the Aswan High Dam, annual flooding replenished the rich agricultural soils and counteracted land losses from coastal erosion.
4. ( ) Local bedrock, mechanically abraded to clay-size gouge, obstructs groundwater flow, resulting in local, shallow, groundwater resources.
5. ( ) Lava flows add new land to counteract losses by coastal landslides and erosion.

6. Which **two** can be forecast with the higher degrees of accuracy and certainty?
a. time and location of an eruption in an active volcanic center such as the Mammoth Lakes area, CA
b. arrival times and maximum expected water surface elevations of a flood moving downstream on a major river such as the Mississippi
c. arrival time in the Hawaiian Islands of a tsunami that originates during an earthquake along the coastal regions of Alaska
d. arrival time, landfall location, and maximum wind speeds expected for a hurricane still a few hundred miles "out to sea"

7. Which philosophy or attitude is the better one for scientists and public officials involved in conveying concerns and warnings of impending natural hazards to the press and the public?
a. Inaccurate, sensationalized media accounts will probably trigger irrationality and panic; thus to reduce potential damage and casualties, public and private organizations ought to initiate quiet preparations for the possible event while minimizing the need for public vigilance and preparedness.
b. A well-informed media and public encourage preparation and readiness for dealing with consequences of likely natural hazards. Advance planning and training enhance a community's ability to respond appropriately, quickly, and efficiently to such events.

Questions 8-9; Matching: Disaster Recovery Responses, Events, and Localities
a. Anchorage, Alaska, 1964 earthquake   b. Rapid City, South Dakota, 1972 flood

8. ( ) A rapid influx of federal and insurance company funds shortened the time period for restoration and reconstruction without much concern for mitigating effects from a similar event in the future.

9. ( ) land-use planning integrated with patient restoration and reconstruction activities have greatly reduced the anticipated dollar costs and causalities expected from a similar event in the future.

Questions 10-13; Do you agree (A) or disagree (D) with the following statements? For the "disagree" statements, suggest a better, more responsible course of action.

10. ( ) Following natural disasters such as floods, fires, and earthquakes, federal disaster funds should be poured into the effected area immediately to shorten the restoration and reconstruction phases of recovery. Otherwise valuable economic assets will be underutilized or lost.

11. ( ) The federal government should not provide public funds for restoration and reconstruction following natural disasters. This responsibility should be left to private companies and local and state governments. Why should Californians pay for flood damages in easily identified, flood-prone areas of Missouri and Illinois and why should midwesterners pay for earthquake damage in Los Angeles and other known earthquake prone areas in California? Such a policy would greatly reduce damages from natural hazards by encouraging appropriate, local, anticipatory responses, controls on development, and land-use planning.

12. ( ) The federal government should provide low-cost insurance in categories of natural hazards, such as coastal areas subjected to repeated storm damages and flooding, where known risks are too high for private companies to profitably market hazard insurance. Without such insurance, entrepreneurial real estate developments could be slowed or stopped entirely, resulting in large potential losses for the local economy.

13. ( ) Channelization, levees, and dams guarantee minimum risks of flooding, especially from rare, high-magnitude floods. Thus artificial flood control projects eventually pay for themselves in the form of increased land and real estate values in areas formerly vulnerable to repeated flooding.

14. Which is the one incorrect statement concerning the Mexico City area?
a. High population density increases its vulnerability for casualties and damage from earthquake shaking.
b. Earthquakes along Mexico's west coast pose little groundshaking hazard to Mexico City.
c. Many buildings are already tilted and or structurally weakened due to land subsidence associated with excessive withdrawals of groundwater.
d. The city is built on water-bearing lake beds; shaking effects from distant earthquakes may be greatly amplified.

15. Which person would probably be better informed concerning the given local hazard? Why?
a. home owner, Santa Monica, CA; brush/forest fire risks
b. blackjack dealer, Las Vegas, NV; earthquake risks

16. Which one is not associated with strong El Niño events? See Figure 4.8.
a. droughts in southern Africa and eastern Australia
b. heavy rains in normally very dry areas of Bolivia, Peru, and Ecuador
c. unusually warm, equatorial surface waters in the Atlantic Ocean
d. temporarily weakened eastern Pacific trade winds

17. Which two statements concerning hazards associated with the volcano Nevado del Ruiz, Colombia, are correct?
a. An 1845 mudflow resulted in thousands of fatalities; subsequently, the townsite of Armero was moved to a safe location well above the Lagunilla River floodplain.
b. Improved forecasting, monitoring, and timely warnings limited casualties from the 1985 mudflows to a few unlucky people working in agricultural fields along the Lagunilla River.
c. The townsite of Armero, on the floodplain of the Lagunilla River, was developed around the rich agricultural soils laid down by the 1845 mudflow.
d. The 1985 mudflow event showed that in the absence of prudent planning related to a well-known natural hazard, increased population density translated directly into increased fatalities.

18. On the provided world map, accurately plot and label the following countries, features, or locations.

a. Bangladesh	b. Kobe, Japan	c. Hawaiian Islands
d. Mount Rainier, WA	e. Rapid City, SD	f. Northridge, CA
g. Peru	h. Colombia	j. eastern Australia
	i. Mexico City, Valley of Mexico	

Environmental Geology Quiz • Chapter 4

1. Which statement better characterizes the magnitude-frequency concept of natural hazards and resulting property damage?
a. In general, the more powerful the natural event, the longer its recurrence interval. These rare, very powerful, natural hazard events seldom produce any long-lasting consequences.
b. In general, lower-magnitude, less powerful, natural hazard events occur more often than higher-magnitude events. Thus in many parts of the United States, expansive soils and frost heaving can account for more property damage during a given year than earthquakes or floods.

Questions 2-3; Matching: Natural Hazards and Long-Term Consequences
a. Island of Hawaii   b. northern Egypt

2. ( ) Lava flows add new land to counteract losses by coastal landslides and erosion.

3. ( ) Annual flooding used to replenished rich agricultural soils and counteract losses of land to coastal erosion.

4. Which **one** can be forecast with the highest degrees of accuracy and certainty?
a. arrival time, landfall location, and maximum wind speeds expected for a hurricane still a few hundred miles "out to sea"
b. arrival time in the Hawaiian Islands of a tsunami that originates during an earthquake along the coastal regions of Alaska
c. time and location of an eruption in an active volcanic center such as the Mammoth Lakes area, CA

5. Which **one** best characterized financial and environmental responses to the Rapid City, SD, flash floods of 1972?
a. Land-use planning integrated with patient restoration and reconstruction activities have greatly reduced the anticipated dollar costs and causalities expected from a similar event in the future.
b. A rapid influx of federal and insurance company funds shortened the time period for restoration and reconstruction without much concern for mitigating effects from a similar event in the future.

6. Which **one** statement concerning Mexico City and the Valley of Mexico is true?
a. A low-density, dispersed population decreases the area's vulnerability to earthquake shaking and volcanic eruptions.
b. Distant earthquakes off Mexico's west coast pose little threat to the area.
c. The city is built on water-bearing lake beds; thus earthquake shaking effects may be greatly amplified.

7. Which **two** are **not** associated with strong El Niño events?
a. droughts in southern Africa and eastern Australia
b. heavy rains in normally very dry areas of Bolivia, Peru, and Ecuador
c. unusually cool, equatorial surface waters in the Pacific off Ecuador and Peru
d. strengthened southeast trade winds in the North Atlantic

8. Which **one** statement concerning Amero and Nevado del Ruiz, Colombia, is correct?
a. The 1985 floods and mudflows were caused by catastrophic failure of a flood-control dam upstream from the village of Amero.
b. Improved forecasting, monitoring, and timely warnings limited casualties from the 1985 mudflows and floods to a few unlucky people working in agricultural fields along creeks flowing off Nevado del Ruiz.
c. Armero, on the floodplain of the Lagunilla River, was developed around rich agricultural soils laid down by volcanic mudflows from Nevado del Ruiz in 1845.

Questions 9-13; Do you agree (A) or disagree (D) with the following statements? For the "disagree" statements, suggest a better, more responsible course of action.

9. ( ) Official concerns and warnings of possibly impending natural hazards usually trigger inaccurate, sensationalized press and media coverage and substantial economic losses in the affected communities. Thus public and private organizations ought to initiate quiet preparations for the possible event while assuring the public that everything is fine.

10. ( ) The federal government should provide low-cost insurance in categories of natural hazards, such as coastal areas subjected to repeated damaging storms and floods, where known risks are too high for private companies to profitably market hazard insurance. Immediately after these events, federal funds should be poured into the affected area to shorten the restoration and reconstruction phases of recovery.

11. On the world map, plot and label the following countries, areas, or locations.
a. Bangladesh   b. Kobe, Japan   c. Colombia
d. Mount Rainier, WA   e. Peru   f. Island of Hawaii

Answers; Home Study Assignment
Natural Hazards: An Overview • Chapter 4

1. b	2. b, d	3. b	4. a	5. c
6. b, c	7. b	8. a	9. b	10. A or D*
11. A or D*	12. D*	13. D*	14. b	15. a*
16. c	17. c, d			

*
10. In this statement, the short-term economic viewpoint is overstated at the expense of prudent, long-term planning and common sense. Rapid recovery is a fine goal, but not if the process involves no consideration of how to prevent casualties and reduce damage costs from a similar event in the future.

11. This is largely a political and social issue. It's hard to argue in favor of public policies that encourage risky investments in disaster-prone areas. On the other hand, it's hard to argue that state and federal governments should leave disaster victims totally on their own with a self-righteous admonition that they should never have been in such an unsafe area in the first place. What's needed is a compassionate program without cost subsidies designed to 1) help disaster victims and communities when needed; and 2) to encourage land-use decisions, construction practices, and public works investments that will reduce damages and casualties should a potential disaster, anticipated or unforeseen, become a reality.

12. The federal government probably shouldn't provide risky, coastal flood insurance in areas where the damage risks are so high that the insurance amounts to a taxpayer financed subsidy to individuals who build in dangerous areas. Real estate developers certainly don't deserve any special consideration just because they risk losing their development to a coastal storm before the individual units are sold. Once sold, the flood liability rests with the new owners.

13. The most controversial issue in this statement is the notion that flood-control structures are justified costwise by increased property values in flood-prone areas. The flood control structures perform well for smaller, more frequent, flood events but can be overwhelmed with disastrous results by the rarer, exceptionally large-magnitude floods.

15. The Santa Monica homeowner probably has experienced at least one serious wildfire event and knows that the risks rise with vegetation mass and dry, hot weather. The Vegas blackjack dealer may not even be aware of the risks posed by much less frequent seismic events in the Las Vegas area.

Answers: Environmental Geology Quiz • Chapter 4

| 1. b | 2. a | 3. b | 4. b | 5. a |
| 6. c | 7. c, d | 8. c | 9. A or D* | 10. A or D* |

*
9. Official concerns and warnings of potential disasters may, in fact, produce some inaccurate and sensationalized media coverage and economic losses. However, using this as justification for withholding vital information from the public is not acceptable behavior and cannot be tolerated in a free society. A well-informed, well-educated, and well-prepared public is light-years ahead of one that is ignorant of, unaware of, and unprepared for some potential natural disaster.

10. This is largely a political and social issue. It's hard to argue in favor of public policies that encourage risky investments in disaster-prone areas. On the other hand, it's hard to argue that state and federal governments should leave disaster victims totally on their own with a self-righteous admonition that they should never have been in such an unsafe area in the first place. What's needed is a compassionate program without cost subsidies designed to 1) help disaster victims and communities when needed; and 2) to encourage land-use decisions, construction practices, and public works investments that will reduce damages and casualties should a potential disaster, anticipated or unforeseen, become a reality.

In the second part of the statement, short-term, economic considerations are overstated at the expense of prudent, long-term planning and common sense. Rapid recovery is a fine goal, but not if the process involves no consideration of how to prevent casualties and reduce damage costs from a similar event in the future.

Commentary/Media Watch

Ideas and Notions

Chapter 4 is a short chapter that introduces students to natural hazards. In geological and ecological senses, natural disasters and calamities are "bad news-good news" situations. Short intervals of drastic physical and ecological disruption are followed by gradual recovery, generation of new habitats, and repopulation. Humans may or may not be part of the scene. Recurrence intervals, times between successive, similar events in the same area, and prediction accuracy vary widely among different events and from area to area.

Some, like summer monsoon flooding on the Indian subcontinent, come at regular, one-year intervals and arrival times are fairly predictable, especially once the rains start falling. Relatively rare, high magnitude flooding events such as the hundred-year flood on a river or stream are notable for their lack of predictability; on the basis of extended, hydrologic records, we can estimate the discharge and stage of the 100-year flood and feel confident, in a statistical sense, that such an event will come once every hundred years. Its arrival time, however, is not predictable; weather-related events such as floods, droughts, and hurricanes are more or less random events and the expected recurrence interval has little relevance to event prediction.

On the other hand, regional seismicity or eruptions of a specific volcano may exhibit historically- or scientifically-documented recurrence intervals that help with evaluation of possible future activity. Evaluation and predictability of natural risks are important facets of environmental geology. A well-educated, informed public will be a prepared public. In stark contrast, indifference and ignorance translate directly into lack of preparation and increased property damage and casualties. The tragic story of the volcano Nevado del Ruiz and the town of Armero, Colombia, 1985 (p. 98 and Chapter 8), demonstrate the need for competent risk assessment and preparedness.

Areas subjected to essentially randomly timed, hazardous events can easily be identified in many cases. Lands on river floodplains will be flooded sooner or later; mudflows will ravage stream and river valleys on the lower flanks of stratovolcanoes such as Mt. Rainier and Nevado del Ruiz; earthquakes will strike along active fault zones; and tropical storms will strike the Atlantic and Gulf Coasts. To varying degrees of certainty, we know where the disasters will strike; we just don't know when. Once in motion, the aforementioned events do however, possess lead times. The arrival of a flood crest or storm surge can be established from a few hours to a few days in advance. Anomalous strain rates, increased radon levels in wells, and changes in bedrock geophysical parameters may indicate an impending earthquake. Advance warning time is precious; it can be devoted to preparation, and preparation reduces damages and casualties. Unannounced, sudden events present the much greater threat.

Public Behavior and Educating the Public

To the extent that scientific studies assisted by modern technology can improve our ability to forecast natural hazards, public education, preparedness, and common sense land use regulations and building codes can greatly reduce damage costs and casualties. In addition to scientific and technological content, natural hazards forecasting and damage mitigation have very significant social, political, and economic components. Thus we in the earth sciences need to communicate willingly and effectively with the public and with responsible officials in public agencies and private enterprise.

Satellite observations and the nationwide Doppler radar network have greatly improved weather forecasting, and increasingly, the American public is better prepared for potentially hazardous, weather-related events such as tornadoes, hurricanes, and flash floods. On-the-spot coverage of hurricanes and other potentially dangerous storms and documentaries of past, severe, storm events on The Weather Channel are routinely accompanied by commentary from expert meteorologists. Although flash flood victims still profess amazement at how the flood waters seem to come suddenly out of nowhere, flash flood warnings based on Doppler radar rainfall rates and cumulative totals are routinely given hours in advance. River stages and flood peak arrival times of larger-scale, lower basin flooding events are predicted well in advance, providing valuable time for reducing property damage and saving lives. In less affluent countries, inadequate forecasting capabilities and communications and transportation networks make it more difficult to warn people of impending danger and to have them moved out of harm's way. For example in Bangladesh, even if public authorities know a powerful tropical cyclone is approaching, little can be done to reduce damage and casualties.

In more affluent countries, advanced seismic networks, scientific studies of past seismic activity, up-to-date land use zoning regulations, and strict building codes drastically reduce earthquake damages and casualties. Experience gained from previous earthquakes is constantly being used to upgrade building codes, to recommend ways to strengthen older buildings and infrastructure, and to increase overall earthquake preparedness. These activities all pay big dividends in lowered damage costs and lower casualties. Collapsed buildings are mostly to blame for the appalling death tolls often associated with moderate magnitude earthquakes in less affluent countries. Relatively new, modern, top-heavy, concrete-slab apartment buildings in Russia and Mexico and traditional stone and mud homes in North Africa, Iran, and India are equally dangerous during earthquakes.

Supplementary References

Multiple Authors, 1994, Environmental risks and hazards: Cutter, S., editor, Prentice Hall, Upper Saddle River, NJ, 413 p.

This book is a collection of articles plus supplementary educational materials that examine natural hazards from the diverse perspectives of different social sciences.

Siegel, F. R., 1996, Natural and anthropogenic hazards in development planning: Academic Press, Inc., Orlando, FL and London, U. K., 200 p.

Rivers and Flooding • Chapter 5; Home Study Assignment

1. Which is the **one** incorrect statement concerning sediments carried by streams and rivers?
a. Common constituents of the dissolved sediment load are bicarbonate (HCO_3^{1-}), sulfate (SO_4^{2-}), and calcium (Ca^{2+}) ions.
b. Bed load usually exceeds suspended load in quantity; bed load sediments include mainly sand and gravel.
c. Suspended load particles include clays and silt; turbulence in the moving water keeps the sediment particles suspended.
d. Bed load particles move by rolling, bouncing, and skipping along the channel bottom.

Questions 2-7; True (T) or False (F): Terms Related to Streams and Rivers

2. ( ) A stream's drainage basin is the total land area contributing runoff to the stream. Watershed is another term for drainage basin.

3. ( ) Stream power depends directly on the area of the stream's drainage basin.

4. ( ) Longitudinal stream profiles usually show steeper gradients in upstream channels and lower, less steep gradients in downstream locations where the stream is approaching its base level.

5. ( ) A stream's gradient is the vertical drop in elevation between two points on the stream surface divided by the distance the water flows between the same two points.

6. ( ) Streams have lower slopes or gradients than longitudinal, downvalley, land surface profiles of their associated valleys.

7. ( ) Between successive flood events, streams and rivers with wide, well-developed floodplains are vigorously downcutting and deepening their channels.

8. What is meant by the term **reach** in regard to rivers?
a. simple; the distance you walk until you reach the river
b. describes a specific length or segment of a stream
c. refers to the distance a river flows from its headwaters to its mouth
d. refers to the drainage basin that supplies runoff and sediment to a river

9. Which **one** defines the average velocity of water flowing in a stream channel?
a. alluvial capacity divided by the stream competency
b. stream's gradient times its average depth   c. stream's capacity times its gradient
d. stream's discharge divided by its cross-sectional area

10. Which statement is essentially correct?
a. Alluvial fans are deposited by highly competent, ephemeral streams where gradients decrease abruptly, usually at the mouths of canyons that drain steep, mountainous terrain.
b. As low-competency, high gradient streams approach sea level or a lake that functions as a temporary base level, they slow down and deposit their dissolved loads and suspended loads of sand and gravel in the form of a delta.

11. In general, how do streams respond to locally increased sediment loads?
a. Initially, deposition of sediments raises the elevation of the channel bottom and steepens the local gradient, raising the stream's capacity enough to transport the additional mass of sediment.
b. The additional sediment load raises the stream's capacity and stream power, thus promoting erosion of the stream's banks and bottom.

12. Which land-use change will normally be associated with locally increased sediments yields in a drainage basin?
a. conversion of forest land to agriculture   b. reforestation of agricultural lands

13. Historical and scientific studies show that a river once flowed over a bedrock channel, then deposited 5 meters of sediment in its channel, and has since downcut about 2 meters into its channel bottom sediments. Which land-use scenario fits best with the observed stream history? Study Figure 5.5, p. 104.
a. Initially the basin was forested; the area was clear-cut for timber, then converted to corn and soybean production.
b. Initially, the basin was forested; the area was cleared and converted to agriculture, then converted back to forests.

Questions 14-18; Matching: Channel and Sediment Load Characteristics
a. braided channel   b. meandering channel

14. ( ) characterizes a stream with excessively large quantities of coarse bed load, such as one flowing away from the snout of a glacier
15. ( ) gradients are steeper, channels are wider and shallower
16. ( ) gradients are less steep; channels are sinuous, deeper, and not so wide
17. ( ) stream characterized by numerous smaller channel segments that divide and recombine in a complex pattern
18. ( ) bed load is sand; typically deposited as point bars

19. Which **one** correctly describes how a stream's sediment budget responds to construction and filling of a dam and reservoir? Study Figure 5.6, p. 105.
a. Sediments are deposited in the reservoir and on the channel bottom upstream from the reservoir; bottom erosion and downcutting affect the channel immediately downstream from the dam.
b. Sediments are deposited in the reservoir and downstream from the dam, raising the elevation of the channel bottom. Bottom erosion and scouring lower the elevation of the channel bottom upstream from the reservoir.

Questions 20-23; True (T) or False (F): See Figures 5.8 and 5.9, p. 106 and 107.

20. ( ) Meander scrolls on a floodplain mark past locations of river channels.
21. ( ) Lateral bank erosion and maximum depths occur in straight channel reaches associated with riffles.
22. ( ) Old point bars and oxbow lakes represent former locations of river channels.
23. ( ) Lateral migration of a meandering channel is accomplished by deposition of point bars and lateral erosion of the bank opposite the point bar.

24. Refer to Figure 5.12c, p. 108. What discharge corresponds to a stage of 1.0 meter?
a. 10 cubic meters/sec   b. 20 cubic meters/sec   c. 50 cubic meters/sec

25. Which **two** responses correctly characterize the pool and riffle channel pattern?
a. Different depths and velocities associated with pool and riffle channel patterns provide ecological diversity in stream habitats.
b. Pools tend to be scoured and deepened during floods.
c. Riffles usually have finer size channel-bottom sediments than associated pools.
d. At low discharges, average water velocities and turbulence in the pools exceed those of water moving through the riffles.

Questions 26-30; Matching: Physical Units and Hydrologic Quantities

26. ( ) feet or meters   a. stage
27. ( ) square feet or square meters   b. discharge
28. ( ) feet/mile or meters/kilometer   c. water velocity
29. ( ) feet/sec or meters/sec   d. cross-sectional area of a stream
30. ( ) cubic feet/sec or cubic meters/sec   e. stream gradient

31. Which is the correct relationship to determine how often a given magnitude flood is expected to occur?
a. recurrence interval = (number of years of records + 1)/rank of the discharge
b. recurrence interval = (rank of the discharge + 1)/number of years of records

32. Which is the correct interpretation of the hundred-year flood?
a. The corresponding discharge is expected to occur once during a hundred-year interval; the probability of this discharge occurring in any one year is 1/100.
b. The discharge corresponding to the 100-year flood is 100 times that associated with the average yearly flood-stage discharge.

Questions 33-38; Matching: Disastrous Flood Events

a. Front Range, Colorado   b. West Virginia   c. Las Vegas, Nevada   d. eastern Ohio
e. Tucson, Arizona   f. coastal southern California, north of Los Angeles

33. ( ) In 1995 and 1992, flood waters swept through an RV park built on a normally-dry distributary channel of the Ventura River. Historical records show numerous, severe flooding events associated with this location in the past.

34. ( ) In 1983 and again in 1993, severe bank erosion and channel migration caused extensive property damage to homes and other structures built on previously undeveloped land along Rillito Creek.

35. ( ) Following 14 cm (5.5 inches) of rainfall in less than 4 hours, sudden failure of debris dams piled-up behind bridges on Pipe and Wegee Creeks unleashed 5 meter-high walls of water that devastated downstream areas.

36. ( ) City and county officials bet heavily on storm-water retention basins, artificial channels, pipelines, and culverts to reduce flooding damages from heavy, local, summer thunderstorms.

37. ( ) Failure of a coal-waste dam on Buffalo Creek unleashed a flood wave with a discharge equivalent to 40 times that of the 50-year flood.

38. ( ) In the summer of 1976, severe flash floods unleashed by unusually heavy, local thunderstorms resulted in more than one hundred fatalities and severe damage to homes, bridges, and roads along Big Thompson Canyon.

39. Which statements correctly describe passage of the peak flood discharge down the Chattooga and Savannah Rivers between Clayton, GA, Calhoun Falls, SC, and Clyo, GA? Refer to Figure 5.15, p. 110.
a. The peak discharge at Clyo was about 2000 cubic meters/second.
b. The river crested at Clyo 3 days after cresting at Calhoun Falls.
c. At Calhoun Falls, the river receded to normal levels on day 13 after cresting.
d. At all three localities, discharge declined from peak flow to normal flow conditions in less time than was required for the stage to rise from normal flow to peak discharge conditions.
e. Average downstream speed of the flood peak between Clayton and Clyo was about 60 kilometers per day.

40. Which **two** statements correctly describe the hydrologic characteristics of an urbanized drainage basin as compared to an equivalent natural drainage basin?
a. Impervious land cover in an urbanized basin results in increased runoff and less infiltration for a given storm event.
b. In extensively urbanized basins, peak discharges associated with short recurrence interval storm events (2 to 10 years) are significantly raised.
c. Extensive urbanization results in longer lag times between storm precipitation and peak discharges.
d. Urbanized streams generally have more appealing aesthetic qualities and more diverse aquatic habitats than natural streams.

41. Which **one** statement is false? Refer to Figures 5.16, 5.17, and 5.18, p. 110-112.
a. With 40 % impervious cover and area served by storm sewers, bankfull discharge events on the main stream in a small drainage basin can be expected to occur roughly 3 times more frequently than before urbanization.
b. Peak discharges associated with the 2.3 year recurrence interval storm in a natural basin increase rapidly with the first 20 % of impervious cover and area served by storms sewer. Additional impervious cover and storm sewers in the basin have little effect on peak discharges associated with this same storm event.
c. Peak discharges associated with 50-year and longer recurrence interval storms are relatively insensitive to percentages of impervious cover in the drainage basins.

42. Which **one** correctly describes the hydrologic effects of a well-managed, stormwater retention pond? See Figure 5.22, p. 116.
a. decreases lag times and increases peak downstream discharges
b. increases lag times and reduces peak downstream discharges

Question 43-44; Matching: Contrasting Responses to Flood Hazards
a. prevention   b. adjustment

43. ( ) Land use activities in flood prone areas are regulated to limit damage costs from future flooding events; is especially important in reducing damages expected from very rare, long recurrence interval flooding events.

44. ( ) Hard structures such as dams, levees, floodwalls, and diversion channels are constructed to eliminate flooding in some otherwise floodprone areas. A false sense of security and unrestrained development may greatly inflate damage costs incurred by rare, very severe, long recurrence interval flooding events.

45. Which explanation correctly describes the long-term natural hazard associated with the lower Mississippi River and Mississippi Delta?
a. Diversion of the Mississippi into the Red-Atchafalaya River system could happen during a major flood on the lower Mississippi. If so, the present day channel from Baton Rouge to New Orleans and the delta lands would be abandoned.
b. Diversion of the Red-Atchafalaya River system into the present-day Mississippi River during a major, lower Red River flood event would increase discharges and flood hazards for Baton Rouge, New Orleans, and the delta lands.

46. A specific reach of the naturally meandering Blackwater River in Missouri was channelized in 1910. Which two statements accurately characterize the effects of this action. See the *case history*, p. 126.
a. The modified, straightened stream channel has a lower gradient than the meandering, former, natural channel.
b. Bank erosion and channel widening, initiated following channelization, evidently are still continuing today.
c. Flooding frequencies along the modified channel have decreased; flooding frequencies along the downstream, unmodified, natural channel have increased.
d. The channelized reach supports larger fish populations and more diverse aquatic environments than the natural, unmodified reaches of the river.

47. What is a pilot channel?
a. a manmade meandering channel simulating a natural stream that is incorporated into a much larger, floodway channel project designed to protect against the 100-year flood event
b. a smooth, concrete-lined stream channel designed to guide storm runoff discharges downstream without attendant flooding

48. Which two responses describe environmentally sound aspects of urban stream channel restoration projects?
a. In designed meanders, riprap can be used to stabilize the higher, steeper banks of an asymmetrical channel while sand or gravel bars develop along the opposite, shallower side of the channel.
b. Removing trees from a floodplain spurs growth of aquatic organisms and enhances aquatic habitats because more sunlight can now strike the stream bottom.
c. Straight smooth channels and open ditches provide for more aesthetically pleasing landscapes than natural streams because their geometric regularity bears witness to Man's superior capacity for rationality and reason.
d. Meandering channels with pool and riffle patterns provide better aquatic habits than straight channels without pools, especially during periods of low discharge.

49. Which river, associated with the Lake Okeechobee-Everglades National Park, Florida hydrologic system is presently being "restored" to a simulated natural state after having been "straightened" for land drainage and flood control?
a. Atchafalaya   b. Savannah   c. Kissimmee   d. Suwannee

Questions 50-53; True (T) or False (F): Mississippi/Missouri River Flooding Issues

50. ( ) Reservoirs constructed on the Missouri River in Montana and the Dakotas allowed for storage of high discharges associated with the 1973 and 1993 flood events and resulted in reduced, downstream peak discharges and flood damage.

51. ( ) Long, extensive levee systems effectively narrowed portions of the Mississippi and Missouri River floodplains, and raised peak water levels associated with floods of given discharges.

52. ( ) Levees constructed by federal agencies failed far more often than those built and maintained by private organizations, showing again that big government rarely competes with private enterprise in cost efficiency and quality.

53. ( ) The 1993 upper Mississippi Basin floods were unusually severe but of short duration; they resulted from a week-long storm system that stalled over the upper Midwest and melting of an extensive snowpack in the same area.

54. The table lists maximum discharges for each of ten consecutive years for Fictitious River, a crystal-clear, fish-filled, whitewater river known from the dreams of rafters, kayakers, and fly fisherman. Complete the rank order and recurrence interval columns. Refer to the *special feature* Magnitude and Frequency of Floods, p. 108-109.

Year	Maximum Discharge; cubic feet/sec	Rank Order	Recurrence Interval; yrs
1	660		
2	1700		
3	2050		
4	1200		
5	900		
6	740		
7	1500		
8	800		
9	1050		
10	520		

55. Graph the data on probability paper; estimate the 50-year flood discharge.

56. What is the probability of a 50-year flood occuring in any 25-year interval?

57. On the U. S. map provided, plot and label these areas, locations, or features.
   a. Upper Mississippi River Valley, St. Louis, MO to Minneapolis, MN
b. mouth of Ventura River, Pacific Coast, CA   c. Savannah River, SC and GA
d. Blackwater River, west-central Missouri   e. Mississippi delta lands, LA
f. Las Vegas, NV   g. Tucson, AZ   h. Kissimmee River, FL
   i. Austin, TX, and the basin of the Colorado River, Texas

Environmental Geology Quiz • Chapter 5

1. Which is the **one** incorrect statement concerning sediments carried by streams?
a. Suspended load particles include clays, silt, and fine-size sand; turbulence in the moving water keeps the sediment particles suspended.
b. Bed load usually exceeds suspended load in quantity; bed load sediments include mainly sand and gravel.
c. Common constituents of the dissolved sediment load are bicarbonate (HCO_3^{1-}), sulfate (SO_4^{2-}), and calcium (Ca^{2+}) ions.

Questions 2-6; True (T) or False (F)

2. ( ) A stream's drainage basin is the total land area contributing runoff to the stream. Watershed is another term for drainage basin.
3. ( ) Stream gradient is the vertical drop in elevation between two points on the stream surface divided by the distance the water flows between the two points.
4. ( ) Lateral bank erosion and maximum depths occur in straight channel reaches associated with riffles.
5. ( ) Old point bars, oxbow lakes, and meander scrolls represent former locations of braided river channels.
6. ( ) Between successive flood events, streams with wide, well-developed floodplains are vigorously downcutting and deepening their channels.

7. Which statement is essentially correct?
a. Alluvial fans are deposited by competent, ephemeral streams where gradients decrease abruptly, usually at mouths of canyons that drain mountainous terrain.
b. As low-competency, high gradient streams approach sea level or a lake, they slow down and deposit their dissolved and suspended loads in the form of a delta.

8. Assume that detailed field studies show that a river once flowed in a bedrock channel, then deposited 5 meters of sediment, and has since downcut about 2 meters into the sediments. Which land-use scenario best fits this history?
a. the basin was forested; cleared and used for agriculture; converted back to forests
b. the basin was forested; clear-cut for timber; converted to agriculture

Questions 9-11; Matching; Stream Channels   a. braided   b. meandering

9. ( ) characterizes a stream with excessively large quantities of coarse bed load, such as one flowing away from the snout of a glacier
10. ( ) gradients are lower; channels are sinuous, deeper, and not so wide
11. ( ) stream characterized by numerous smaller channel segments that divide and recombine in a complex pattern

12. Which **one** correctly describes how a stream's sediment budget responds to construction and filling of a dam and reservoir?
a. deposition in the reservoir and upstream channel; erosion and downcutting of the channel immediately downstream from the dam
b. deposition in the reservoir and channel downstream from the dam; bottom erosion and scouring of the channel upstream from the reservoir

13. Which is the correct relationship to determine how often a given magnitude flood is expected to occur?
a. recurrence interval = (rank of the discharge + 1)/number of years of records
b. recurrence interval = (number of years of records + 1)/rank of the discharge

14. Which **two** responses correctly characterize the pool and riffle channel pattern?
a. Riffles usually have finer size channel-bottom sediments than associated pools.
b. At low discharges, average water velocities and turbulence in the pools exceed those of water moving through the riffles.
c. Pool and riffle channel patterns provide ecological diversity in stream habitats.
d. Pools tend to be scoured and deepened during floods.

Questions 15-17; Matching: Physical Units and Hydrologic Quantities

15. ( ) cubic feet/sec or cubic meters/sec   a. gradient
16. ( ) feet/mile or meters/kilometer   b. velocity
17. ( ) feet/sec or meters/sec   c. discharge

Questions 18-20; Matching: Disastrous Flood Events

a. Big Thompson Canyon, Front Range, Colorado   b. Rillito Creek, Tucson, Arizona
c. Ventura River, southern California, north of LA   d. Buffalo Creek, West Virginia

18. ( ) Failure of a coal-waste dam unleashed a flood wave with a discharge equivalent to 40 times that of the 50-year flood.
19. ( ) In the summer of 1976, severe flash floods unleashed by unusually heavy, local thunderstorms resulted in more than one hundred fatalities and severe damage to homes, bridges, and roads.
20. ( ) In 1995 and 1992, flood waters swept through an RV park. Historical records show numerous, other, past, severe flooding events at this same location.

21. Which **one** statement correctly describes hydrologic characteristics of urbanized and/or equivalent natural drainage basins?
a. Peak discharges associated with 50-year and longer recurrence interval storms are relatively insensitive to percentages of impervious cover.
b. Extensive urbanization results in longer lag times between storm precipitation and peak discharges.
c. Urbanized streams generally have more appealing aesthetic qualities and more diverse aquatic habitats than natural streams.

22. Which statement concerning the 1910 channelization of the Blackwater River, Missouri, is inaccurate and factually incorrect?
a. The channelized reach supports larger fish populations and more diverse aquatic environments than natural, unmodified reaches of the river.
b. Bank erosion and channel widening, initiated following channelization, evidently are still continuing today.

23. Which **one** describes the effects of a well-managed, storm-water retention pond?
a. decreases lag times and increases peak downstream discharges
b. increases lag times and reduces peak downstream discharges

24. What is a pilot channel?
a. manmade meandering channel, simulating a natural channel, included in a large floodway channel designed to protect against the 100-year flood event
b. a smooth, concrete-lined stream channel designed to guide storm runoff discharges downstream without attendant flooding

25. Which one describes an adjustment to the flooding threat, not flood prevention?
a. construction of dams, levees, floodwalls, and diversion channels
b. regulated land-use activities to limit damage costs from future floods

26. Which river in southern Florida, straightened" for land drainage and flood control, is presently being "restored" to simulate a natural stream?
a. Savannah   b. Suwannee   c. Atchafalaya   d. Kissimmee

27. Which one defines the average velocity of water flowing in a stream channel?
a. alluvial capacity divided by competency   b. gradient times average depth
c. stream's capacity times its gradient   d. discharge divided by cross-sectional area

Questions 28-31; Agree (A) or Disagree (D); Briefly explain your disagreements.
28. ( ) For designed meanders incorporated into urban stream channel restoration, riprap can stabilize the higher, steeper bank of the asymmetrical channel; sand or gravel bars will develop along the opposite, shallower side of the channel.

29. ( ) The 1993 upper Mississippi Basin floods were unusually severe but of short duration; the floods resulted from a week-long storm system that stalled over the upper Midwest and melting of an extensive snowpack in the same area.

30. ( ) During the 1993 floods, extensive levees effectively narrowed parts of the Mississippi and Missouri River floodplains, resulting in steeper gradients, higher average velocities, and lower peak water levels associated with a given discharge.

31. On the map of the U. S. provided, accurately plot and label the following.
   a. Upper Mississippi River Valley, St. Louis, MO to Minneapolis, MN
b. mouth of Ventura River, Pacific Coast, CA   c. Tucson, AZ
d. Blackwater River, west-central Missouri   e. Kissimmee River, FL

Answers; Home Study Assignment
Rivers and Flooding • Chapter 5

1. b	2. T	3. F	4. T	5. T
6. T	7. F	8. b	9. d	10. a
11. a	12. a	13. b	14. a	15. a
16. b	17. a	18. b	19. a	20. T
21. F	22. T	23. T	24. b	25. a, b
26. a	27. d	28. e	29. c	30. b
31. a	32. a	33. f	34. e	35. d
36. c	37. b	38. a	39. a, b, e	40. a, b
41. c	42. b	43. b	44. a	45. a
46. b, c	47. a	48. a, d	49. c	50. F
51. T	52. F	53. F		

54.

Year	Max Q cfs	Rank Order	Recurrence Interval; yrs	Year	Max Q cfs	Rank Order	Recurrence Interval; yrs
1	660	9	1.22	6	740	8	1.38
2	1700	2	5.50	7	1500	3	3.67
3	2050	1	11.00	8	800	7	1.57
4	1200	4	2.75	9	1050	5	2.20
5	900	6	1.83	10	520	10	1.10

55. about 3700 cfs (cubic feet per second)   56. one-half (1/2) or 50 %

Answers: Environmental Geology Quiz • Chapter 5

1. b	2. T	3. T	4. F	5. F
6. F	7. a	8. a	9. a	10. b
11. a	12. a	13. b	14. c, d	15. c
16. a	17. b	18. d	19. a	20. c
21. a	22. a	23. b	24. a	25. b
26. d	27. d	28. A	29. D*	30. D*

*

29. The 1993 Upper Mississippi Basin floods were unusually severe, but hardly of short duration. The dangerous floodstage conditions lasted on and off for most of May and June. Meteorological conditions favorable for heavy rains persisted in different parts of the basin for nearly two months. The conditions themselves were not unusual, but their continued, persistent repetition over two months was indeed a vary rare event.

30. This statement is essentially correct except for the reference to "lower peak water levels associated with a given discharge". Effectively constricting and narrowing the floodplain by building levees, dikes, or floodwalls raised river stages associated with given discharges above those observed for similar discharges before the floodplain was modified.

Commentary/Media Watch

Introductory Comments

After numerous congressional threats, government shutdowns, vetoes, and negotiations, President Clinton finally signed the 1996 Budget Bill in April. Buried deeply in the billions of dollars of future government spending were $200 million for environmental protection and restoration of Everglades National Park.

Once the early returns assured a Clinton victory in the 1996 presidential race, election-night television news coverage shifted to congressional races and interesting state-wide ballot initiatives. Florida's ballot included a proposal to levy a one-cent per pound tax on Florida's sugar production with the proceeds dedicated to hydrologic restoration and water quality improvements. The initiative was defeated by a modest margin, but the results showed that almost one half of Florida's voters were willing to pay slightly higher prices for sugar and sugary sweets in return for environmental protection in South Florida and the Everglades.

May through December, 1996, when most of this manuscript was written, was a banner period for media coverage of floods and flood-related events, both domestic and international. Noteworthy 1996 flooding events not included in the following discussions are the record-breaking, early October floods in southern Maine, the worst flooding since 1916 to strike the upper Broad River basin and Lake Lure area in west-central North Carolina, disastrous, early December flooding in northern Greece, and two episodes of flash flooding that struck Bucks County, Pennsylvania, and adjacent areas just north of Philadelphia.

Riparian Habitat Restoration, the Grand Canyon

As described in the *case history* The Grand Canyon (p. 284) riparian habitats in the Grand Canyon have undergone significant changes, many of them detrimental, since the Glen Canyon Dam (Lake Powell) was completed in 1963. Annual spring floods no longer affect the downstream reaches of the river; sandy beaches and shoreline sand bars had gradually disappeared, exotic brush and trees took hold at the expense of native vegetation adjusted to the annual flooding cycle, and coarse, bouldery debris dams dumped at tributary mouths by flash floods were left in place, forming upstream "pools" and downstream "rifles". In addition, native fish species adjusted to high suspended sediment loads and seasonal changes in water temperature, discharge, and river-bottom, spawning habitats have fared poorly in competition with introduced species such as rainbow trout.

In the unusually wet year of 1983, forced, large-volume releases of water from Glen Canyon Dam showed that a "manmade flood" could reverse some of the observed habitat deterioration. Subsequently, environmental groups, the National Park Service, and the Bureau of Reclamation finally agreed to a manmade flood in the spring of 1996 and extensive follow-up studies and analyses to determine if such flooding on a regular schedule would be environmentally beneficial and "economically feasible".

The beginning of the spring flood event in March attracted some media coverage. CNN showed brief footage of the powerful, massive, water jets spurting for exit conduits at the base of the dam and brief articles appeared in many newspapers, As you might expect, media interest quickly shifted elsewhere, but a few latter references to the project did appear. On October 9, Secretary of the Interior Bruce Babbitt announced that henceforth, Glen Canyon Dam will be operated so as to better protect riparian ecosystems in Grand Canyon National Park, suggesting that other manmade floods can be expected in the future. Initial results of the follow-up scientific studies indicate that the proscribed flooding did achieve some of the hoped-for modifications in riparian and riverbed habitats. Complete results of these studies will be presented in the special session "Controlled Floods, Natural Floods, and Riverine Processes" at the December (1966), Fall Meeting of the American Geophysical Union in San Francisco.

Anonymous, 1996, Grand Canyon flooded: in News Update, National Parks, v. 70, no. 5-6, p. 23
This news brief notes the March 26, 1996 opening of the Glen Canyon floodgates to simulate a week-long, spring flood.

Carlowicz, M., 1996, Controlled flood of Colorado River creates stream of data: EOS. v. 77, no. 24, p. 225 & 230
The article gives some basic hydraulic data for the '96 "spring flood" and presents a generally positive assessment of channel scouring and riparian beach rebuilding during the flood. Also noted were some results of an experiment utilizing dye injection to study mixing, turbulence, and eddy behavior. Velocity measurements will be used to improve roughness estimates for different reaches in the canyon.

Anonymous, 1996, Restoring habitat and clean air in the Grand Canyon: EDF Letter, v. XXVII, no. 4, July, p. 1 & 3
This note suggested that the flood had beneficial effects on some of the major issues of concern, such as spawning habitat for native fishes, beach restoration, channel scouring, and restoration of damaged riparian habitats. The Environmental Defense Fund, EDF, has devised a computer model that analyzes how regional electricity demand and production at Glen Canyon are combined with hydrologic impacts on the river and its riparian system; progressive peak shaving/load management policies evidently can allow for continued annual releases of water to simulate spring floods below the dam.

Flash Floods, south-central Pennsylvania, Gettysburg area, and Maryland panhandle near Hagerstown, MD
Media coverage, 6/19/96 and 6/20/96, included CNN, The Weather Channel, and newspapers. In some areas, up to 12 inches of rain fell in 6 hours; this is more than enough precipitation to trigger dangerous flash flooding. At least four fatalities were reported. Flood-impacted residents interviewed on CNN expressed surprise and amazement that the floods came "so quickly and without warning". This reaction is hard to believe since The Weather Channel and local television channels had already broadcast flash flood warnings for the area, based on the unusually heavy rains forecast for the region and Doppler radar returns that confirmed the torrential rains. Evidently, local public officials did not provide

adequate warnings or the warnings were ignored. Neither situation is in the best interests of public safety.

Buffalo Creek, Colorado, Front Range, southeast of Denver, CO

Media coverage on CNN and in newspapers briefly described damage from flash floods and debris flows, 7/13/14/96, triggered by a strong, nearly stationary, thunderstorm that settled in over the Buffalo Creek watershed.

Seven western Pennsylvania counties north of Pittsburgh, PA

CNN, 7/20/96, reported flash flooding in a seven-county-wide area of western Pennsylvania. Heavy precipitation was widespread enough to merit concern that flooding of the Allegheny River at Pittsburgh was possible in the next few days.

Hurricane Fran

In early September, 1996, Hurricane Fran made landfall just west of Wilmington, NC, and moved to the north-northwest across North Carolina, Virginia, and western Maryland. Rainfall totals in the Shenandoah and Potomac River basins were especially heavy, resulting in near-record crests on the Shenandoah River at Harpers Ferry, WV, at the confluence of the two rivers. Video footage on CNN showed flooding of low-lying areas along the Shenandoah in Harpers Ferry and focused on the threat to historical features and National Park Service facilities in the area. Railroad bridges across both rivers were loaded with railroad cars to add weight and help stabilize them. In nicely illustrated graphics, the flood crest was routed down the Potomac to Washington, DC, alerting the public to expected arrival times and stages at downstream locations.

Predicted heavy flooding occurred along most of the major rivers in the Coastal Plain and central Piedmont regions of North Carolina. Record stages were evidently recorded in Goldsboro, Kinston, and other North Carolina towns along the Neuse River. Large "forced" releases from Falls Lake Reservoir, dangerously filled to the brim, and additional heavy rains unrelated to Fran aggravated the situation and added to flooding woes.

Falls Lake is the main water supply reservoir for the City of Raleigh, NC, and other surrounding areas. The reservoir, managed by the Army Corps of Engineers, has some flood-control function, but its major purpose is water supply, and the reservoir is managed accordingly. Following the floods, managers of the Falls Lake Reservoir were accused of worsening or even causing the near-record crests at Goldsboro and Kinston. Local politicians openly made such accusations on television news broadcasts and in the print media; little mention was made of the 6 to 12 inches of basin-wide precipitation produced by the storm! Local residents described how the floods arrived "without warning". In regard to advance notice, this writer viewed many warnings and announcements of the approaching flood crest days before its arrival, and public officials in Kinston were featured on local television news tying pink flagging to trees and telephone poles to show the predicted height of the crest. In self-defense, the Corps of Engineers countered with television clips and news announcements pointing out that forced releases were necessary to "save the dam" and that no one, not even the Corps, can predict the future. Neither side in the controversy, however, is addressing the major issue of how much water was stored in the reservoir when the hurricane hit. Being managed as a water supply reservoir, we can expect that the Corps did not allow the pool level to decline to low-enough levels to intercept and store "theoretical", upper basin, flood-stage runoff from some future, unpredictable hurricane. Needless to say, for flood control, a reservoir is only as useful as its available storage capacity; a filled reservoir offers no help with flood control and may worsen the situation.

This story did have its amusing moments though. The anchor for the 11:00 News (Channel 9, Greenville, NC, September, 26, 1996) in reporting on the controversy, stated the discharge from the reservoir in units of *cubic square feet per second.* Let's hope he got that wording from a staff writer, not from the engineers managing the dam!

Two additional items related to the storm and flooding were of public concern, releases of untreated sewage during power outages and "hog lagoons". Trees uprooted by the storm caused widespread power outages in the basin; in some areas electricity was not restored to residential customers for ten days or more. Most towns and cities in the Neuse River Basin do not have onsite, backup, emergency sources of electricity for their sewage treatment plants. Thus while the power is off, raw, untreated sewage is released into the rivers.

North Carolina has a firm grip on second place in U. S. swine production, one step behind Iowa but closing fast. Recent growth has been "explosive" and mainly concentrated in the eastern one-third of the state. Hog-raising "palaces" are regarded as agricultural enterprises rather than as industrial facilities and thus are subject to minimal regulation. Animal wastes are directed to "hog lagoons" where natural bacterial and biochemical processes decompose the waste. When overtopped by floodwaters or filled to overflowing by heavy rains, such as happened in June, 1995, the contents of the lagoons are released into streams and rivers. The lagoon waters are loaded with organic matter, nitrates, and phosphates, and represent huge nutrient loads and oxygen demands, even under floodstage conditions. In 1995, lagoon wastes caused significant deterioration of water quality in the state's sounds (estuaries) and coastal waters (see Chapter 11, this manual for additional information on these issues).

Damage estimates from Fran's winds and flooding topped $6 billion, and state officials are now describing Fran as "the worst storm" in North Carolina history; it will almost certainly be the most expensive in the state's history.

Hurricane Hortense followed a week after Fran; its track passed through the Virgin Islands, eastern Puerto Rica and western Dominican Republic. Video on CNN (September 13 and 14) showed raging flash floods in Puerto Rico and the Dominican Republic and wind damage in the San Juan area. Rainfall totals from 12 to 23 inches were reported from St. Croix, VI, and parts of eastern Puerto Rico.

Tropical Storm/Hurricane/Extra-Tropical Storm Josephine

Tropical cyclonic system Josephine formed in the western Gulf of Mexico in early October and, as a tropical storm, minimal hurricane, and extra-tropical storm, followed a path across northern Florida into the Atlantic and northward along the East Coast. Torrential rains, totaling 18 inches in some parts of

Massachusetts and southern Maine caused record flood stages in many rivers and streams.

Rains associated with Josephine hit eastern North Carolina during the evening hours of October 6 and carried over until about 11:00 A. M. the next morning. This was a region-wide storm and produced 5-6 inches of precipitation over the entire drainage basin of Green Mill Run, a small creek flowing through the city of Greenville, NC. National Weather Service Doppler radar images, carried on the local Utilities Commission television channel and fixed location rain gauges showed heaviest rainfall amounts, about 1 in/hr, fell during the period between 7:30 and 10:30 A. M. on October 7; thus about one-half the total precipitation fell during the last two and one-half hours of the storm.

Green Mill Run is a classic example of a small stream in a rapidly urbanizing drainage basin. In the past few years, new apartment complexes, shopping centers, parking lots, housing subdivisions, new roads, and widened streets have added significantly to the percentage of impervious cover in the basin. Storm sewers and drainage ditches carry the runoff quickly and directly to the creek. Over the years, the channel has been rerouted and straightened; an old meander scar is evident in a creekside city park and, prior to construction of a new classroom building, another used to be visible on the East Carolina University campus. Steep to vertical channel walls are supported by a tough marine clay unit; the channel has not been hardened by concrete walls or riprap. As with other such streams, the pre-urban pool and riffle character is largely gone, having been replaced by a smooth, sand-covered bottom. Much of the floodplain is forested or city parkland and most nearby homes and commercial structures are well above the elevation expected for the 100-year flood.

An exception is a student parking lot on the East Carolina University campus that has seen occasional slight flooding in the past. Swollen by Josephine's heavy rains that morning, the creek rose rapidly, inundating the parking lot to a depth of almost three feet. Unfortunately, the lot was full; cars were locked; and the owners, unaware of the flash flood threat, were scattered all over campus. Once the threat was recognized, University personnel made a valiant effort to get the cars out of danger, but about 90 vehicles were partly submerged and sustained serious water damage.

Flash flood warnings for eastern North Carolina were regularly broadcast on The Weather Channel, CNN, and local television channels. These were initially based on predicated rainfall totals and later on radar-detected totals. Thus the flooding came with "advance warning". The warnings were either 1) not heard, 2) ignored, or 3) felt not deserving of any special precautions or protective actions. The last possibility is most likely. A high-ranking University official characterized the flooding as an act of God, a legal phrase meant to imply that the University was immune from liability because the flooding event was completely unexpected and unforeseen. The following day, the Pitt County Emergency Services Coordinator, as quoted in a local newspaper, attributed the severity of the flooding to ground still saturated from heavy rains of Hurricane Fran. Thus owners of the flooded cars might be eligible for FEMA relief funds designated for victims of Fran.

His statement was incorrect. Rainfall totals from Fran were not that heavy; our backyard rain gauge showed 3.5 inches. About 5 inches of rain fell during the four weeks since Fran, most during the two weeks following the storm; since then, rainfall totals in the Green Mill Run basin have been insignificant and known areas of throughflow-seepage had dried. The unusually high flood crest in Green Mill Run resulted from two factors; 1) the large increase in area of impervious cover in the basin in the past few years; and 2) the fact that the heaviest rains fell late in the storm when retention ponds were nearly full and surface soils were thoroughly wetted and probably saturated or close to saturation. These combined to produce very high percentages of runoff during the last three hours of the storm.

If the flooding was an act of God, it certainly was aided and abetted by Man in the form of urbanization. Although its basic causes, embodied in Figure 5.20, p. 114, are well-understood among professionals, such flooding still comes as a "complete surprise" to many people, even to otherwise well-educated individuals. As teachers and educators, all we can do is to keep trying; perhaps some day, sound planning based on earth and environmental science concepts will be as important to communities as new convention centers, new sports arenas, and lower taxes.

National Flood Insurance Program (NFIP)

Advertisements for the National Flood Insurance Program are regularly included in publications such as TIME Magazine and aired on CNN and other television channels. The adds point out that year after year, floods are far and away the leading category of natural disasters in the country and the most expensive. However, most families and businesses are not insured against flood damage; in fact, most homeowner's insurance policies specifically exclude coverage for flood damage, as opposed to, for example, water damage caused by a broken hot water heater or leaking, storm-damaged roof. Flood insurance availability and rates are based on flood-risk maps, specifically with reference to the area expected to be inundated by the 100-year flood.

FEMA is a bureaucratic stepchild of the Cold War. Its original mission was to plan for and deal with consequences of potential nuclear bomb attacks. As the nuclear threat has receded, the agency has been mandated to take a leading role in providing assistance and relief from natural disasters such as earthquakes and storms. For information on the National Flood Insurance Program, contact either FEMA (Federal Emergency Management Agency) or NFIP (National Flood Insurance Program) as listed below.

FEMA/MSC, BOX 1038, Jessup, MD 20797-9408; 1-800-611-6123 ext. 616

NFIP, 500 C Street SW, Washington, DC 20472; 1-800-427-5593; http://www.fema.gov

Flash Flooding in the Tuscany Region, Italy

Very damaging flash floods struck the Tuscany region of northwest Italy during the third/fourth week of June, 1996. Media coverage included articles in U. S. newspapers and video footage on CNN and The Weather Channel. Very powerful video images showed muddy torrents with impressive standing waves roaring along a narrow, quaint, village street lined with old stone buildings. Damage was heavy; Italian officials place the total at $30 million (U. S.). The heavy rains also triggered numerous mudslides and debris avalanches. The most severe flash flooding may have resulted from failure of debris dams that temporarily impounded flood waters in upstream canyons and valleys.

The Italian environmental minister was quoted as saying that deforestation, excessive development, and unauthorized stream diversion, some of which may have been unlawful, contributed to the severity of the flooding. Similar comments concerning the Tuscany floods were restated in NBCTV's program Dateline, 6/25/96. In this program, video images and panelists contrasted the Tuscany devastation (very localized, upper basin flash flooding) with the 1993 upper Mississippi Basin flooding (much slower, longer lasting, and more widespread flooding of a major river and its tributaries); in addition, the panelists discussed how flood control policies, zoning regulations, and land use planning can reduce fatalities and property damage from flooding.

It will be interesting to see if any Italian governmental officials and/or private citizens embroiled in the disaster will actually be indicted and tried for these alleged violations of environmental regulations and laws. Following the Vaiont Dam disaster in 1963, some Italian governmental officials were tried, convicted, and sent to prison for their negligence and complicity in the catastrophe. About 2600 lives were lost in the Vaiont Dam floods (see *case history*, Vaiont Dam, p. 143); fatalities in the Tuscany flooding were substantially lower.

Tragic Flash Floods, northern Spain

In early August, 1996, flash floods wiped out a crowded, family, vacation campground nestled in a narrow canyon in the foothills of the Pyrenees Mountains near the town of Biescas, northern Spain. Most of the campers were Spanish nationals. Some were in tents; others were in minivans, light trucks with camper tops, and other kinds of recreational vehicles. Most were asleep when the floodwaters hit. About seventy bodies were recovered in the first three days following the tragedy. Others were still being found up to three weeks later; some had been carried downstream over 20 miles from the campsite. The final victim count will probably approach ninety.

On the morning of August 8, CNN showed sand bars and coarse, angular gravel deposits strewn along a main street of Biescas that parallels the creek channel. Smashed vans and campers, some upside-down, were mixed in with the flood debris. Subsequent Associated Press articles (August 9, 10, and 11) provided the following information.

In addition to flash floods, the torrential rains resulted in numerous debris slides in the area. Survivors stated that two successive flood waves up to 12 feet high had surged along the creek channel and through the campground. The August 11 article stated that a "concrete walkway", built across the channel, had collapsed, unleashing large amounts of water impounded behind the structure.

This mechanism is not very convincing. Basically, it might explain the origin of the first wave, but not the second. Also the reported 12-foot height of the flood waves required sudden releases from much deeper, impounded pools than would accumulate behind a walkway. More likely, the waves were triggered by collapse of temporary dams that formed when debris slide deposits, including broken tree trunks and other vegetation, blocked the channel and bottom of the canyon. Debris slide dams could have impound large volumes of water and provided sources for the thick deposits of sand and very coarse, angular gravel encasing smashed vehicles farther downstream. Release of water impounded behind dams of organic debris, such as tree trunks, branches, and brush, piled up at bridges and other channel obstructions, is another possibility. This mechanism was identified as the causative factor in the eastern Ohio flash floods of June, 1990 (see *case history*, Flash Floods in Eastern Ohio, p. 115).

Evidently, no advance warnings were issued. Local and regional officials seemed surprised or indifferent to the tragedy, despite the fact that flash floods are evidently fairly common in this area. For example in one of the Associated Press articles, it was pointed out that flash floods near Yebra the previous summer had claimed eleven victims. The Green Group, a small, Spanish, environmental political party, accused the Aragon provincial government of criminal negligence in permitting the campground site. Government representatives argued that torrential rains and flash floods are impossible to predict; thus camp owners and public planners could not have foreseen the risk.

This is hogwash! Any student reading Dr. Keller's book will see that the summer flash flood threat and the campground location presented a dangerous combination, one that cried out for denial of the campground permit, or in the least, that demanded prudent planning and preparation, including a strategy for issuing warnings and evacuation orders. A half-hour advance warning based on knowledge of the torrential rainfall event probably could have saved all the campers. This was evidently a privately-owned camp licensed by the provincial government so this event has put campground owners and government officials "on the defensive".

Floods and Flash Flooding, Chicoutimi-Lac St. Jean Area, Quebec, Canada

An extended period of flooding and flash flooding accompanied two, successive storm systems that produced record-breaking combined rainfall totals over parts of the upper Saguenay River basin in Quebec Province, north of Quebec City. Video clips of the peak-stage floodwaters were shown on CNN and The Weather Channel (7/21-23). Images of the floodwaters surging down the streets of Chicoutimi were particularly impressive. The steep gradients, bedrock channels, and huge discharges produced record water velocities, powerful turbulence, and impressive standing waves and hydraulic jumps at channel obstructions. The following information was gleaned from an article written for the Boston Globe (C. Nickerson) that also appeared in the Raleigh, NC, News & Observer (7/23/96).

The floods struck the upper Saguenay River basin near Chicoutimi and Lac St. Jean. Ten fatalities were documented, 10,000 people were made homeless, and property damage to provincial infrastructure only, not including damage to private property, was estimated at $30 million dollars. At the height of the flooding, dikes and small dams burst or were threatening to burst, adding to the danger. Some areas received over 11 inches of rain in one 24-hour period; this deluge followed prolonged, heavy rains during the preceding two weeks.

Common knowledge and verbal reports from a reliable, Canadian source indicate that Hydro-Québec may have played a part in the flooding. Hydro-Québec is a huge hydroelectric corporation with massive investments in reservoir construction and drainage modifications in northern Quebec. Its specific projects and overall vision of converting vast regions of northern Quebec into a huge, interconnected, "hydroelectric pool" have previously come under intense scrutiny from environmental groups and native North Americans living in the affected areas.

Expected benefits from dam and reservoir projects include income from the sale of electricity; also projected savings from flood control are typically cited. Under normal conditions, all or a portion of an upstream floodstage discharge can be stored in a reservoir, producing two important benefits; first, potential floodstage discharges below the reservoir can be regulated and reduced, and second, the stored water can be released later to generate electricity at times when usage is normally high and stream levels are normally low. However, the flood control value of a reservoir is only as great as its unfilled storage capacity; thus a reservoir's flood control value is maximized when it's empty! A completely filled reservoir has no flood control value and may, in fact, compound or magnify the downstream flooding danger. For example, a brim-full reservoir may threaten the integrity of the dam itself. Earthen dams can be shoved aside or breached; spillways and other overflow discharge conduits may not work as expected, and unusually large, rapid releases may be required to "save the dam and stabilize the situation". Thus, the flooding impacts downstream may actually exceed those expected on an equivalent undammed, unregulated river for the same sequence of storm events.

For example, Granite Dam and reservoir on the Verde River above Phoenix, AZ, has very rarely filled and been opened with disastrous results downstream along the normally dry course of the Salt River through the Phoenix metropolitan area. Should the reservoirs on the Salt River upstream from Phoenix ever fill completely in some future, record-shattering wet year, managers of the Salt River project will be faced with a similar, no-win situation.

Pool level regulation involves a bet on the future. A good bet means that no additional heavy storm events are imminent, and thus water can be stored and held in reserve for late summer-fall electricity generation; a bad bet means that heavy runoff from an unexpected, unanticipated storm event hits just after reservoir storage capacity has been augmented by runoff from an earlier storm event. In the later case, the dam managers have no recourse but to "open the floodgates" and hope for the best. Forced releases from overfilled dams may have contributed to the worst of the Chicoutimi flooding. Justified or not, initial public anger toward Hydro-Québec was evident in the flooded communities, and the company's record on environmental and regional social issues has made it an easy target for public hostility.

North Korea and the Korean Demilitarized Zone

In the last week of July, torrential rains, totaling up to 20 inches over a two-day period in some areas, struck the hilly terrain along both sides of the demilitarized zone separating North and South Korea. The rains set off flash floods, mudflows, and debris avalanches. South Korean authorities reported that 24 South Korean soldiers were buried by mass wastage debris that swept over their defensive positions along the border; an estimated 50,000 South Korean civilians were evacuated from the devastated areas. The situation was probably much the same north of the demilitarized zone, but as usual, little information was forthcoming from North Korean officials.

Sparse information leaking from the north was subject matter for an informative article in USAToday (7/30/96) and an in-depth news feature on National Public Radio (8/6/96). The severe, demilitarized-zone flooding event was the last in a series of heavy summer rains that had heavily damaged North Korea's infrastructure (roads, bridges, dams, electrical power grids, etc.) and wiped out almost all the 1996 rice crop. This flooding followed equally heavy flooding last summer (1995). Given two successive years of widespread crop losses and damage to the nation's transportation and hydroelectric networks, the country was sinking into paralysis. By mid-August, the North Korean Government was quietly requesting food and other assistance from the international community in a desperate effort to head off social chaos and famine.

By early October, news sources in Seoul, South Korea, reported that about 1000 North Koreans per day were dying of starvation and that normally well-fed North Korean military personnel were facing food shortages. North Korea has no foreign currency reserves for food purchases and requests for international humanitarian aid have essentially been ignored. Given the present situation, a major winter catastrophe is looming for the North, and as usual, the regime is silent. The unusually heavy, summer rains of 1995 and 1996 may be major contributors to eventual political changes on the Korean peninsula. North Korea, as we've known it since the end of World War II, is on the verge of collapse.

Fujian and Hubnei Provinces, China

In early July, television news stories and brief newspaper articles appeared describing the massive flooding along the Yangtze River and its tributaries in Hubnei Province, China. Rainfall totals from June 27 to July 4 were quoted as from 13 to 22 inches. Damage estimates easily topped $2 billion and 3.2 million acres of croplands were inundated. As of late July, known flood-related fatalities had risen to 1500. To compound the misery, tropical storms/typhoons struck the Chinese coast in early August, dumping huge quantities of additional rain on the stricken areas. Officials in the city of Wuhan, Hubnei Province, characterized the situation as equivalent to the disastrous flooding of 1954. As of mid August, the flooding was subsiding, but officials were deeply concerned about cholera and other water-born infections; needless to say, the flooding had temporarily destroyed any semblance of water purification and sewage treatment in the region; thus sanitary conditions were expected to deteriorate rapidly as the floodwaters gradually subsided. Sandbagging and other flood-prevention activities were further hampered by unusually warm daytime temperatures. Day after day at the height of the flooding, workers along the 200-milelong floodwall enclosing Wuhan toiled in 100° F temperatures. It remains to be seen if China can sustain these massive agricultural losses without having to initiate major imports of grains and other basic foodstuffs.

Large-scale public works projects such as flood control walls and dikes, canal building, and interbasin river transportation, etc. go way back in Chinese history. In contrast, because of obvious differences in settlement patterns, large-scale public works projects in North America are essentially restricted to the last two hundred years. The extensive historical records that probably exist concerning flooding in areas such as the middle and lower Yangtze Basin, could provide raw materials for scholarly projects in historical hydrology and the historical evolution of the public works idea. By way of comparison, historical records have contributed greatly to studies of earthquake patterns in China over the past two thousand years or so.

Monsoon-Related Flooding, Indian Subcontinent

Heavy rains in late June and early July had produced floodstage discharges in the lower Ganges River basin and other major rivers in Bangladesh. A July 8 Associated Press article noted that flood waters were threatening the capital city of Dhaka; over 100,000 residents had already been displaced from their homes. Newspaper reports of July 24 indicated that Bangladesh was one-third inundated, and that 5 to 6 million people had been directly affected by the flooding. Most of the country is only a few feet above sea level; as bad as the river flooding is, the real threat to Bangladesh is from storm surge and flooding associated with tropical cyclones (hurricanes) coming ashore from the Indian Ocean/Bay of Bengal. One such storm, November, 1970, accompanied by an 18-foot storm surge and powerful winds, accounted for an estimated 300,000 fatalities and $63 million in crop damage (p. 229). A November, 1996 "cyclone" struck the east coast of India, causing severe wind damage, flooding, and probably hundreds of fatalities.

Future storm-driven disasters are inevitable. The appalling death counts could be reduced by better advanced warnings, improved transportation, and forced evacuations. Unfortunately, such prudent and costly improvements are not likely in the world's poorer countries in the near future.

Late July heavy rains in the Bombay region, India, caused flash flooding and a small number of fatalities; heavy rains in northern India and the Himalayan Mountains caused severe flash flooding and numerous landslides. In Nepal alone, flooding and mass wastage events had displaced over 30,000 people from their homes. In late August, CNN reported that up to 19 inches of rain from a single storm had fallen around Lahore, in the Punjab region, Pakistan. This area is part of the main Indus River drainage basin south of the Himalayan front.

Strongly seasonal, rainy and dry seasons, associated with the Indian monsoon climate, effect huge, seasonal variations in river discharges. One beneficial effect of wet-season flooding is to temporarily flush out the massive load of industrial and animal-related pollutants that concentrate in rivers, such as the Ganges, under low-discharge conditions toward the end of the dry season (see Hazarika, S., 1994, Plan to clean up India's Holy River fails to stem the tide of filth: in Themes of the Times; The Changing Earth, Fall, 1995; an annual collection of contemporary articles from The New York Times, distributed exclusively by Prentice Hall).

Other International Flash-Flooding and Flooding Events, 1996

In late July, tropical storm Caesar moved along the Caribbean coast of Colombia, setting off heavy rains, flash floods, and slope failures. One particularly, tragic, slope failure near Bogota resulted in ten fatalities, including eight children from a single family. The storm then turned to the north and drenched parts of Costa Rica and eastern Nicaragua, causing slope failures in hilly terrain and severe flooding.

Typhoon Gloria swept through the northern Philippines in late July, triggering slope failures and flooding. Areas around Mt. Pinatubo were of particular concern since unconsolidated ash and other volcanic debris from the 1991 eruption is still highly susceptible to debris avalanches and other rapid slope failures that generate mud flows in the affected streams and rivers.

In late July, tropical storm Frankie battered the Red River delta area of northern Vietnam, one of the republic's major agricultural areas. News reports cited severe flooding, extensive crop damage, and at least seventeen fatalities from the storm. Later, in early October, floods in the Mekong River delta of southern Vietnam had inundated 25,000 acres of croplands and were threatening to destroy 600,000 tons of harvested rice; more than 200,000 homes were evacuated and twenty-one fatalities were reported.

In mid-September, 1996, hurricane Fausto struck southern Baja California and parts of Sonora and Chihuahua. Remnants of Pacific hurricanes like these can trigger unusually heavy rains and flash flooding in the normally semiarid areas of northwestern Mexico and southwestern United States.

Supplemental References

Brakenridge, G. R., Knox, J. C., Paylor, E. D., and Magilligan, F. J., 1994, Radar remote sensing aids study of the Great Flood of 1993: EOS, v. 75, no. 45, November, p. 521 & 526-527

Laenen, A. and Dunnette, D., editors, 1996, River quality: Dynamics and restoration: Lewis Publishers, Boca Raton, FL, 480 p.

Mairson, A., 1994; The great flood of '93: National Geographic Magazine, v. 185, no. 1, January, p. 42-81

Manning. J. C., 1997, Applied principles of hydrology, third edition: Prentice Hall, Upper Saddle River, NJ, 276 p.

Landslides and Related Phenomena • Chapter 6; Home Study Assignment

Questions 1-5; True (T) or False (F): Slope Elements; Refer to Figure 6.1, p. 133.

1. ( ) In general, soils and unconsolidated, weathered, rock debris overlying bedrock are thickest on the upper convex and lower concave portions of the slope and thinnest where slope angles are steepest.
2. ( ) The angle of repose is the steepest slope angle sustainable in unconsolidated granular materials such as dry sand and even-sized, crushed rock.
3. ( ) Wash slope and convex slope refer to the same upper portion or crest of a slope.
4. ( ) Rockfalls are to be expected from bedrock cliffs above debris slopes.
5. ( ) In general, bedrock cliff faces and debris accumulations are more common in humid regions than in drier, arid and semiarid regions.

Questions 6-9; Matching: Slope/Land Failures, Deformations, and Movements

a. flowage   b. subsidence   c. sliding   d. falling

6. ( ) Land surface sinks in association with collapse of underground openings such as caverns, mine openings, and pore space in dewatered, unconsolidated sedimentary deposits.
7. ( ) A relatively coherent, intact block or slab of bedrock or very cohesive soil moves downslope along a fracture, bedding plane, or other well-defined slip surface.
8. ( ) Detached bedrock blocks freely drop, bounce, and roll downhill; blocks accumulate as a debris slope at the base of a steep bedrock slope or cliff.
9. ( ) Downhill movement of weakly to moderately cohesive surface soils is accompanied by internal deformation of the moving material.

10. Which response correctly characterizes mass wasting movements illustrated in Figure 6.3, p. 133?
a. Slip movements characterize the upper portion of the landslide; flowage dominates movements in the lower portion.
b. Flowage movements characterize the upper portion of the landslide; slip movements dominate in the lower portion.

Questions 11-13; Matching: Classification of Landslides; See Table 6.1, p. 134.

a. debris avalanche   b. rotational slide/slump   c. soil creep

11. ( ) very slow, downslope flowage of unconsolidated, unsaturated soil
12. ( ) an unsaturated, cohesive block of soil/colluvium moves downhill along a curved slip surface
13. ( ) extremely rapid, catastrophic, downslope movement of water-saturated, unconsolidated soil and colluvium

14. A relatively low-gradient slope is covered with a deep snowpack. Which situation has the higher risk of rapid slope failure as the snowpack melts?
a. the area is heavily forested   b. the area was clear-cut within the past few years

15. Which **two** statements concerning slope stability are correct?
a. Driving forces involved in slope stability are gravitational; they increase with the steepness of the slope and include mainly the weight of the slope-forming materials such as soil and colluvium.
b. Fractures, clay partings in sedimentary rocks, and foliation planes in metamorphic rocks offer weak surfaces for potential soil slips, especially if oriented parallel or nearly parallel to the slope.
c. A stable slope has a safety factor well in excess of 1; that is, gravitational forces pulling slope material toward the base of the slope greatly exceed the shear strength and internal cohesion of the slope material.
d. Excavation of a steep roadcut reduces the total weight of any potential slide mass and raises the slope's safety factor.

16. What is wrong, explicitly or implicitly, with the two incorrect statements in Question 15? How can each be rephrased/rewritten to be correct?

Questions 17-20; Matching: Slope Failures/Mass Movements
a. soil slip   b. slump   c. slab failure   d. shale terrain

17. ( ) This term describes an irregular, hummocky (lumpy), slope topography associated with slope instability and mass movements.
18. ( ) This term describes a rotational slide in soils/colluvium or weak bedrock; slip surface is curved and surface of slipped, rotated block may show a slight inclination in the opposite direction to the overall slope inclination.
19. ( ) A thin mass of soil slides along shallow, planar, slip planes in colluvium or shallow, deeply weathered bedrock; the movement is translational.
20. ( ) A resistant caprock, such as a well-cemented sandstone, is undercut by erosion of softer, weaker, underlying shale beds; blocks of the resistant caprock break away from the outcrop and fall or slide down to the base of the slope.

21. Which **two** statements concerning the 1995 La Conchita, CA, slide are reasonably correct and consistent with our knowledge of this landslide? See Figure 6.8.
a. This was a large, complex landslide; rotational block slumps in the upper part graded into earthflows at the toe or downslope portion of the slide.
b. Local oversteepening was caused by stream erosion of the concave lower portion of the slope. Clear-cutting of old-growth forest reduced the resisting forces on the slope and increased the driving forces.
c. The slide broke away from the crest of an old seacliff that had been free of slope failures for the past 40,000 years.
d. Water infiltrating from heavy rains and irrigation lowered the shear strength of portions of the slope-forming material.

22. Consider a relatively steep vegetated slope area with numerous steep ravines. Which areas are most susceptible to soil slips? When are they most susceptible?
a. Colluvium and soils in the ravines are temporarily water-saturated during a sustained, heavy rainfall event.
b. Soil and colluvium on the higher areas between the ravines are covered with thick, brushy vegetation that hasn't burned in over 50 years.

23. Which description more accurately characterizes the scars of recent soil slips?
a. long, narrow, shallow, slip surfaces coincident with steep ravines and gullies
b. broad, short, deep, curved, slip surfaces associated with rotated slump blocks

24. Given similar topography and equal storm events, which has the higher potential for developing a temporary, perched water table?
a. shallow, sandy soil and colluvium overlying relatively massive, unfractured granite or granitic gneiss
b. deep, clay-rich soil and colluvium overlying highly fractured, steeply foliated gneiss and schist

25. Slope failures are common in formerly submerged, unconsolidated material as flood waters recede or as the water level in a reservoir is rapidly lowered. Why?
a. High pore pressures destabilize newly-exposed, saturated, bank material.
b. Spontaneous liquefaction, associated with dewatering, increases the shear strength of newly-exposed, slope-forming material.

Questions 26-37; Matching: Infamous Slope Failures and Land Subsidence Events

a. Winter Park, FL, 1981
b. southern Central Valley, CA
c. The December Giant; 1972, near Montevallo, AL
d. Vaiont Canyon, northeastern Italy, 1963
e. Portuguese Bend, southern California
f. Thistle, UT, 1983
g. Yungay and Ranrahirca, Peru, 1970
h. Lake Peigneur, LA, 1980
i. Lehigh Valley, eastern PA, 1986
j. Rio de Janeiro, Brazil
k. Los Angeles, southern California
l. Juneau, AK

26. ( ) A misdirected, oil exploration drill hole penetrated an abandoned portion of an extensive, underground salt mine, triggering a bizarre, catastrophic, event.

27. ( ) A long-forgotten, filled sinkhole in an area of limestone bedrock suddenly opened, threatening nearby residential developments.

28. ( ) After months of creep and other evidences for dangerously unstable slope conditions, a huge catastrophic landslide plunged into a deep reservoir, triggering a disastrous flood in the downstream river valley.

29. ( ) Natural and manmade increases in water infiltration, road construction, and residential development on a coastal terrace atop a seacliff evidently destabilized part of a much larger, older, inactive landslide mass.

30. ( ) Over three days, cars, buildings, part of a highway, and part of a community swimming pool complex dropped into a newly-formed, enlarging sinkhole.

31. ( ) Unconsolidated subsurface clays and surface soils suddenly collapsed into a large underground cavern formed by dissolution of limestone bedrock.

32. ( ) A high-speed debris avalanche of broken rock, snow, and ice, shaken off the north peak of Nevado Huascaran by an earthquake, buried everything in its path.
33. ( ) Residential subdivisions and dock facilities have been constructed in areas struck by dangerous snow avalanches during the previous 100 years.
34. ( ) Sprawling hillside urbanization and many damaging landslides have led to an increased knowledge concerning slope processes and strict codes to regulate hydrologic and engineering modifications of hillslopes.
35. ( ) A known, occasionally unstable, massive landslide mass was reactivated by infiltrating water associated with an extended period of melting snows and heavy precipitation; a town, railroad, and major highway were flooded by water impounded behind a 60 meter-high landslide debris dam.
36. ( ) Deforestation, roadcuts, excavated terraces, uncontrolled stormwater runoff, steep slopes, and shallow soils over impermeable granite bedrock contribute to dangerous landslides during unusually intense rainfall events.
37. ( ) General land subsidence and local surface fissures have been caused by excessive withdrawals of groundwater and lowering of the water table.

Questions 38-40; True (T) or False (F): For false statements, note what is incorrect.
38. ( ) Slight additions of weight to the surface of the snowpack or slight reductions in shear strength at depth in the snowpack can trigger a snow avalanche.

39. ( ) Surface drains, cutoff trenches, and impermeable surface coverings reduce amounts of infiltrating surface water accompanying precipitation events, lowering pore pressures and maintaining or increasing a slope's stability factor.

40. ( ) Installing drains to lower pore pressures will sometimes stabilize otherwise slowly moving and potentially dangerous landslide masses.

41. On the U. S. map provided, plot and label these areas, locations, or features.
a. Santa Barbara, CA   b. La Conchita, CA   c. Winter Park, FL; near Orlando
d. Portuguese Bend area, California   e. Thistle townsite, near Spanish Fork, UT
f. Cascade Range, Washington, Oregon, and northern California
g. central Virginia; area of debris avalanches triggered by remnants of Camille, 1969
h. Montevallo, AL; home of the December Giant, 1972; central Alabama

42. On the world map provided, plot and label these localities or features.
a. Anchorage, AK   b. Rio de Janeiro, Brazil   c. Juneau, AK
d. Nevado Huascaran, west-central Peru   e. Vaiont Dam and River Valley, Italy

Environmental Geology Quiz • Chapter 6

Questions 1-3; True (T) or False (F)

1. ( ) In general, soils and unconsolidated, weathered, rock debris overlying bedrock are thickest on the upper convex and lower concave portions of the slope and thinnest where slope angles are steepest.

2. ( ) The angle of repose is the steepest slope angle sustainable in unconsolidated granular materials such as dry sand and even-sized, crushed rock.

3. ( ) Rockfalls are to be expected from bedrock cliffs above debris slopes.

Questions 4-5; Matching: Slope/Land Failures

a. flowage   b. subsidence   c. sliding

4. ( ) Sinking of the land surface in association with collapse of underground openings and closure of pore space in unconsolidated sediments.

5. ( ) A relatively coherent, intact, block of bedrock or very cohesive soil moves downslope along a fracture, bedding plane, or other well-defined slip surface.

Questions 6-7; Matching: Slope Failures/Landslides

a. debris avalanche   b. rotational slide/slump   c. soil creep

6. ( ) an unsaturated, cohesive block of soil/colluvium moves downhill along a curved slip surface

7. ( ) extremely rapid, catastrophic, downslope movement of water-saturated, unconsolidated soil and colluvium

8. A relatively low-gradient slope is covered with a deep snowpack. Which situation has the higher risk of rapid slope failure as the snowpack melts?
a. the area was clear-cut within the past few years   b. the area is heavily forested

9. Which **one** statement concerning slope stability is incorrect?
a. Driving forces involved in slope stability are gravitational; they increase with the steepness of the slope and include mainly the weight of the slope-forming materials such as soil and colluvium.
b. Fractures, clay partings in sedimentary rocks, and foliation planes in metamorphic rocks offer weak, potential slip surfaces, especially if oriented parallel or nearly parallel to the slope.
c. Excavation of a steep roadcut reduces the total weight of any potential slide mass and raises the modified slope's safety factor.

10. Rewrite/change the incorrect statement in the previous question to be correct?

11. Which description more accurately characterizes the scars of recent soil slips?
a. long, narrow, shallow, slip surfaces coincident with steep ravines and gullies
b. broad, short, deep, curved, slip surfaces associated with rotated slump blocks

12. Which term describes a rotational slide in soils/colluvium or weak bedrock; slip surface is curved and surface of slipped, rotated block may show a slight inclination in the opposite direction to the overall slope inclination.
a. slab failure   b. slope creep   c. soil slip   d. slump

13. Which **one** statements concerning the 1995 La Conchita, CA, slide is incorrect and not consistent with our knowledge of this landslide?
a. Stream erosion oversteepened the concave lower portion of the slope and clear-cutting of old-growth forest reduced the resisting forces high on the slope.
b. This was a large, complex landslide; rotational block slumps in the upper part graded into earthflows at the toe or downslope portion of the slide.
c. Water infiltrating from heavy rains and irrigation lowered the shear strength of portions of the slope-forming material.

14. Given similar topography and equal storm events, which has the higher potential for developing a temporary, perched water table?
a. deep, clay-rich soil and colluvium overlying highly fractured, steeply foliated gneiss and schist
b. shallow, sandy soil and colluvium overlying relatively massive, unfractured granite or granitic gneiss

15. Why do slope failures occur in formerly submerged, unconsolidated material as flood waters recede or the water level in a reservoir is rapidly lowered?
a. Spontaneous liquefaction, associated with dewatering, increases the shear strength of newly-exposed, slope-forming material.
b. High pore pressures destabilize newly-exposed, still-saturated, bank material.

Questions 16-21; Matching: Infamous Slope Failures and Land Subsidence Events
a. Lehigh Valley, eastern PA, 1986   b. southern Central Valley, CA
c. Juneau, AK   d. Vaiont Canyon, Italy, 1963
e. Portuguese Bend, southern California f. Thistle, UT, 1983
g. Yungay and Ranrahirca, Peru, 1970   h. Rio de Janeiro, Brazil

16. ( ) General land subsidence and local surface fissures have been caused by excessive withdrawals of groundwater and lowering of the water table.

17. ( ) Residential subdivisions and dock facilities have been constructed in areas struck by dangerous snow avalanches during the previous 100 years.

18. ( ) A high-speed debris avalanche comprised of broken rock, snow, and ice, shaken off Nevado Huascaran by an earthquake, buried everything in its path.

19. ( ) Natural and manmade increases in water infiltration, road construction, and residential development on a coastal terrace atop a seacliff evidently destabilized part of a much larger, older, inactive landslide mass.

20. ( ) After months of creep and other evidences for dangerously unstable slope conditions, a huge catastrophic landslide plunged into a deep reservoir, triggering a disastrous flood in the downstream river valley.

21. ( ) A long-forgotten, filled sinkhole in an area of limestone bedrock suddenly opened, threatening nearby residential developments.

Questions 22-23; True (T) or False (F): For false statements, note what is incorrect.
22. ( ) Slight additions of weight to the surface of the snowpack or slight reductions in shear strength at depth in the snowpack can trigger a snow avalanche.

23. ( ) Surface drains, cutoff trenches, and impermeable surface coverings reduce amounts of infiltrating surface water accompanying precipitation events, raising pore pressures and lowering a slope's stability factor.

24. On the attached world map, locate, plot, and label the following localities or features. You may need to estimate some locations.
a. Anchorage, AK   b. Rio de Janeiro, Brazil   c. Juneau, AK
d. Nevado Huascaran, west-central Peru   e. Vaiont Dam and River Valley, Italy
    f. Cascade Range, Washington, Oregon, and northern California
g. Portuguese Bend area, California   h. Thistle townsite, near Spanish Fork, UT

Answers; Home Study Assignment
Landslides and Related Phenomena • Chapter 6

1. T	2. T	3. F	4. T	5. F
6. b	7. c	8. d	9. a	10. a
11. c	12. b	13. a	14. b	15. a, b*
16. *	17. d	18. b	19. a	20. c
21. a, d	22. a	23. a	24. a	25. a
26. h	27. i	28. d	29. e	30. a
31. c	32. g	33. l	34. k	35. f
36. j	37. b	38. T	39. T	40. T

*
16. response c, question 15; The definition of the slope safety factor is reversed. For a safety factor well in excess of 1, the shear strength and internal cohesion of the slope-forming material exceed the gravitational force acting to push the soil downward along the slope.

response d, question 15; The excavation does reduce the total weight of a potential slide mass, but the increased slope angle more than compensates for the weight reduction to raise the gravitational forces acting on the slope. Thus the safety factor is decreased and the slope is less stable and more likely to fail.

Answers: Environmental Geology Quiz • Chapter 6

1. T	2. T	3. T	4. b	5. c
6. b	7. a	8. a	9. c	10. *
11. a	12. d	13. a	14. b	15. b
16. b	17. c	18. g	19. e	20. d
21. a	22. T	23. F*		

*

10. Response c in question 9 is incorrect. Excavation of a road cut does reduce the total weight of slope material and as a result, does reduce the mass component of the downslope, gravitational force. However the slope safety factor is decreased, not raised, because the steeper slope angle more than compensates for the reduced mass, increasing the downward gravitational force over that consistent with the unmodified slope. Removal of the roadcut material also lowers the slope's safety factor because the upslope material is deprived of the stabilizing, buttressing effect formerly exerted by the now-excavated, downslope material.

23. The first part of the statement is correct; surface drains, trenches, and impermeable coverings do reduce infiltration of precipitation. However, less infiltration reduces, not raises, pore pressures, and reduced pore pressures stabilize the slope. Increased pore pressures reduce a slope's safety factor.

Commentary/Media Watch

Ideas and Notions

Mass wastage events, although often complex and to some extent individually unique, can be reduced conceptually to two factors; 1) the gravitational force acting to pull/push the slope material downhill along the slopes, and 2) the strength of the slope material and its ability to resist the gravitational force. The gravitational force varies directly with the slope angle, the steeper the slope, the larger the component of the gravitational acceleration acting parallel to the slope. One point that students often miss is that gravity is relentless. It acts "all the time"; the gravitational force isn't on sometimes and off at other times. Rock exposed on vertical cliff slopes has to resist the maximum gravitational acceleration.

Natural slopes adjust to the local climatic conditions, vegetation, soils, and bedrock; thus similar slope angles are maintained "indefinitely" as weathering, erosion, and mass wasting proceed. Gravitational forces pushing downward parallel to the slope are resisted and/or exceeded by the internal cohesion of the slope-forming materials. On oversteepened slopes the gravitational forces increase while the resisting forces are reduced or unchanged. Thus overall slope stability is reduced and the oversteepenend slope has an increased potential for slope failures. Some oversteepening is natural; for example, wave erosion may undercut a coastal cliff or stream erosion may undercut the toe of a slope. Excavations for highways and other facilities constitute manmade oversteepening.

Strengths of massive plutonic, metamorphic, and strongly-lithified sedimentary rocks are mainly determined by fracture spacing, area, and orientation. Bedding surfaces can also be planes of weakness in sedimentary strata. Any fracture or bedding surface represents a potential slip surface and weakens the rock. Potential slip surfaces inclined in the same direction as the slope are of particularly concern, since they represent potential zones of easy, downhill slip. Weathering along fractures and along weak strata, such as shales interbedded with sandstone, can further reduce the strength of the bulk rock. Water adds weight (adds to the downslope gravitational acceleration); elevated pore pressures reduce frictional forces resisting movement and lower the internal cohesion (strength) of clay-bearing bedrock. Clay-bearing bedrock such as shales and mudstones, especially if pyritic, are considerably weakened by moisture and weathering. Shale terrains (p. 135) have a high potential for landslides and problems related to soil creep.

In weakly lithified to nonlithified materials, internal strength arises from binding forces between individual particles. Cementation and compaction, the two, major processes in lithification, are absent or only weakly developed. Granular aggregates are materials, such as dry sand, that have essentially no grain-to-grain binding; the steepest possible slope sustainable in granular aggregates is the angle of repose, from 33° to about 35°, depending on particle sizes of the aggregate. Oversteepening of granular aggregates will trigger downhill movements that "re-establish" a slope equal to the angle of repose.

Major contents of silt- and clay-size particles contribute significantly to grain-to-grain binding forces in nonlithified surface materials such as soils and colluvium. Thus clayey soils, gravels and sands have, under ordinary, "dry" conditions, much greater internal cohesion than well-sorted aggregates such as clean sand and gravel. As with fractured bedrock, excessive moisture and water-saturation significantly reduce internal cohesion between particles. Elevated pore pressures and water saturation have particularly large, destablizing effects on slopes underlain by clay-bearing soils and colluvium. Thus it's no secret that mass wasting movements are triggered, accelerated, and renewed by heavy infiltration (rainfall or melting snow) and that adequate subsurface drainage is a major engineering weapon in preventing and managing mass movements.

Rock avalanches (p. 142-143) are very dangerous, catastrophic mass movements triggered by earthquakes. These begin as freely-falling debris masses shaken loose in steep mountainous terrain. The falling debris mass traps and compresses air between itself and the land surface, and moves down canyon on a virtually frictionless "compressed-air slip plane". These avalanches can travel great distances and are serious threats to canyon-bottom residents and facilities.

Soils and other unconsolidated foundation materials, especially if saturated with water, are subject to vibration-induced liquefaction. Land failures during earthquakes are common in such areas, even if relief and slopes are gentle.

The Miramar Sinkhole and the Yosemite Valley Rockfall

Most media-noted mass wastage events are integrated with reports of heavy rains and flash floods (see Commentary/Media Watch, Chapter 5). Newsworthy mass wasting events not directly related to storms and heavy rains included a sinkhole opening in Miramar, FL (Headline News, 6/10/96 & 6/11/96), and a rockfall from Glacier Point in Yosemite Valley, CA (CNN, 7/11/96).

The Miramar sinkhole opened in a fairly new subdivision. The original contractor had evidently used a pre-existing sinkhole pond as a disposal site for construction debris and woody waste such as tree trunks and stumps; then the site was covered over and "landscaped". Needless to say, homes around the sink are now in jeopardy, property values have plunged, and lawyers' phones are ringing! The situation is similar to that portrayed in the Lehigh Valley, PA, *case history* (p. 156); however in that case, the failed sinkhole had been filled, covered, and forgotten for many years before a subdivision was built.

At least two other so-called sinkholes were shown on CNN and other news networks, one in the San Francisco Bay area and the other in Oregon. The Bay area "sink" developed above a ruptured, storm water runoff drain and the other formed during heavy rains and flooding associated with the strong mid-November, 1996, storm. However, both so-called "sinkholes" were actually slope failures. Evidently television news writers feel that a tractor-trailer rig dropping into a suddenly opened "sinkhole" is more newsworthy than one tipping into a slope-failure depression.

Rockfalls and rock avalanches constitute a serious threat to Yosemite Park visitors. Yosemite Valley, the most heavily-visited area in the Park, is particularly at risk, so Park Service officials must follow prudent policies regarding siting of campgrounds and other visitor-dedicated facilities. For example, The Slide, an extensive canyon-wall scar and rock avalanche deposit in the upper reaches of the Tuolumne River drainage, is known to have formed during the great Owens Valley, CA, earthquake of 1872. The Glacier Point rockfall, however, was not triggered by an earthquake or by any other external event so far as anyone knows. Trees and buildings on the valley floor were damaged by powerful winds generated as the falling rock displaced air; at least one fatality was confirmed. The rock debris itself impacted higher up on the canyon walls and did not reach the valley floor.

Joint spacing and orientation play dominant roles in determining slope angles in Yosemite Valley (Huber, 1989). Talus accumulations and less-steep valley walls develop from strongly-jointed, small, diorite bodies. Granodiorites and granites with widely-spaced joints hold up the steeper, near-vertical, rock walls.

Supplemental References

Huber, N. K., 1989, The geologic story of Yosemite National Park: U. S. Geological Survey Bulletin 1595; reprinted by the Yosemite Association, Yosemite National Park, CA 95389

Robinson, G., 1988, The forest and the trees; A guide to excellent forestry: Island Press, Covelo, CA 95428, 257 p.

This book offers an informative introduction to forestry and the forest products industry. Management decisions affecting land use and environmental issues, such as insect control, infiltration, runoff, mass wastage, soil erosion, soil fertility, and stream sedimentation, are strongly impacted by economic factors.

Sidle, R. C., Pearce, A. J., and O'Loughlin, C. L., 1985, Hillslope stability and land use: Water Resources Monograph 11, American Geophysical Union, Washington, DC, 140 p.

Earthquakes and Related Phenomena • Chapter 7
Home Study Assignment

1. The terms **capable fault** and **maximum credible earthquake** relate to stringent regulatory requirements regarding siting of what class of facilities?
a. large dams for flood control and water supply
b. large, roofed shopping malls   c. nuclear power plants
d. coastal seawalls to protect against tsunamis and storm wave erosion

Questions 2-4; Matching: Recurrence Intervals, Slip Rates, and Earthquake Magnitudes; see Figure 7.33, p. 193.

a. once every 10 years   b. once every 10,000 years   c. once every 1200 years

2. ( ) 0.1 mm/year slip rate; magnitude 6.0
3. ( ) 100 mm/year slip rate; magnitude 7.0
4. ( ) 1 mm/year slip rate; magnitude 8.6

5. Which **two** responses relating to the Parkfield segment of the San Andreas fault in California are reasonably correct?
a. It has a relatively high conditional probability of rupturing to produce a major earthquake in the next 25 years or so.
b. It is the only segment of the fault characterized mainly by past, dip-slip motion.
c. The 1989 Loma Prieta earthquake, associated with the Santa Cruz Mountains segment, relieved much of its pent-up strain energy.
d. Based on historical records beginning with the 1857 event, this segment shows an average recurrence interval of about 22 years for major earthquakes.

6. Which **two** responses concerning the Pallett Creek locality on the San Andreas fault are reasonably correct?
a. This site, north of San Francisco, California, is important for paleoseismic studies pertaining to the North Coast segment of the San Andreas fault.
b. Over the past twelve centuries, four clusters of major earthquakes have occurred.
c. Radiocarbon dating of organic matter, deposited in swamps and bogs along the fault trace, has been very helpful in working out the chronology of prehistoric, major earthquakes on this segment.
d. Ten major earthquakes, evenly spaced over time and dating back to the seventh century A. D., have occurred along this segment.

7. Which states show the lower probability that over the next 50 years, earthquake-induced, horizontal ground accelerations may reach 0.05 g? See Figure 7.27.
a. North Dakota, Minnesota, and Wisconsin
b. Vermont, New Hampshire, and Massachusetts.

8. Which area is associated with higher earthquake risks? See Figure 7.27
a. New Madrid fault zone, central Mississippi Valley
b. Black Hills area, southwestern South Dakota

9. Which area shows the higher seismic risk? See Figure 7.27.
a. Yellowstone National Park area   b. Charleston, SC

10. Which **two** responses are consistent with the dilatancy-diffusion model of the earthquake cycle? See Figure 7.22.
a. Earth movements and radon emissions stabilize at above average values for some time prior to the earthquake, then abruptly decrease to much lower values following the earthquake.
b. Electrical resistivity drops sharply to a minimum and seismic velocities increase to their normal, long-term values just before the earthquake.
c. During the precursor stage just preceding an earthquake, seismic velocities gradually increase to higher-than-normal values as water flows into the rock and pore pressures rise.
d. Pore pressures and numbers of seismic events reach maximums at the same time that seismic velocities are lowest.

11. Which earthquake magnitude scale depends directly on the area of the fault rupture plane, the average displacement, and the shear strength of the rock along the rupture trace?
a. Richter   b. Modified Mercalli   c. seismic moment   d. superimposed slip

12. The 1988 Armenian (southern part of the former Soviet Union) and 1995 Sakhalin Island (easternmost Siberia, Russia) earthquakes resulted in an appalling loss of life. Why?
a. Both areas were densely populated and at the epicenters of magnitude 8.5 quakes.
b. The quakes hit at rush hour and thousands of motorists were killed when freeways collapsed.
c. Poorly constructed, top heavy, concrete slab, apartment buildings collapsed.
d. Buildings in both areas were constructed on unconsolidated, water-saturated, lake beds and foundation failures were common.

13. Which is the maximum possible damage designation on the Mercalli scale?
a. 10   b. 12   c. X   d. XII

14. Which response best describes seismic gaps?
a. inactive faults cutting anticlines   b. segments of active faults with high creep rates
c. unusually quiet zones along otherwise active faults
d. slices of land bounded by active, strike-slip faults

15. The Marina District in San Francisco was heavily damaged in the 1906 and 1989 quakes. Why?
a. The epicenters of both quakes were right under the district.
b. The area is built on consolidated bedrock, causing the shaking to be amplified.
c. Liquefaction and foundation failures were common.
d. Shaking was no more extensive than elsewhere in the city, but the whole district burned following each quake.

16. What foundation material produces the most amplification of groundshaking during an earthquake?
a. unfractured igneous rock   b. well-cemented sandstone
c. water-saturated silt and clay   d. well compacted, dry soil and colluvium

17. Which **two** responses are appropriate for the 1985 Mexico City quake?
a. A shallow-focus, Richter magnitude quake of 8.1 struck right under the city.
b. Earthquake engineering codes ignored the importance of high-frequency vibrations associated with nearby earthquakes.
c. Much of the city is built on filled-in, shallow lakes and marshes.
d. High, top-heavy, concrete-slab buildings collapsed when shaken by amplified, low frequency vibrations from a distant earthquake.

18. Which **two** statements concerning earthquake waves are correct?
a. Rayleigh waves are surface waves with mainly vertical motion; they have slower speeds than P waves.
b. Love waves often account for the most damaging, horizontal, longer period shaking associated with seismic waves from a distant epicenter.
c. In general, longer period, surface waves are attenuated faster than higher frequency seismic vibrations, such as P and S waves.
d. In general, low buildings with lower frequency natural vibrations are more susceptible to damage from distant earthquakes than high-rise buildings with higher frequency natural vibrations.

Questions 19-27; Matching: Important U. S. Earthquakes

a. Northridge, CA, 1994   b. Landers, CA, 1992   c. Charleston, SC, 1886
d. San Francisco, CA, 1906 e. New Madrid, MO, 1811-12 f. Loma Prieta, CA, 1989
g. Denver area, CO, 1962-65 h. San Fernando Valley, CA, 1971 i. NV and AZ, late 1930s

19. ( ) unexpected, shallow-focus, magnitude 7.5 quake; new, strike slip ruptures suggest that a possible reorientation of the transform fault boundary between the North American and Pacific plates is in progress

20. ( ) magnitude 8+ earthquake on the San Andreas fault; fire damage greatly exceeded damage from groundshaking

21. ( ) series of small earthquakes related to filling of the Lake Mead (Hoover Dam) reservoir

22. ( ) magnitude 7.1 quake; caused collapsed freeways and localized damage in parts of the San Francisco Bay area

23. ( ) series of "mysterious" earthquakes caused by deep-well injection of liquid wastes

24. ( ) three major quakes in four months; caused subsidence, surface failures, and other topographic changes over a large area underlain by soft, water-saturated soils and sediments

25. ( ) major intraplate quake; caused much damage near the epicenter and localized damage due to material amplification and liquefaction as far away as northern Ohio

26. ( ) major, "unannounced" quake in a densely populated urban area; caused by displacement along a "blind thrust"

27. ( ) magnitude 6.6 quake struck a then, rapidly urbanizing area; recorded, peak, horizontal ground accelerations of 1.15 g forced rethinking of design criteria based on much smaller accelerations assumed for a quake of this magnitude

28. Which **two** statements referring to the Kobe, Japan, 1995 earthquake are reasonably correct?
a. struck a heavily industrialized, densely populated, modern urban area
b. caused over 50,000 fatalities and $100 billion worth of physical damage
c. localized on a fault close to the common boundary point of the Pacific, Philippine, and Eurasian plates
d. a high degree of preparedness and planning resulted in insignificant alteration of lifestyles and only minor damage to infrastructure

Questions 29-32; Matching: Faults and Fault Displacements
a. normal fault   b. reverse fault   c. thrust fault   d. strike slip fault

29. ( ) dip slip motion; horizontal compression and crustal shortening
30. ( ) dominantly horizontal displacement along a vertical fault
31. ( ) dip slip motion; horizontal tension and crustal stretching
32. ( ) low-angle, dip slip displacement at depth may merge into an uplifted anticlinal fold

33. For most zoning and planning purposes, a fault is classed as active if fault displacements have occurred during what **two** time periods? **two answers**
a. last 100,000 years   b. last 10,000 years   c. Pleistocene Epoch   d. Holocene Epoch

34. Which **one** statement concerning foci and epicenters is correct?
a. The focus is the faulted point on the surface directly above the epicenter.
b. The fault first cracks at the epicenter and breaks to the surface at the focus.
c. The epicenter is at the surface directly above the focus where the rupture initiates.
d. The earthquake rupture starts at the focus and extends downward to the epicenter.

35. Which **one** of the following statements is correct?
a. P waves travel through solids; S waves do not.
b. P waves and S waves travel through liquids; P waves do not travel through solids.
c. S waves travel through solids, and P waves travel through liquids.
d. P and S waves travel through liquids, but S waves do not travel through solids.

36. Which **two** statements concerning tsunamis are true?
a. They travel as deep water waves at speeds greater than surface, seismic waves but slower than S waves.
b. Wave heights increase and wavelengths decrease as they move into shallower water.
c. The waves are started by fault-induced, horizontal shifts in the seafloor that suddenly propel great masses of water in opposite directions.
d. In the open ocean, wavelengths are many miles or kilometers; wave heights are only a few feet.

37. Which have the highest velocities in rock?
a. primary waves   b. secondary waves   c. surface waves

38. Which **one** is a direct measure of the distance from a seismic receiving station to the focus of a distant earthquake?
a. the time interval between the first P- and S-wave arrivals
b. the ground accelerations induced by surface waves passing a receiving station
c. the time elapsed between the first P-wave arrivals from the strongest aftershocks

39. Which **two** statements are correct?
a. for S waves, materials vibrate parallel to the path of energy transmission
b. for P waves, materials vibrate parallel to the path of energy transmission
c. for P waves, materials vibrate at right angles to the path of energy transmission
d. for S waves, materials vibrate at right angles to the path of energy transmission

40. What are the smaller magnitude quakes that follow a major earthquake?
a. exoshocks   b. aftershocks   c. hyposhocks   d. epishocks

41. Which **one** response includes the instrument used to record groundshaking and the inventor of the first widely-used earthquake-magnitude scale?
a. oscilloscope - Mercalli   b. vibrograph - Prector
c. polygraph - Freud   d. seismograph - Richter

42. What is the approximate travel time of a tsunami, generated by a coastal earthquake in central Chile, South America, to Honolulu, Hawaii? See Figure 7.26, p. 185.   a. 4 hours   b. 1 day   c. 15 hours   d. 55 minutes

43. Consider the Niigata, Japan, 1964 earthquake of magnitude 7.5. Which **two** responses accurately characterize elevation changes in the epicentral region between 1900 and 1970? See Figure 7.31, p. 190-191.
a. Survey stations along the Sea of Japan coastline north and south of Niigata showed continuous, gradual, uplift or subsidence between 1900 and 1940.
b. At four survey stations nearest the epicenter, gradual subsidence since 1900 was followed by more rapid subsidence in the years between 1945 and the earthquake.
c. Ten-year intervals of little or no vertical motion were observed at all survey stations southwest of the epicenter.
d. The survey stations all recorded uplift, some at anomalously rapid rates, during the ten or so years prior to the earthquake.

   Questions 44-47; Matching: Earthquake Prediction/Hazard Evaluation
a. Borah Peak, ID,   b. Coalinga, CA,   c. Tanshan, China,   d. Haicheng, China,
   1983   1983   1976   1975

44. ( ) An earthquake, associated with heavy property damage and over one hundred thousand fatalities, struck without warning.

45. ( ) During the year preceding the major event, an annular-shaped area of low-magnitude earthquakes evidently outlined the epicentral zone of the impending major earthquake.

46. ( ) Interpretation of the foreshock pattern and observations of anomalous animal behavior contributed to successfully predicting a major earthquake.

47. ( ) Prior to the earthquake, recognition and careful mapping of prehistoric fault scarps identified an active fault and a significant seismic hazard.

48. Which **one** should be taken as serious evidence that a major earthquake might occur within the next few hours or so?
a. gradual reduction in regional seismicity over the preceding five years
b. another quiet day on a conspicuous seismic gap
c. very gradual, continuous uplift related to tectonic creep along a buried fault
d. sudden, coastal uplift of a meter or so

49. An operating earthquake warning system could give what kind of advance notice of groundshaking triggered by a major quake?
a. about 24 hours   b. a minute at most   c. about one hour   d. five to ten minutes

50. On the U. S. map provided, accurately plot and label the following areas, locations, or features.

a. Denver, CO   b. New Madrid, MO   c. Charleston, SC
d. San Francisco Bay region, California   e. Hoover Dam/Lake Mead, NV and AZ
f. San Fernando Valley, California   g. Los Angeles basin area, California

51. On the provided world map, accurately plot and label the following countries, features, or locations.

a. Aleutian Islands   b. Honolulu, HI   c. Niigata, Japan
d. Sakhalin Island, Pacific Coast, Russia   e. coastal region of central Chile
      f. Armenian Republic and Caucasus Mountains (former USSR)

Environmental Geology Quiz • Chapter 7

1. The terms **capable fault** and **maximum credible earthquake** relate to stringent regulatory requirements regarding siting of what class of facilities?
a. large dams for flood control and water supply
b. large, roofed shopping malls   c. nuclear power plants
d. coastal seawalls to protect against tsunamis and storm wave erosion

2. Which slip rate would result in the longer earthquake recurrence interval for magnitude 7 earthquakes on an active fault?
   a. 10 mm/year   b. 1 mm/year

3. Which **one** response relating to the Parkfield segment of the San Andreas fault, California, is incorrect?
a. Based on historical records beginning with the 1857 event, this segment shows a fairly regular, average recurrence interval of about 22 years for major earthquakes.
b. This segment of the fault is characterized mainly by past, dip-slip motion.
c. It has a relatively high conditional probability of rupturing to produce a major earthquake in the next 25 years or so.

4. Which **one** response concerning the Pallett Creek locality on the San Andreas fault, California, is reasonably correct?
a. This site, north of San Francisco, California, is important for paleoseismic studies pertaining to the North Coast segment of the San Andreas fault.
b. Carbon-14 dating of organic matter deposited in swamps and bogs along the fault has identified four clusters of major earthquakes over the past twelve centuries.
c. Since the seventh century A. D., major earthquakes along this segment have occurred at evenly-spaced time intervals.

5. Which is associated with higher earthquake risks?
a. Black Hills area, SD   b. central Mississippi Valley area, AR, TN, and MO

6. Which response is consistent with the dilatancy-diffusion earthquake model?
a. Earth movements and radon emissions stabilize at above-average values prior to the earthquake, then decrease to much lower values following the earthquake.
b. Before the earthquake, seismic velocities gradually increase to higher-than-normal values as water flows into the strained rock and pore pressures rise.

7. Which earthquake magnitude scale depends directly on the area of the fault rupture, the average displacement, and the shear strength of the rock?
a. seismic moment   b. superimposed slip   c. Richter   d. Modified Mercalli

8. The 1988 Armenian (southern part of the former Soviet Union) and 1995 Sakhalin Island (easternmost Siberia, Russia) earthquakes resulted in appalling losses of life. Why?
a. Both areas were densely populated and at the epicenters of magnitude 8.5 quakes.
b. Buildings in both areas were constructed on unconsolidated, water-saturated, lake beds and foundation failures were common.
c. Poorly constructed, top heavy, concrete slab, apartment buildings collapsed.

9. Which is the maximum possible damage designation on the Mercalli scale?
   a. 10   b. XII   c. X   d. 12

10. The Marina District, San Francisco, was heavily damaged in the 1906 and 1989 quakes. Why?
a. Shaking was no more extensive than elsewhere in the city, but the whole district burned following each quake.
b. Liquefaction and foundation failures were common.
c. The area is built on consolidated bedrock, causing the shaking to be amplified.
d. The epicenters of both quakes were right under the district.

11. What foundation material produces the most amplification of earthquake groundshaking?   a. unfractured igneous rock   b. well-cemented sandstone
c. water-saturated silt and clay   d. well compacted, dry soil and colluvium

12. Which response is appropriate for the 1985 Mexico City earthquake?
a. Top-heavy, concrete-slab buildings collapsed when shaken by amplified, low frequency vibrations from a distant earthquake.
b. Earthquake engineering codes ignored the possibility of amplified high-frequency vibrations generated by a nearby earthquake.

13. Which **one** statement concerning earthquake waves is incorrect?
a. Rayleigh waves are surface waves with mainly vertical motion; they have slower speeds than P waves.
b. Love waves often account for the most damaging, horizontal, longer period shaking associated with seismic waves from a distant epicenter.
c. In general, longer period, surface waves are attenuated faster than higher frequency seismic vibrations, such as P and S waves.

Questions 14-18; Matching: Important U. S. Earthquakes

a. Charleston, SC   b. Landers, CA   c. Denver, CO
d. Northridge, CA   e. New Madrid, MO

14. ( ) 1994; major, "unannounced" quake in a densely populated urban area; caused by displacement along a "blind thrust"

15. ( ) 1962-65; earthquakes caused by deep-well injection of liquid wastes

16. ( ) 1886; major intraplate quake; caused much damage near the epicenter and distant localized damage due to material amplification and liquefaction

17. ( ) 1811-12; three major quakes in four months; caused subsidence and surface failures over a large area underlain by soft, water-saturated soils and sediments

18. ( ) 1992; unexpected, shallow-focus, magnitude 7.5 quake; new, strike-slip ruptures suggest that a possible reorientation of the transform fault boundary between the North American and Pacific plates is in progress

19. Which **one** statement referring to the Kobe, Japan, 1995 earthquake is correct?
a. a high degree of preparedness and planning resulted in insignificant alteration of lifestyles and only minor damage to infrastructure
b. struck a heavily industrialized, densely populated, modern urban area
c. caused over 50,000 fatalities and $100 billion worth of physical damage

20. Which is associated with dip slip motion, horizontal tension, and crustal stretching?   a. normal fault   b. reverse fault   c. thrust fault   d. strike slip fault

21. Which faulting is associated with dominantly horizontal displacement along a vertical rupture?   a. normal   b. reverse   c. thrust   d. strike slip

22. Which **two** of the following statements are correct?
a. For most zoning and planning purposes, a fault is classed as active if fault displacements have occurred during the past 10,000 years.
b. P waves and S waves travel through solids; P waves do not travel through liquids.
c. Fault rupture starts at the focus; the epicenter is the surface point above the focus.
d. P and S waves travel through liquids; S waves do not travel through solids.

23. Which **one** statement concerning tsunamis is incorrect?
a. Heights increase and wavelengths decrease as they move into shallower water.
b. The waves are started by fault-induced, horizontal shifts in the seafloor that suddenly propel great masses of water in opposite directions.
c. In the open ocean, wavelengths are many miles; wave heights are only a few feet.

24. Which **one** is a direct measure of the distance from a seismic receiving station to the focus of a distant earthquake?
a. the time interval between the first P- and S-wave arrivals
b. ground accelerations of long-period surface waves 100 kms from the epicenter
c. the time elapsed between the first P-wave arrivals from the strongest aftershocks

25. Which **one** response includes the instrument used to record groundshaking and the inventor of the first widely-used earthquake-magnitude scale?
a vibrograph - Mercalli   b. polygraph - Freud   c. seismograph - Richter

Questions 26-28; Matching: Earthquake Prediction/Hazard Evaluation
a. Coalinga, CA, 1983   b. Borah Peak, ID, 1983   c. Haicheng, China, 1975

26. ( ) During the preceding year, an annular-shaped area of low-magnitude events evidently outlined the epicentral zone of the impending major earthquake.

27. ( ) Interpretation of the foreshock pattern and observations of anomalous animal behavior contributed to successfully predicting a major earthquake.

28. ( ) Prior to the earthquake, recognition and careful mapping of prehistoric fault scarps identified an active fault and a significant seismic hazard.

29. Which **one** indicates that a major earthquake might be coming very soon?
a. gradual reduction in regional seismicity on a conspicuous seismic gap
b. continued gradual uplift related to tectonic creep along a nearby buried fault
c. apparent lowering or rise in relative sea level in a few hours

30. On the map/maps provided, plot and label the following locations or areas
a. Kobe, Japan   b. New Madrid, MO   c. Charleston, SC
d. San Francisco Bay region, California   e. Hoover Dam/Lake Mead, NV and AZ
f. Sakhalin Island, Pacific Coast, Russia   g. Los Angeles Basin, California
h. Aleutian Islands   i. coastal central Chile

Answers; Home Study Assignment
Earthquakes and Related Phenomena • Chapter 7

1. c	2. c	3. a	4. b	5. a, d
6. b, c	7. a	8. a	9. a	10. a, b
11. c	12. c	13. d	14. c	15. c
16. c	17. c, d	18. a, b	19. b	20. d
21. i	22. f	23. g	24. e	25. c
26. a	27. h	28. a, c	29. b	30. d
31. a	32. c	33. b, d	34. c	35. c
36. b, d	37. a	38. a	39. b, d	40. b
41. d	42. c	43. a, c	44. c	45. b
46. d	47. a	48. d	49. b	

Answers: Environmental Geology Quiz • Chapter 7

1. c	2. b	3. b	4. b	5. b
6. a	7. a	8. c	9. b	10. b
11. c	12. a	13. c	14. d	15. c
16. a	17. e	18. b	19. b	20. a
21. d	22. a, c	23. b	24. a	25. c
26. a	27. c	28. b	29. c	

Commentary/Media Watch

Ideas and Notions

Each earthquake teaches us much about how to improve our existing construction codes, earthquake safety regulations, and overall earthquake preparedness. Densely populated urban areas in technologically-advanced countries such as Japan and the United States come through major shocks quite well. Severe infrastructure damage tends to be localized and spotty; and although numerous lives are suddenly disrupted and altered, fatalities and injuries tend to be relatively low. Building collapse is probably the single most important issue in earthquake safety. If a structure collapses, its occupants have little chance to survive; survival chances are greatly enhanced if a structure "hangs together", even if it's seriously damaged.

Structural performance often depends directly on a building's age and real estate evaluation. Older buildings often fare poorly, unless they are unusually massive, well-constructed, and have foundations in bedrock. Many such buildings survived the 1906 San Francisco earthquake, although some were destroyed in the ensuing fire. Gleaming, newer, expensive buildings, designed according to the latest earthquake safety considerations, usually come through with "flying colors". This contrast is most evident in "third world" and "newly developing " countries where heavy damage in older and poorer neighborhoods contrasts sharply with relatively light damage to the modern, high-rise hotels and office buildings. Top-heavy, concrete-slab buildings, no matter when they were built, are particularly dangerous and subject to collapse. Traditional, weakly cemented mud and stone wall structures, such as those in the Balkan countries, North Africa, parts of India, and elsewhere, are death traps. Appalling losses of life can result even from moderate magnitude earthquakes.

Compare, for example, the numbers of fatalities associated with the Kobe, Japan (p. 162; Reid, 1995, and Holzer, 1995) and Northridge, CA quakes (p. 162; *special feature* The Northridge Earthquake and Buried Faults, p. 166-167) to the appallingly high numbers of deaths from the 1985 Mexico City quake (p. 162; p. 173-174; Rial and others, 1992) and from earthquakes in recent years striking in Russia and the former republics of the Soviet Union (Leith, 1995). Building design, materials, workmanship, and foundation stability are the major factors. Maximum ground accelerations and duration and frequency of shaking are "acts of God". Such comparisons convincingly demonstrate that rigorous building codes and common-sense zoning regulations can greatly reduce structural damages and casualties from earthquakes without excessively raising costs or unduly infringing on private property rights.

Seismic Events in the Media and Earthquake Preparedness

Earthquakes, especially the destructive ones, are big media events. For example, after the 1994 Northridge quake, CNN showed nearly continuous coverage of the epicentral area during most of the day. Likewise, the 1995 Kobe, Japan earthquake received plenty of media attention. Initial coverage details "life-and-death" issues such as casualties, infrastructure damage, rescue efforts, dangers from aftershocks, broken water mains, broken gas lines, hazardous spills, and fire. In succeeding days and weeks, media coverage diminishes in quantity and shifts toward efficacy of relief efforts, damage assessment, debris removal, repairs, and rebuilding.

Monday editions of the Raleigh, NC, News & Observer have a feature called "Earthweek: Diary of a Planet". Among other events of interest to earth science students, it includes a listing of the previous week's moderate to strong earthquakes from around the world; most of these are ignored by the domestic media or given one-line attention at best. This feature for August 12, 1996, included a prediction by a Chinese geophysicist that at least one magnitude 6 earthquake would probably accompany filling of the huge, controversial, Three Gorges dam now under construction on the Yangtze River in western China. The Science section of Time, September 16, 1996, featured an article titled "The Quake That Wasn't" written by Jeffrey Kluger and taken from a paper in Nature by A. T. Linde and others. Based on detailed observations and monitoring, Linde's research team documented slip along a central California segment of the San Andreas fault that had it occurred in a time frame of a few minutes, should have produced a moment magnitude 4.8 earthquake. However, this slip event persisted for about a week and thus did not produce any earthquakes! High, pore pressures in rock along the fault and low residual stresses were suggested as possible explanations for such episodes of accelerated aseismic slip.

Supplemental References

Bolt, B., 1993, Earthquakes: W. H. Freeman and Company, New York, NY, 331 p.

Camelbeeck, T. and Meghraoui, M., 1996, Large earthquakes in Northern Europe more likely than once thought: EOS, v. 77, no. 42, p. 406 & 409

This well-illustrated article presents current results of historical and paleoseismological studies that point toward higher-than-previously believed seismic risks in the northwest part of the Rhine graben in The Netherlands, Belgium, and northwestern Germany. These are ominous findings for an area partly below sea level and so dependent on manmade structures for protection against river flooding and North Sea storm surges. Near-surface water tables and unconsolidated or weakly consolidated surface materials over most of this area also raise the specter of extensive coseismic liquefaction and other soft-sediment deformations during an earthquake.

DeMets, C. and others, 1995, Anticipating the successor to Mexico's largest historical earthquake: EOS, v. 76, no. 2, October, p. 417 & 424

Fuis, G. S. and numerous other authors, 1996, Images of crust beneath southern California will aid study of earthquakes and their effects: EOS, v. 77, no. 18, p. 173 & 176

Gore, R., 1995, Living with California's faults: National Geographic Magazine, v. 187, no. 4, April, p. 2-35

Holzer, T. L., 1995, The 1995 Hanshin-Awaji (Kobe), Japan, earthquake: GSA Today, v. 5, no. 8, August, p. 153-156 & 165

Holzer, T. L. and five other authors, 1996, Seismograms offer insight into Oklahoma City bombing: EOS, v. 77, no. 41, p. 393 & 398-399

Keller, E. and Pinter, N., 1995, Active tectonics: Earthquakes, uplift and landscape: Prentice Hall, Upper Saddle River, NJ, 348 p.

Leith, W., 1995, Sakhalin earthquake renews concerns about seismic safety in the former Soviet Union: EOS, v. 76, no. 26, June, p. 257-258

Penick, J. Jr., 1976, The New Madrid earthquakes of 1811-1812: Univ. of Missouri Press, Columbia, MO 65201, 181 p.

Reid, T. R., 1995, Kobe awakes to a nightmare: National Geographic Magazine, v. 188, no. 1, July, p. 112-136

Rial, J. A., Saltzman, N. G., and Ling, H., 1992, Earthquake-induced resonance in sedimentary basins: American Scientist, v. 80, no. 6 (Nov.-Dec.), p. 566-578

Sheppard, P. R. and White, L. O., 1995, Tree-ring responses to the 1978 earthquake at Stephens Pass, northeastern California: Geology, v. 23, p. 109-112

Thurber, C., Roecker, S., Lutter, W., and Ellsworth, W, 1996, Imaging the San Andreas fault with explosion and earthquake sources: EOS, v. 77, no. 6, February, p. 45 & 57-58

Volcanic Activity • Chapter 8; Home Study Assignment

1. Which Cascade Range volcano erupted in 1980? It was the first volcanic eruption in the United States, excluding Alaska and Hawaii, since Mt. Lassen, California, erupted in 1915-16.   a. Rainier   b. Shasta   c. Kilauea   d. St. Helens

2. Which statement is true?
a. A high viscosity lava will flow more readily and easily than a low viscosity lava.
b. A low viscosity lava will flow more readily and easily than a high viscosity lava.

3. In general, which magma will be more likely to erupt explosively?
a. a highly viscous magma such as rhyolite
b. a relatively lower viscosity magma such as basalt

4. Which one commonly forms lava flows with smooth, locally ropy surfaces?
a. basalt erupted from a shield volcano
b. rhyolite erupted explosively from a rising lava dome

5. Which is the overall, most abundant, volatile or gaseous constituent dissolved in magmas and released during volcanic eruptions?
a. water vapor   b. carbon monoxide   c. hydrogen chloride   d. nitrogen

Questions 6-10: Matching; Varieties of Volcanoes
a. shield volcanoes   b. composite or stratovolcanoes   c. cinder cones

6. (   ) massive, gently sloping volcanoes built of successive, basaltic lava flows
7. (   ) large, fairly steep-sided cones composed of lavas and pyroclastic layers
8. (   ) small basaltic cones built during one, short, eruptive episode
9. (   ) volcanoes of southwest Alaska and the Aleutian Islands
10. (   ) volcanoes of Hawaii

11. Which response describes a caldera rather than a crater?
a. a small, steep-sided, volcanic depression bored out by an eruptive plume
b. a large volcanic depression caused by collapse after a large, explosive eruption

12. Which volcanic feature has its slope angles determined by the angle of repose?
a. cinder cone   b. lava dome   c. shield volcano   d. stratovolcano

13. Which is the common lava erupted from the active volcanoes of Hawaii and Iceland?   a. basalt   b. rhyolite   c. andesite

14. Which ocean basin is rimmed by the most subduction zones and active stratovolcanoes?   a. Atlantic   b. Indian   c. Pacific   d. Arctic

15. Did Earth's climate temporarily warm up or cool off in the few years following the major 1991 eruption of Mt. Pinatubo?   a. cooled off   b. warmed up

16. Which volcanic centers are associated with deep mantle hot spots? **one answer**
a. Vesuvius and the other volcanoes of Italy
b. the volcanoes of Hawaii and the Yellowstone National Park area
c. the very young basaltic cinder cones scattered along the Ring of Fire

17. Which was the largest, most powerful, explosive volcanic eruption of historic time?   a. Mt. Pelée, Martinique, 1902   b. Tambora, Indonesia, 1815
c. Vesuvius, Italy, 79 A. D.   d. Nevado del Ruiz, Colombia, 1985

18. Which phenomenon occurs for a few years after major explosive volcanic eruptions such as Tambora and Pinatubo? **one answer**
a. heavy falls of volcanic ash within 100 km of the volcano
b. unusually warm weather in the tropics and subtropics
c. a world-wide rise in sea level of one to three centimeters
d. brilliantly colored sunrises and sunsets

19. Which **one** statement about the May, 1980, eruption of Mount St. Helens is false?
a. During the 1980 eruptive cycle, the mountain peak was substantially built up by lava flows and pyroclastic debris.
b. An eruptive plume of ash rose high into the atmosphere.
c. Mudflows accompanied the volcanic activity.
d. The most powerful explosive eruption was preceded by a massive landslide.

20. Generally, which are the two, most abundant gases emitted during volcanism?
a. chlorine, sodium   b. oxygen, nitrogen   c. helium, neon   d. water, carbon dioxide

21. In 1902, what volcanic event destroyed the city of St. Pierre, Martinique?
a. mudflows   b. slow-moving, basaltic lava flows   c. a heavy ashfall
d. a fast-moving nuée ardente generated from a pyroclastic flow

22. Which **one** statement concerning cinder cones is not true?
a. They are small volcanoes with fairly steep sides.
b. They are built mostly or entirely during one eruptive cycle.
c. The cinders and other tephra are tightly welded into consolidated rock.
d. The cinders and other pyroclastic particles are most commonly basalt.

23. Which kind of eruptive activity is most likely to be highly explosive?
a. lava flows from a shield volcano   b. basaltic lavas erupted from fissures
c. cinder cone activity   d. summit eruptions of composite cones or stratovolcanoes

24. Which kind of volcanism is typical of oceanic ridge systems?
a. explosive, composite cones   b. andesite lava flows and pyroclastic eruptions
c. submarine, basaltic lava flows   d. rhyolitic, pyroclastic-flow activity

25. Kilauea and Mauna Loa are what kind of volcanoes?
a. explosive, rhyolitic volcanoes   b. basaltic shield volcanoes
c. andesitic stratovolcanoes   d. small, basaltic cinder cones

26. Large, composite cones or stratovolcanoes have average compositions similar to which lava?   a. basaltic   b. larharic   c. andesitic   d. rhyolitic

27. How are the Iceland volcanoes related to Earth's tectonic plates?
a. They lie on a spreading center where two plates are converging.
b. They lie on a subduction zone where two plates are converging.
c. They lie on a spreading center where two plates are moving apart.
d. They lie along a subduction zone where two plates are diverging.

28. Which one best describes volcanism in the Cascade Range, northwestern U. S.?
a. related to a hot spot   b. related to an oceanic ridge system
c. related to a subduction zone   d. related to deep, transform faults

29. The big Hawaiian volcanoes are best described how?
a. are directly above a transform plate boundary that cuts deeply into the mantle
b. are directly above an active subduction zone
c. lie along the crest of an oceanic ridge or spreading center
d. are in the interior of the Pacific plate above a hot spot deep in the mantle

30. Which one precedes a summit eruption of the volcano Kilauea on Hawaii?
a. inflation and uplift of the summit area
b. violent, submarine eruptions occur just offshore
c. emission of volcanic gases stops completely for a few hours
d. the summit caldera sinks rapidly and tilts toward the ocean

31. Note the ash cloud path and pattern of airfall ash deposits associated with the May 18, 1980 explosive eruption of Mount St. Helens, WA. See Figure 8.29, p. 222. Which response best characterizes the atmospheric wind patterns on the day of the eruption and on the succeeding few days?
a. weak to absent, low altitude winds; high altitude winds blowing to the west at low velocities
b. fairly strong low altitude winds blowing toward the east; consistent, strong, upper altitude jetstream winds blowing to the east-southeast as far as the middle Mississippi River Valley, then curving east-northeast over New England
c. low and high altitude winds were very strong but blowing in opposite directions, low altitude winds toward the east and jetstream winds toward the west
d. strong low altitude winds directed toward the north; high altitude jetstream winds, highly variable in direction, spread the ash over a broad area extending from the Gulf of Mexico to the southern prairie provinces of Canada

Questions 32-37; Matching: Volcanoes and Historical Eruptions

a. Mt. Unzen   b. Mt. Vesuvius   c. Heimaey Island
d. Krakotoa   e. Nevado del Ruiz   f. Loihi

32. ( ) This volcano was dormant for many centuries prior to a major eruption that buried the Roman city of Pompeii. Italy, 79 A. D.

33. ( ) Mudflows, triggered by hot ash and pyroclastic flows erupted onto snow and glacial ice in the summit region, devastated riverbank villages on the lower flanks of the volcano. Colombia, 1985

34. ( ) A tsunami, caused by a massive pyroclastic flow hurtling into the sea, and a nuée ardente, detached from the same pyroclastic flow, devastated nearby coastal areas. Indonesia, 1883

35. ( ) Cinders, ash, and lava flows from a newly formed basaltic vent threatened to overwhelm a normally quiet, peaceful, commercial fishing village. Iceland, 1973

36. ( ) This active, submarine volcano is growing off the southern coast of Hawaii.

37. ( ) Renewed explosive volcanic activity produced damaging mudflows, some lava flows, and numerous, very dangerous, pyroclastic flows. Japan, 1991

38. What caused the brilliantly colored sunrises and sunsets seen in the five or so years following the 1991 eruption of Mt. Pinatubo, Philippine Islands?
a. The eruption added large amounts of carbon dioxide to the atmosphere.
b. The eruptive cloud destroyed parts of the Earth's protective ozone layer.
c. Radioactive uranium atoms blown into the atmosphere glowed red as they decayed.
d. Sulfur dioxide and other erupted gases formed aerosols in the lower stratosphere.

Questions 39-47; Matching: Localities and Volcanic Events

a. Lake Nios, Cameroon
b. Mount Rainier
c. Mammoth Lakes, CA
d. Heimaey Island, Iceland
e. Spirit Lake
f. St. Pierre, Martinique
g. Yellowstone Nat. Park
h. Kalapana
i. Lake Taal

39. ( ) Very large magnitude, Pleistocene and earlier explosive eruptions produced extensive rhyolitic, ashflow tuff deposits, a huge caldera, and thick, voluminous, caldera-filling, rhyolite lava flows. It is the world's largest geothermal field.

40. ( ) This area was virtually filled and recontoured by debris from the major landslide that decapitated the laterally bulging, summit region magma chamber and triggered the catastrophic, May 18 explosive eruption.

41. ( ) This village on Hawaii was destroyed by lava flows in 1990; subsequent geologic studies showed this area was highly vulnerable to such a fate.

42. ( ) Thousands of people and their livestock were asphyxiated by carbon dioxide emitted from a normally tranquil, density-stratified water body occupying an extinct volcanic crater.

43. ( ) In recent years, this popular ski resort has been beset with swarms of earthquakes, resurgence, and increased geothermal activity, all indicative of a rising subsurface magma body and possible renewed explosive volcanic activity.

44. ( ) A fast moving nuée ardente, unleashed during the major, explosive, pyroclastic flow-forming event on Mt. Pelée, wiped out this coastal city of 30,000 inhabitants; it was all over in a "heartbeat".

45. ( ) Water flooding was successfully utilized to deflect, reroute, and stop basaltic lava flows threatening to overrun a coastal fishing port and its harbor.

46. ( ) Fast moving, surge clouds of superheated steam and fine-sized rock particles erupted from a stratovolcano of the same name are an ever-present threat to local residents.

47. ( ) The gigantic, presettlement, Electron and Osceola mudflows originated here; similar events today would cause extensive casualties and damage in river valleys of the Puyallup River basin and in the Puget Sound lowlands.

48. Which two statements concerning the Long Valley Caldera, California, are reasonably correct? The other two contain serious factual errors.
a. The caldera is elliptically shaped; its long diameter is about 25 km, and most of the most recent lava domes and dome vents are clustered along the western margin of the caldera.
b. The roughly circular, 10-km diameter resurgent area in the west central part of the caldera is believed to overlie a still, partially molten, magma body with its top about 9 km below the surface.
c. The rhyolite domes and vents in the Mono and Inyo Craters areas were localized along faults that cut the caldera-fill rock deposits. However the Mono and Inyo Craters activity predated eruption of the Bishop Tuff and collapse of the caldera.
d. The Bishop Tuff was erupted 100,000 years ago during the major, caldera-forming explosive event. The youngest volcanic activity in the area consisted of basaltic lava flows erupted from vents centered in the area of the 1980-82 uplift.

Questions 49-52; Do you agree (A) or disagree (D) with the following statements? Briefly explain any disagreements.

49. ( ) The volcanic mudflows that buried Amero and other river valley communities on the lower flanks of Nevado del Ruiz were sudden and unexpected events that could not have been anticipated in any rational way.

50. ( ) Locally increased seismic activity, new areas of heated ground, and increased emissions of sulfur dioxide may signal that a new batch of magma is rising beneath a volcanic vent or crater.

51. ( ) Increased slope tilting and inflation accompany the gradual rise of magma into the summit region of the Hawaiian volcano Kilauea. The same summit region subsides once the eruption is in progress.

52. ( ) Heavy volcanic ashfalls can cause total darkness, roofs to collapse, extensive damage to communications and electronic equipment, and severe respiratory stress in humans and animals.

53. On the provided world map, accurately plot and label the following countries, features, or locations.
a. Iceland   b. Philippine Islands   c. Cascade Range volcanoes
d. Island of Martinique, West Indies   e. Yellowstone National Park
    f. Long Valley caldera; Mammoth Lakes area, California
        g. Krakatoa; between Java and Sumatra, Indonesia
            h. Lake Nios, Cameroon Republic, Africa

Environmental Geology Quiz • Chapter 8

1. Which Cascade Range volcanic eruption, 1980, was the first in the United States, excluding Alaska and Hawaii, since Mt. Lassen, California, erupted in 1915-16.
   a. Shasta   b. St. Helens   c. Rainier   d. Kilauea

2. Which statement is true?
a. A low viscosity lava will flow more readily and easily than a high viscosity lava.
b. A high viscosity lava will flow more readily and easily than a low viscosity lava.

3. In general, which magma will be more likely to erupt explosively?
a. a relatively lower viscosity magma such as basalt
b. a highly viscous magma such as rhyolite

4. Which is the overall, most abundant, volatile or gaseous constituent dissolved in magmas and released during volcanic eruptions?
a. nitrogen   b. carbon monoxide   c. hydrogen chloride   d. water vapor

Questions 5-7: Matching; Varieties of Volcanoes
   a. cinder cones   b. composite or stratovolcanoes   c. shield volcanoes

5. ( ) massive, gently sloping volcanoes built of successive, basaltic lava flows; volcanoes of Hawaii

6. ( ) large, fairly steep-sided cones composed of lavas and pyroclastic layers; volcanoes of the Cascade Range, southwest Alaska, and the Aleutian Islands

7. ( ) small, unconsolidated, basaltic volcanic structures built during one, short, eruptive episode; Sunset Crater, AZ and Helgafell, Iceland

8. Which response describes a caldera rather than a crater?
a. a small, steep-sided, volcanic depression bored out by an eruptive plume
b. a large volcanic depression caused by collapse after a large, explosive eruption

9. Which volcanic feature has its slope angles determined entirely by the angle of repose?   a. cinder cone   b. lava dome   c. shield volcano   d. stratovolcano

10. Which is the common lava erupted from the active volcanoes of Hawaii and Iceland?   a. andesite   b. basalt   c. rhyolite

11. Which was the largest, most powerful, explosive volcanic eruption of historic time?   a. Mt. Pelée, Martinique, 1902   b. Tambora, Indonesia, 1815
c. Vesuvius, Italy, 79 A. D.   d. Nevado del Ruiz, Colombia, 1985

12. Which phenomenon occurs for a few years after major explosive volcanic eruptions such as Tambora and Pinatubo? **one answer**
a. brilliantly colored sunrises and sunsets
b. a world-wide rise in sea level of one to three centimeters
c. unusually warm weather in the tropics and subtropics
d. heavy falls of volcanic ash within 100 km of the volcano

13. In 1902, what volcanic event destroyed the city of St. Pierre, Martinique?
a. a heavy ashfall   b. slow-moving, basaltic lava flows   c. mudflows
d. a fast-moving nuée ardente generated from a pyroclastic flow

14. Which kind of eruptive activity is most likely to be highly explosive?
a. lava flows from a shield volcano   b. basaltic lavas erupted from fissures
c. cinder cone activity   d. summit eruptions of composite cones or stratovolcanoes

15. Kilauea and Mauna Loa are what kind of volcanoes?
a. andesitic stratovolcanoes   b. small, basaltic cinder cones
c. explosive, rhyolitic volcanoes   d. basaltic shield volcanoes

16. Large, composite cones or stratovolcanoes have average compositions similar to which lava?   a. basaltic   b. rhyolitic   c. andesitic

17. Which **two** statements concerning plate tectonics and volcanism are correct?
a. The volcanoes of Iceland lie on a spreading center; two plates are moving apart.
b. The Cascade Range volcanoes lie on a spreading center; two plates are diverging.
c. The Cascade range volcanoes lie above a subduction zone; two plates are converging and one is sinking beneath the other.
d. The volcanoes of Iceland lie above a subduction zone; two plates are converging and one is sinking beneath the other.

18. Which **one** precedes a summit eruption of the volcano Kilauea, Hawaii?
a. the summit caldera sinks rapidly and tilts toward the ocean
b. violent, submarine eruptions occur just offshore
c. emission of volcanic gases stops completely for a few hours
d. inflation and uplift of the summit area

Questions 19-24; Matching: Volcanoes and Historical Eruptions

a. Krakatoa   b. Nevado del Ruiz   c. Loihi
d. Heimaey Island   e. Mt. Vesuvius   f. Mt. Unzen

19. ( ) This volcano was dormant for many centuries prior to a major eruption that buried the Roman city of Pompeii. Italy, 79 A. D.

20. ( ) Mudflows, triggered by hot ash and pyroclastic flows erupted onto snow and glacial ice in the summit region, devastated riverbank villages on the lower flanks of the volcano. Colombia, 1985

21. ( ) A tsunami, caused by a massive pyroclastic flow hurtling into the sea, and a nuée ardente, detached from the same pyroclastic flow, devastated nearby coastal areas. Indonesia, 1883

22. ( ) Cinders, ash, and lava flows from a newly formed basaltic vent threatened to overwhelm a normally quiet, peaceful, commercial fishing village. Iceland, 1973

23. ( ) This active, submarine volcano is growing off the southern coast of Hawaii.

24. ( ) Renewed explosive volcanic activity produced damaging mudflows, some lava flows, and numerous, very dangerous, pyroclastic flows. Japan, 1991

25. What caused the brilliantly colored sunrises and sunsets seen in the five or so years following the 1991 eruption of Mt. Pinatubo, Philippine Islands?
a. Radioactive uranium atoms blown into the atmosphere glowed as they decayed.
b. Sulfur dioxide and other erupted gases formed aerosols in the lower stratosphere.
c. The eruptive cloud destroyed parts of the Earth's protective ozone layer.
d. The eruption added large amounts of carbon dioxide to the atmosphere.

Questions 26-31; Matching: Localities and Volcanic Events

a. Rainier
b. Heimaey
c. Mammoth Lakes
d. Taal
e. Yellowstone
f. Nios or Nyos

26. ( ) national park, U. S.; Very large magnitude, Pleistocene and earlier explosive eruptions produced extensive rhyolitic, ashflow tuff deposits, a huge caldera, and thick, voluminous, caldera-filling, rhyolite lava flows.

27. ( ) lake in Cameroon; Thousands of people and their livestock were asphyxiated by carbon dioxide emitted from this normally tranquil, density-stratified water body occupying an extinct volcanic crater.

28. ( ) town and resort area in California; In recent years, this popular ski resort has been beset with swarms of earthquakes, caldera resurgence, and increased geothermal activity, all indicative of a rising subsurface magma body and suggesting possible renewed explosive volcanic activity.

29. ( ) island cluster, Iceland; Water flooding was successfully utilized to deflect, reroute, and stop basaltic lava flows threatening to overrun a coastal fishing port.

30. ( ) lake and active volcano, Philippine Islands; Fast moving, surge clouds of superheated steam and fine-sized rock particles erupted from the stratovolcano are an ever-present threat to local residents.

31. ( ) volcano in the Pacific Northwest, U. S.; The Electron and Osceola mudflows originated here; similar mudflows today would cause extensive casualties and damage in river valleys and heavily populated lowlands west of the mountain.

Questions 32-33; Do you agree (A) or disagree (D) with the following statements? Briefly explain the reasons for any disagreements.

32. ( ) Locally increased seismic activity, new areas of heated ground, and increased emissions of sulfur dioxide may signal that a new batch of magma is rising beneath a volcanic vent or crater.

33. ( ) Subsidence of the summit caldera region accompanies the gradual rise of magma in the Hawaiian volcano Kilauea. The same summit region rebounds back to its pre-eruption elevation once the eruption is over.

34. On the map provided, plot and label these countries, areas, or features.
a. Iceland
b. Philippine Islands
c. Long Valley caldera
d. Island of Martinique, West Indies
e. Yellowstone National Park

Answers; Home Study Assignment
Volcanic Activity • Chapter 8

1. d	2. b	3. a	4. a	5. a
6. a	7. b	8. c	9. b	10. a
11. b	12. a	13. a	14. c	15. a
16. b	17. b	18. d	19. a	20. d
21. d	22. c	23. d	24. c	25. b
26. c	27. c	28. c	29. d	30. a
31. b	32. b	33. e	34. d	35. c
36. f	37. a	38. d	39. g	40. e
41. h	42. a	43. c	44. f	45. d
46. i	47. b	48. a, b	49. D*	50. A
51. A	52. A			

*

49. The mudflows were predictable on the basis of the explosive volcanic activity, ice, and snow in the summit region. Summit eruptions had triggered previous mudflows in the same downstream localities. Warnings and evacuation orders should have been given. The disastrous results were simply due to a communications breakdown or criminal negligence, depending on your point of view.

Answers: Environmental Geology Quiz • Chapter 8

1. b	2. a	3. b	4. d	5. c
6. b	7. a	8. b	9. a	10. b
11. b	12. a	13. d	14. d	15. d
16. c	17. a, c	18. d	19. e	20. b
21. a	22. d	23. c	24. f	25. b
26. e	27. f	28. c	29. b	30. d
31. a	32. A	33. D*		

*

33. The rise of magma into the Kilauea volcano is accompanied by inflation, swelling and outward tilting, of the summit region. Following the summit eruption, often coincident with a flank eruption, the summit region deflates, shrinks and tilts inward, and is restored to its pre-eruption configuration.

Commentary/Media Watch

Pyroclastic Flows

Pyroclastic flows, also called glowing avalanches, originate in at least three ways; 1) mass wastage failure of growing summit lava domes or spines, 2) fluidized vent material containing abundant suspended fragments overflowing the crater rim, or 3) collapse of an eruption column. The first two usually produce small-volume flows that seldom reach the lower flanks of the volcano. Large-volume pyroclastic flows originate via collapse of particle-rich eruption columns.

Pyroclastic flows are particle-rich, ground-hugging avalanches that follow canyons and valleys down the flanks of the volcano. Flows unleashed during large caldera-forming eruptions, such as those from Long Valley caldera that deposited the Bishop Tuff, can run out great distances from their ring-fracture vents and spread thick, nonstratified, pyroclastic-flow (ashflow) deposits over vast areas.

As they move, pyroclastic flows generate highly turbulent, roiling clouds of heated gases with dilute suspensions of fine-size particles. Clouds hot enough to radiate visible light, usually dark red, are known as nuée ardentes. Rapid heating of entrained air and low particle density impart the cloud with extreme fluidity and violent convective turbulence; net forward speeds match those of the basal avalanche. Although damage due to hurricane-force winds and strong turbulence may be extreme, cloud deposits typically consist only of thin mantles of silt- and dust-size ash, seldom more than a foot or so in thickness. These highly-fluid, roiling clouds can become separated from their parental pyroclastic flows and continue moving more or less as independent entities. Needless to say, they are highly mobile and extremely dangerous. Two of the most disastrous explosive volcanic events in history, the destruction of St. Pierre, Martinique, 1902, and the heavy casualties and devastation along the coast of Sumatra during the Krakatoa explosive event, 1883, involved such independently-moving, detached clouds.

Pyroclastic flows unleashed during the devastating Mount Pelée eruption followed river valleys and canyons, entering the sea well beyond the city limits of St. Pierre. However, as the major pyroclastic flow followed around an abrupt bend in a river valley, the associated cloud detached itself, jumped the valley divide, and surged in a straight-line path directly toward St. Pierre. It was all over in a matter of minutes; there were two survivors (p. 210 in Keller; Bullard, 1979). The city was essentially leveled by powerful hurricane-plus force winds and the silent ruins were covered with a thin mantle of fine-grained volcanic ash.

During the most violent phase of the Krakatoa eruption, two different, eruption-related entities devastated nearby coastal areas of Sumatra. The first was a hot, roiling cloud similar to the one the struck St. Pierre. As voluminous, fast-moving, subaerial, pyroclastic flows on Krakatoa entered the sea, the associated clouds were detached and continued on across the 30 or so miles of open water to strike the coast. In addition, the pyroclastic flows entered the sea at such high rates of speed and in such large volumes that enough seawater was displaced fast enough to generate tsunamis; these struck the same coastal areas previously devastated by the hot, roiling clouds.

Volcanic Surges

Volcanic surges, such as those that make Taal (p. 210) such a dangerous volcano, are much like the roiling clouds associated with pyroclastic flows, but they lack a distinct, high-particle-density basal avalanche. Most of the surge is a fast-moving, highly turbulent cloud with dilute concentrations of fine-grained particles. At ground level, sand-size and larger particles may be entrained. Surge dynamics simulate movements of widely-spaced particles entrained in a strong wind, not of densely-concentrated particles in a pyroclastic flow. Surge deposits characteristically display good particle size sorting and stratification, either laterally extensive planar beds or dune-style cross stratification. Super, hurricane-force wind speeds and extreme turbulence result in leveling of forests and structures impacted by the surge.

Surges are generated by collapse of unstable eruption columns. Whereas pyroclastic flows are generated by the continuing collapse of "stable" eruption columns, surges result when exit velocities and particle densities at the vent mouth fluctuate widely, usually during the initial stages of an explosive eruption. Surge particles may be mostly lithic or fragmented new magma. Surges generated by steam explosions, such as those noted at Taal Volcano, are known as phreatic surges; those generated partly by steam explosions and partly by magma fragmentation are designated as phreatomagmatic surges. The former are fairly cool (just a few hundred degrees centigrade) and carry mainly lithic particles; the latter may be much hotter and carry both lithic particles and fragmented particles of new magma. The main, 1980, Mount St. Helens eruption was evidently a phreatomagmatic surge generated when rock saturated with pressurized, superheated water and gas-charged, near-surface magma were suddenly decompressed to atmospheric pressure by the massive Goat Island slope failure.

Volcanological History of Mount Vesuvius

The famous 79 A. D. eruption of Mount Vesuvius, near Naples, Italy (Table 8.1, p. 203; Bullard, 1979) makes for an informative, entertaining classroom lecture and/or project on volcanism and volcanic hazards. The volcano had evidently been dormant for hundreds of years prior to the eruption, and its volcanic origin was evidently unrecognized or only dimly perceived by local residents. The cone had been largely destroyed in an earlier eruption. Only a flat-topped, truncated cone, heavily vegetated with scrub trees and vines and a large central depression, probably a summit caldera, remained; one hundred and fifty years earlier, these brush-covered slopes and summit depression were exploited as a nearly impregnable hideout by Spartacus and his rebellious band of escaped gladiators and other slaves.

The cities of Pompeii and Herculaneum had grown and prospered on the broad, lower flanks of the volcano. Pompeii in particular, had a notorious reputation as a center for "holiday relaxation and sinful enterprises". Both cities were buried during the 79 A. D. event. Well-stratified ash and pumice beds, "eyewitness" historical accounts, and positions of victims indicate that Pompeii was buried by ashfalls. Some trapped victims evidently managed to stay alive for days, perhaps even a week or so, before succumbing to asphyxiation, dehydration, and starvation. The ashfalls were so heavy that all sunlight was blotted out; those who waited too long to escape were faced with absolute darkness. Only a "lights out" experience in a cavern can give the same awareness of "total darkness". Herculaneum, in contrast, escaped the heaviest airfalls but was buried suddenly, probably by pyroclastic flows or by mudflows or avalanches generated from ash that had accumulated on the higher slopes of the volcano. The subsequent discoveries (mid eighteenth century) and archeological importance of the two buried cities and the "modern" volcanological history of Vesuvius provide a nice wrap-up to the 79 A. D. story (see Bullard, 1979).

Soufrière Hills Volcano, Island of Montserrat, Lesser Antilles

Newspapers and CNN reports, as of September, 1996, documented the continuing, sporadic eruptive activity at Soufrière Hills Volcano on the Island of

Montserrat, Lesser Antilles. The current episode began in July, 1995. During the most recent eruption, buildings in an evacuated area were set afire and the capital city of Plymouth was coated with a dusting of volcanic ash. There was obviously still significant danger of a major, explosive eruption, and forced evacuations are put into motion at the first signs of impending activity. Continued activity, shown in brief reports on CNN and Headline News (December, 1, 1996), had authorities worried about a possible impending catastrophic caldera-forming eruption that also might trigger coastal tsunamis if rapidly-moving, large-volume pyroclastic flows reached the sea. Active volcanoes/fumarole fields named Soufrière or La Soufrière are also present on Saint Vincent and Guadeloupe. Soufrière on St. Vincent most recently erupted in the 1970s; an explosive eruption in 1902 resulted in heavy loss of life. This later eruption, May 7, occurred on the day before the disastrous 1902 eruption of Mt. Pelée, Martinique (p. 210).

Popocatépetl Volcano, Mexico

Sporadic, small-scale, eruptive activity since late 1994 of Popocatépetl, one of the giant stratovolcanoes rimming the Valley of Mexico, caused renewed concern about the vulnerability of the heavily populated metropolis to volcanic activity. CNN featured video of the earlier activity and of the eruption column of October 28 and 29, 1996. Muds flows triggered by ashfalls, snowmelt, and mass wastage failures in the summit region of the volcano and heavy, regional ashfalls present the greatest threat. Siebe and others (1996; see Supplemental References), in a very timely study, worked out the Prehispanic history of the volcano and documented numerous past, eruptive events that, were they to happen today, would produce disastrous consequences in the Valley. Similar hazards, courtesy of the big Cascade Range stratovolcanoes such as Mount Rainier, present ominous threats to parts of Washington, Oregon, and northern California (p. 214-215).

Ruapehu Volcano, New Zealand

Eruptive activity of Ruapehu Volcano, New Zealand, in June and July, 1996, attracted media coverage that included images of the eruptive plumes and interviews with public officials and volcanologists (CNN) and short articles in U. S. newspapers. The eruptions were of particular concern for three reasons: 1) the activity might be a prelude to a possible violent, catastrophic, explosive eruption; 2) the mountain and its surroundings had blossomed into a popular ski resort in recent years and, being at the height of the southern hemisphere winter, the summit had a heavy snowpack and resort hotels were full; and 3) the volcano is close enough to Auckland so that ashfalls presented a real threat to the city. For example, the airport was closed temporarily due to poor visibility, and, of course, heavy ashfalls can reduce visibility to zero, ash is highly detrimental to living organisms, and it is lethal to electronics equipment and delicate machinery (p. 209-210). The renewed activity at Ruapehu prompted a thorough review of volcanic hazards and land use issues associated with the volcano (Bryan and others of the Ruapehu Surveillance Group, 1996; see Supplemental References).

Sub-Glacial Meltwater Bursts, Iceland

In early October, CNN and its sister station Headline News showed short video scenes of steam and ash eruption plumes from the Icelandic volcano Loki. The eruption occurred beneath the Vatnajokull ice cap and glacier, and the eruptive plume had just broken through to the atmosphere. Public safety authorities were very concerned about the potential for a jökulhlaup, a sudden flood triggered by bursting of a subglacial meltwater lake impounded beneath the ice sheet. Such floods are well-known to local authorities and rank among the worst natural disasters to have struck Iceland during historic time. Eventually on November 5, the meltwater did break out in much larger volumes than expected, causing severe flooding and destroying two major highway bridges (see Glacial burst overwhelms Skeidará, EOS, v. 77, no. 47, p. 470). Fortunately, this area is fairly remote and its few inhabitants had long since been evacuated. The danger is not yet over because the eruptive activity was renewed soon after meltwater release.

Supplemental References

Blong, R. J. 1984, Volcanic hazards: A sourcebook on the effects of eruptions: Academic Press; Harcourt Brace Jovanovich Pub., London, U. K., 424 p.
This is a comprehensive reference concerning volcanic hazards.

Bryan, C. and others of the Ruapehu Surveillance Group, 1996, Volcanic eruption at a New Zealand ski resort prompts reevaluation of hazards: EOS, v. 77, no. 20, p. 189-191

Bullard, F. M., 1976, Volcanoes of the Earth: Univ. of Texas Press, Austin, TX, 579 p.
This is a comprehensive work on basic volcanology, historical eruption patterns, and environmental effects of volcanism and volcanic eruptions.

Ladbury, R., 1996, Model sheds light on a tragedy and a new type of eruption: Physics Today, v. 49, no. 5 (May), in Search and Discovery section, p. 20-22
This article is an informative review of the Lake Nyos, Cameroon, eruption event (p. 210 & 214 in Keller).

Larsen, D., 1993, Recovery of Spirit Lake: American Scientist, v. 81, p. 166-177
This article documents the ecological recovery of Spirit Lake following the 1980 eruption of Mount Saint Helens.

Siebe, C., Abrams, M., Macias, J. L., and Obenholzner, J., 1996, Repeated volcanic disasters in Prehispanic time at Popocatépetl, central Mexico: Past key to the future? Geology, v. 24, no. 5, p. 399-402

Torres, R., Self, S., and Punongbayan, R., 1995, Attention focuses on Taal: Decade volcano of the Philippines: EOS, v. 76, no. 24, p. 241, 246-247

Coastal Hazards • Chapter 9; Home Study Assignment

1. Which **two** segments of the U. S. Atlantic and Gulf coasts are most likely to be struck by a hurricane in any given year?
a. Long Island, NY to Cape Cod, MA   b. southern Florida and the Florida Keys
c. Georgia and north Florida   d. the central Texas coastline

2. Where is the Thames barrier and what is its purpose?
a. protect against storm surges coming in from the North Sea; London, England
b. protect against coastal erosion; Normandy coastline, France
c. encourage sand accumulation along resort beaches; Miami, Florida
d. protect against tropical cyclone storm surges; Ganges River Delta, Bangladesh

Questions 3-9; Matching: Waves

a. period   b. fetch   c. swell   d. wave length
   e. height   f. L/T   g. 0.5 L

3. ( ) distance that wave-generating winds blow across open water
4. ( ) horizontal distance between neighboring wave crests
5. ( ) vertical distance from tough to crest of a wave
6. ( ) typical incoming waves at the beach; generated in distant storms at sea
7. ( ) time required for one complete wave form (crest to crest) to pass a fixed point
8. ( ) velocity of deep water waves
9. ( ) minimum water depth for a deep water wave

Questions 10-14; Matching: Beach Morphologic Features; See Figure 9.9, p. 236.

a. longshore trough   b. swash zone   c. longshore bar
d. base of seacliff/dune line   e. berm   f. surf zone

10. ( ) marks highest elevation reached by the strongest storm waves
11. ( ) relatively level to gently sloping, beach surface built by seasonal deposition from summer swell or winter storm waves
12. ( ) ( ) two zones comprising the zone of littoral transport
13. ( ) ( ) submerged, bottom features of the surf and breaker zones
14. ( ) part of the zone of littoral transport subject to the landward and seaward rush of water on a beach

15. Which **two** responses concerning typhoons are correct?
a. severe winter storms in the northern Pacific
b. Southern Hemisphere hurricanes
c. the special name for tropical cyclonic storms in the western Pacific
d. the same type of storms as hurricanes

16. Which **two** responses concerning plungers are reasonably correct?
a. a surfer's bummer   b. favored by steep offshore beach slopes
c. favored by gentle offshore beach slopes   d. a surfer's dream

17. A hurricane is heading due west toward the south Florida coast. Which **two** responses are correct or reasonably correct?
a. The strongest onshore winds and highest storm surge will generally be on the north side of the eye.
b. After crossing Florida, the storm could reform or intensify and threaten the Gulf Coast.
c. This area is largely rural, and as a result, property damages will be relatively slight.
d. Unusually high atmospheric pressures act to temporarily reduce wind speeds and lower the height of the storm surge as the eye passes overhead.

18. What is meant by the term fetch?
a. the beach front area where dogs are trained to retrieve sinking swimmers
b. a large expanse of open water over which the wind blows and generates waves
c. the rotational movements of water particles beneath a passing, surface wave

19. Which **two** statements are true for waves moving from deep waters into shallower waters along a coastline?
a. wave speeds increase and wave lengths decrease
b. wave heights increase and wave speeds decrease
c. wave heights decrease and wave lengths increase
d. wave speeds and wave lengths decrease

20. A deep water wave exists under what conditions? **one answer**
a. the wave length exceeds one-half the wave height
b. the water depth exceeds one-half the wave length
c. the wave length exceeds one-half the water depth
d. the wave height exceeds the water depth

21. Water movement and sand transport parallel to the beach are fundamentally caused by which **one**?   a. a long fetch parallel to the beach
b. deep water waves breaking offshore   c. waves impinging obliquely onto a beach
d. strong, offshore winds creating a pileup of water along the beach front

22. How does refraction affect the crest and trough orientations of incoming waves along a beach?
a. As the waves move into shallower water, the angle between the wave crests (troughs) and the shoreline decreases.
b. As the waves move into shallower water, the angle between the wave crests (troughs) and the shoreline increases.

23. How can crashing, collapsing, storm waves generate explosive forces and stresses on rocky outcrops and manmade structures?
a. oscillating, refractive waves shake the hard materials into small fragments
b. pressurized water and compressed air are driven into cracks and fissures
c. backwash shatters the rock or concrete; rip currents carry the fragments out to deeper water

24. What is a reasonable rate of change for sea level in the past few hundred years?
a. fallen; about 10 inches per century   b. risen; about 30 cm per century
c. risen; about 1 cm per century   d. fallen; about 1 inch per century

25. Which **one**, if its concentration in the atmosphere continues to increase, will eventually produce global warming?
a. volcanic aerosols   b. water vapor   c. ozone   d. carbon dioxide

26. Which **two** structures are built more or less parallel to the beach?
a. seawalls   b. groins   c. jetties   d. breakwaters

27. Ignoring dams and other manmade structures, in which area would rivers and streams supply the larger proportion of sand to the local beach environment?
a. a topographically low coastal area with many large estuaries, such as the Atlantic coastline of North Carolina
b. a topographically rugged, coastal area with steep seacliffs and small estuaries such as the Pacific coastline of California

28. Large estuaries are more common on which coastline?
a. East and Gulf Coasts, U. S.   b. West Coast, U. S.

29. Which area has the smaller, more localized, littoral cells?
a. Atlantic and Gulf Coasts, U. S.   b. Pacific Coast, U. S.

30. When was the Ocean City Inlet cut through the Maryland barrier island system?
a. during hurricane Alexis, 1410   b. during hurricane Hugo, 1989
c. during hurricane Andrew, 1992   d. during a hurricane in 1933

31. The rapidly escalating costs of coastal storm damage in the United States are due mainly to which factor?
a. The number of hurricanes and severe winter storms impacting on U. S. coastlines has risen steadily since about 1900.
b. As coastal development continues, more and more new buildings and other structures become vulnerable to storm damage.

32. Which **one** refers to the broad dome of water moving with the eye and frontal portion of a hurricane?
a. sea dome   b. eyewall tear   c. cyclonic mound   d. storm surge

Questions 33-35; Matching: Beach Preservation Options, Costs, and Environmental Effects

a. Promote coastal development; build massive, hardened structures to stop all but the most powerful storm waves.
b. Prohibit coastal development; build nothing, declare victory over the sea and retreat from the coastline.
c. Allow limited coastal development; make a long-term commitment to beach nourishment.

33. ( ) low initial costs and modest maintenance costs; high potential for storm damage and low to moderate potential for negative environmental effects

34. ( ) high initial costs and eventually high maintenance costs; high potential for storm damage and negative environmental consequences

35. ( ) low initial costs and low maintenance costs; low potential for storm damage and for negative environmental consequences

36. Which two are components of littoral zone transport?
a. beach drift b. storm drift c. longshore drift d. tidal drift

37. Which response correctly describes water movements associated with a deep water wave?
a. water particles move in elliptical paths with increasing diameters downward
b. water particles move in circular paths with decreasing diameters downward
c. water particles move in elliptical paths with decreasing diameters downward
d. water particles move in circular paths with increasing diameters downward

38. Which component of the wave climate in southern California is most responsible for rebuilding sandy beaches during the summer months? See Figure 9.12, p. 238.
a. low wave-height swell originating in far-away, Southern Hemisphere storms
b. powerful, 12 second-period waves that originate from strong storms in the northern Pacific and Gulf of Alaska

Questions 39-46; Agree (A) or Disagree (D)? Briefly explain the basis for any disagreements.

39. ( ) Of the two barrier island segments along the Maryland coast, Assateague Island is the more natural and least developed.

40. ( ) The Ocean City Inlet, Maryland, has been more or less in its present location since before the first white settlers arrived.

41. ( ) Overall sand transport along the Fenwick-Assateague Islands segment of the Atlantic coastal barrier island system is toward the south, but strong storm swell from hurricanes farther south can produce temporary north-flowing longshore currents and sand movements.

42. ( ) Since the Ocean City Inlet jetties, Maryland, were built, sand has accreted on the southeast tip of Fenwick Island; the northeast tip of Assateague Island has experienced sand loss and shoreline erosion.

43. ( ) Storm overwash allows natural, low-profile, barrier islands to gradually migrate landward as sea level rises.

44. ( ) Historical records show variations of about 2 meters in the level of Lake Michigan. Low-stand levels are characterized by wide sandy beaches and low erosion rates; high-stand levels are characterized by narrow beaches, high erosion rates, and major lake front property damage.

45. ( ) Once expensive oceanfront homes, hotels, and condominiums are built, the average taxpayer, through federal and state government, should assume financial liability for storm damages. Otherwise real estate values will fall and the owners might lose money.

46. ( ) Shoreline engineering solutions to beach erosion problems usually perform as planned; additional construction and expenditures of public funds on the same problem beach segment are seldom necessary.

47. The Hatteras Lighthouse, a national historic resource on the Outer Banks of North Carolina, is being seriously threatened by coastal erosion. What is the currently-accepted policy concerning saving this resource?
a. Build a massive concrete seawall around the lighthouse and leave it where it is.
b. Dismantle the lighthouse and refabricate it far enough inland to be reasonably safe for the foreseeable future.
c. Do nothing; allow the lighthouse to eventually be undercut and destroyed by wave erosion.
d. Leave the lighthouse where it is; protect its base with extremely coarse boulder riprap and very large sacks filled with sand.

48. Which explanation is appropriate for the E-10 line and E-60 zone in the FEMA coastal-zone management recommendations?
a. The E-10 line is ten meters back from the mean high tide line; the E-60 zone is the zone nearest to the beach in which permanent buildings can be constructed.
b. The E-10 line shows the expected location of the shoreline in ten years, given current measured erosion rates. Buildings erected in the E-60 zone must be readily moveable.

49. Incoming waves slow down and rotate toward being more nearly parallel to a beachline. What term denotes this process?
a. translation   b. refraction   c. reflection   d. relaxation

50. Which will not be a likely effect of a breakwater?
a. increased sand deposition between the beach and the breakwater
b. increased longshore current velocities between the breakwater and the beach
c. dissipation of storm wave energy on the seaward side of the breakwater
d. increased erosion of the beach on one side of the breakwater

51. Which **one** is not directly related to wave refraction?
a. erosional energy is focused on headland areas along the beach
b. speeds decrease as waves enter shallow water
c. deposition is concentrated in bays and protected waters
d. outgoing tidal currents carry large quantities of sand from estuaries to the sea coast side of a barrier island system

52. Which **two** responses concerning wave refraction along a coastline are reasonably correct? The other two contain serious conceptual or factual errors.
a. Wave normals converge toward resistant headlands or rocky points, thus concentrating wave erosional energy on these same areas.
b. As incoming sea swell waves slow down in shallowing water, the crests and troughs move closer together.
c. Incoming sea swell with crests and troughs oriented to the northeast-southwest will produce north-flowing longshore drift and currents along a west-facing, straight, north-south segment of shoreline.
d. If the crests and troughs of sea swell are perpendicular to a beach, wave refraction is eliminated and longshore currents are not formed.

53. Which manmade, coastal structures are designed to keep tidal inlets from shifting location or filling with sand?
a. groins b. jetties c. breakwaters d. seawalls

54. On the U. S. map provided, accurately plot and label the following features or locations.
a. Florida Keys b. Galveston, TX c. Miami Beach, FL d. Lake Michigan
e. Louisiana coast; Mississippi Delta f. Santa Barbara littoral cell, California
g. Cape Hatteras; Outer Banks, NC h. Fenwick and Assateague Islands, MD
i. Penobscot River/Estuary; Bangor, ME j. Matagorda area, Texas coast

55. On the world map provided, accurately plot and label the following countries, features, or locations.
a. Bay of Bengal b. London, U. K.; Thames River Estuary
c. Trieste, Italy; Adriatic Sea d. Netherlands; the Rhine Delta

Environmental Geology Quiz • Chapter 9

1. Where is the Thames barrier and what is its purpose?
a. protect against tropical cyclone storm surges; Ganges River Delta, Bangladesh
b. protect coastal farmlands from winter storms in the North Sea; The Netherlands
c. protect against storm surges coming in from the North Sea; London, England

Questions 2-5; Matching: Waves

a. period   b. height   c. 0.5 L   d. wave length

2. ( ) horizontal distance between neighboring wave crests
3. ( ) vertical distance from tough to crest of a wave
4. ( ) time required for one complete wave form (crest to crest) to pass a fixed point
5. ( ) minimum water depth for a deep water wave

6. A hurricane is heading due west toward the south Florida coast. Which **one** response is correct or reasonably correct?
a. The strongest onshore winds and highest storm surge will generally be on the north side of the eye.
b. The area is largely rural, and as a result, property damages will be relatively slight.
c. Unusually high atmospheric pressures act to temporarily accelerate eye-wall wind speeds and lower the height of the storm surge as the eye passes overhead.

7. Which **one** response concerning typhoons is correct?
a. severe winter storms in the northern Pacific
b. Southern Hemisphere hurricanes
c. same storms as hurricanes but in the western Pacific

8. Which **two** responses concerning spillers are reasonably correct?
a. a surfer's bummer   b. favored by steep offshore beach slopes
c. favored by gentle offshore beach slopes   d. a surfer's dream

Questions 9-11; Matching: Beach Morphologic Features

a. berm   b. swash zone   c. seacliff base/dune line

9. ( ) marks highest elevation reached by the strongest storm waves
10. ( ) relatively level to gently sloping surface built by seasonal deposition from summer swell or winter storm waves
11. ( ) subjected to the landward and seaward rush of water from breaking waves

12. What is meant by the term fetch?
a. the beach front area where dogs are trained to retrieve sinking swimmers
b. a large expanse of open water over which the wind blows and generates waves
c. the rotational movements of water particles beneath a passing, surface wave

13. Which **one** statement is true for waves moving from deep waters into shallower waters along a coastline?
a. wave speeds increase; wave lengths decrease
b. wave heights increase; wave speeds decrease
c. wave heights decrease; wave lengths increase

14. What is a reasonable rate of change for sea level in the past few hundred years?
a. fallen; about 1 inch per century   b. risen; about 1 cm per century
c. risen; about 30 cm per century   d. fallen; about 10 inches per century

15. How does refraction affect the crest and trough orientations of incoming waves approaching a beach?
a. The acute angle between the wave crests (troughs) and the shoreline decreases.
b. The acute angle between the wave crests (troughs) and the shoreline increases.

16. Water movement and sand transport parallel to the beach are fundamentally caused by which one?   a. deep water waves breaking offshore
b. waves impinging obliquely onto a beach   c. a long fetch parallel to the beach
d. strong, offshore winds creating a pileup of water along the beach front

17. Which two structures are built more or less parallel to the beach?
a. seawalls   b. groins   c. jetties   d. breakwaters

18. In which area would undamed rivers and streams supply the larger proportion of sand to the local beach environment?
a. a topographically low coastal area with many large estuaries, such as the Atlantic coastline of North Carolina
b. a topographically rugged, coastal area with steep seacliffs and small estuaries such as the Pacific coastline of California

19. Which U. S. coastal area has the smaller estuaries and smaller, more localized, littoral cells?   a. Atlantic and Gulf Coasts   b. Pacific Coast

20. Which coastal structures are designed to keep tidal inlets from shifting location or filling with sand?   a. groins   b. jetties   c. breakwaters   d. seawalls

21. Which factor better accounts for the rapidly escalating costs of coastal storm damage in the United States?
a. As coastal development continues, more and more new buildings and other structures become vulnerable to storm damage.
b. The number of hurricanes and severe winter storms impacting on U. S. coastlines has risen steadily since about 1900.

22. Which component of the southern California wave climate is responsible for rebuilding sandy beaches during the summer months?
a. Low wave-height swell originating in far-away, Southern Hemisphere storms.
b. Powerful, 12 second-period waves that originate from strong storms in the northern Pacific and Gulf of Alaska.

23. Which one is the low-cost, low-maintenance strategy for beach management?
a. Promote coastal development; build massive, hardened structures to stop all but the most powerful storm waves.
b. Prohibit coastal development; build nothing, declare victory over the sea and retreat from the coastline.
c. Allow limited coastal development; make a long-term commitment to beach nourishment.

24. Which **one** refers to the broad dome of water moving with the eye and frontal portion of a hurricane?   a. sea mound   b. storm surge   c. cyclone pool

Questions 25-27; Agree (A) or Disagree (D)? Briefly explain any disagreements.

25. (   ) Once expensive, privately-owned oceanfront homes, hotels, and condominiums are built, the average taxpayer, through federal and state government, should assume financial liability for storm damages. Otherwise real estate values will fall and the owners might lose money.

26. (   ) Overall sand transport along the Fenwick-Assateague Islands segment of the Atlantic coastal barrier island system is toward the north, but strong storm swell from winter hurricanes in the North Atlantic can produce temporary south-flowing longshore currents and sand movements.

27. (   ) The Hatteras Lighthouse, a national historic resource on the Outer Banks of North Carolina threatened by coastal erosion, can be saved indefinitely by protecting its base with very coarse boulder riprap and huge sandbags.

28. (   ) Historical records show variations of about 2 meters in the level of Lake Michigan. Low-stand levels are characterized by wide sandy beaches and low erosion rates; high-stand levels are characterized by narrow beaches, high erosion rates, and major lake front property damage.

29. Which response is correct for the E-10 line and E-60 zone in the FEMA coastal-zone management recommendations?
a. The E-10 line is ten meters back from the mean high tide line; the E-60 zone is the zone nearest to the beach in which permanent buildings can be constructed.
b. The E-10 line shows the expected shoreline location in ten years, given current measured erosion rates. Buildings in the E-60 zone must be readily moveable.

30. On the map provided, plot and label the following features or locations.
a. Trieste, Italy; Adriatic Sea			b. Lake Michigan
c. Cape Hatteras; Outer Banks, NC		d. Florida Keys
			e. Ganges Delta, Bangladesh

Answers; Home Study Assignment
Coastal Hazards • Chapter 9

1. b, d	2. a	3. b	4. d	5. e
6. c	7. a	8. f	9. g	10. d
11. e	12. d, f	13. a, c	14. b	15. c, d
16. b, d	17. a, b	18. b	19. b, d	20. b
21. c	22. a	23. b	24. c	25. c
26. a, d	27. b	28. a	29. b	30. d
31. b	32. d	33. c	34. a	35. b
36. a, c	37. b	38. a	39. A	40. D*
41. A	42. A	43. A	44. A	45. D*
46. D*	47. b	48. a	49. b	50. b
51. d	52. a, b	53. b		

*

40. The inlet was cut through the barrier island system in 1933 during a hurricane.

45. Federal and state governments should not encourage risky, capricious, coastal development. The government should "be there" in the face of unforeseen, unpredictable events. Insurance never gave the insured the right to live dangerously and to be protected from foolish risks. Leave beachfront areas natural and uncluttered; look elsewhere for inflating land prices and real estate values.

46. As often as not, shoreline engineering project do not perform as expected or as predicted; always assume that project costs and maintenance costs will escalate over time.

Answers: Environmental Geology Quiz • Chapter 9

1. c	2. d	3. b	4. a	5. c
6. a	7. c	8. a, c	9. c	10. a
11. b	12. b	13. b	14. b	15. a
16. b	17. a, d	18. b	19. b	20. b
21. a	22. a	23. b	24. b	25. D*
26. D*	27. D*	28. A	29. b	

*

25. Federal and state governments should not encourage risky, capricious, coastal development. The taxpayer is not interested in subsidizing foolish decisions and ignorance. However, the government should "be there" in the face of unforeseen, unpredictable events. Insurance never gave the insured the right to live dangerously and to be protected from foolish risks. Leave beachfront areas natural and uncluttered; look elsewhere for inflating land prices and real estate values.

26. The buildup of sand at the south end of Fenwick Island behind the north Ocean City Inlet jetty shows that dominant sand transport is toward the south. This reflects the predominance of incoming waves from strong, North Atlantic,

winter storms that drive southward-moving, longshore currents. Hurricanes, tropical storms, and tropical depressions that persist north of the inlet will also produce net southward, longshore drift. Waves generated by summer and fall, Atlantic, tropical storm systems south of the inlet will impinge from the south, thus causing temporary northward longshore drift.

27. The Hatteras Lighthouse is temporarily being protected with huge, sand-filled plastic sacks. Sooner or later, probably sooner, these will not be sufficient to protect the base and foundation of the lighthouse from shoreline erosion. Long-term plans call for dismantling the lighthouse and refabricating it well back from its present, exposed position. Dismantling and moving the structure present numerous, difficult complications and will be very, very expensive.

Commentary/Media Watch

Brief Introductory Summary

Hurricanes Bertha and Fran and a high-profile controversy concerning a barrier island resort condominium brought coastal environmental issues "front-and-center" during 1996 in North Carolina. Bertha, the smaller and less powerful of the two storms, made landfall just north of Wilmington, NC, in mid-July. On September 6, Fran slammed ashore a short distance west and south of Wilmington, leaving a tangled mess of uprooted trees, downed powerlines, and ruined crops as far north as the Raleigh-Durham area. As noted previously (Commentary/Media Watch, Chapter 5), heavy rains associated with Fran resulted in severe flooding on major river systems in North Carolina and Virginia. Although the two hurricanes dominated media coverage, the disposition of the Shell Island Resort near Wilmington is emerging as a major test of the State's coastal management policies. In addition, the Oregon Inlet jetties project, once presumably scuttled for good, reappeared like a Phoenix from the barrier island sands.

The Oregon Inlet Jetties Project

The Oregon Inlet jetties project, North Carolina Outer Banks, was the subject of Associated Press and other articles carried in North Carolina newspapers in July and August. Governor Hunt (D, NC) and Senator Helms (R, NC) had appeared before a panel of the Senate Committee on Energy and Natural Resources holding hearings on the proposed jetties project. Despite frequently voiced opinions against "wasteful government spending" and " big government", both men strongly endorsed funding for the project; environmental groups, the Department of Interior, the National Park Service, and the Fish and Wildlife Service oppose the project. Maps (Figures 18.9, p. 517 and 18.12, p. 520) and a spacecraft photo (Figure 9.13, p. 238) are useful references concerning the following discussion.

These hearings were just the latest event in a long-running saga; the jetties project has been on and off the federal ledgers for at least 35 years. The modern version was conceived in the early 1960s when state officials decided to commit millions of dollars to 1) a proposed industrial park and fish-distribution center in the Roanoke Island community of Wanchese, 2) a highway bridge spanning Oregon Inlet, and 3) modernization of state highway 12 from the Manteo-Kill Devil Hills area south to Cape Hatteras. To support the increased boat traffic envisioned for the facility at Wanchese, Oregon Inlet had to be deepened and "stabilized".

Oregon inlet opened during a September hurricane in 1846; the only other inlet to the extensive Currituck-Pamlico Sound estuarine system is Hatteras Inlet, fifty miles farther south. Oregon Inlet's modern history is characterized by "shoaling in" and southward migration. Small fishing boats can navigate the inlet, but larger, deeper-draft boats are forced to use Hatteras Inlet, adding over two hundred miles to the Wanchese-North Atlantic round trip. In addition, a migrating inlet would not be compatible with a "static" bridge. Thus the Oregon Inlet jetties project was devised with two main purposes; to support the state's investment in the fish-distribution facility by guaranteeing that fishing boats have permanent, safe passage through the inlet, and to protect the new highway bridge from inlet migration. The new bridge also was destined to support a real estate boom on Hatteras Island, here-to-for accessible only by state-operated ferry, private plane, or boat.

Despite strong support by state officials and the state's congressional delegation, the jetties project initially fell victim to high projected costs, environmental concerns, and lack of broadly-based congressional support. The main environmental opposition came from citizens groups, the National Park Service, entrusted with overall management responsibilities for Hatteras National Seashore, and the U. S. Fish and Wildlife Service. The major issue was and remains the proposed jetties' effects on longshore sand transport. The Pea Island National Wildlife refuge lies just south of the inlet and the barrier island system is very narrow from the Pea Island Refuge south to the village of Avon (Figure 18.12). These narrow sections are frequently overwashed during storms or periods of unusually high surf, necessitating berm building, sand removal, pavement repairs, and other expensive highway maintenance.

Impinging waves generated by winter storms in the North Atlantic generate south-flowing longshore currents, and surf emanating from summer and fall tropical storms farther south in the Atlantic generates north-flowing currents. However, net sand transport is from north to south. Thus sand would be expected to accumulate north of the jetties and erosion would be predicted for oceanfront beaches at Pea Island and the narrow areas farther south. Untested, untried "sand bypass" schemes included in the original proposal substantially raised projected maintenance costs and did nothing to help overcome the general skepticism. In addition, proposed land transfers from agencies of the Interior Department to the Corps of Engineers, required before the project can proceed, have met with repeated legislative failure.

The Oregon Inlet bridge was opened to traffic in November, 1964, and highway improvements were subsequently completed, but the Wanchese fish-distribution facility never got off the ground and slowly faded from public view. This was a project whose time had "not come"! The jetties project sank to a very low budget priority during the austerity years of the Reagan and later administrations and occasional attempts to revive it have met with strong opposition. A declining fishing industry has significantly reduced the political clout of "helping our fishermen". In the meantime, short groins have been installed to stop the inlet from undercutting the south approaches to the bridge and dredging keeps the inlet navigable for shallow-draft boat traffic. The jetties project is "down", but probably not "out".

Hurricanes Bertha and Fran, Summer, 1996

In preparation for an increasingly more likely North Carolina landfall for Hurricane Bertha, extensive media coverage was directed toward the approaching storm. Coastal erosion, the danger from storm surge, efforts to protect homes and boats, and emergency procedures for possible evacuations and activating repair crews were emphasized. On days following the storm, media coverage focused on damaged homes, uprooted trees, electrical outages, and crop damage. Wind and storm surge damage were heaviest in the barrier island community of North Topsail Beach, just north of Wilmington, NC. Newer homes built after implementation of the state's coastal management regulations in 1985 did not qualify for federal disaster relief. An Associated Press article (7/24/96) noted that sea bottom debris from broken piers and other wreckage was ruining shrimper's nets and that a 50 % reduction in shrimp yields could be expected from affected areas. An editorial (7/25/96) in the Greenville, NC, Daily Reflector strongly defended the state's commitment to "soft, coastal management policies and defended FEMA's decision not to make disaster relief funds available to North Topsail residents whose homes were built after 1985 despite repeated warnings of the danger and lack of federal insurance. The editorial writer stressed that taxpayers should not have to subsidize damage costs in areas where warnings of the dangers were well-publicized, widely known, and ignored. That is the common sense American approach.

A special supplement "Hurricane Watch '96 (8/4/96) in Sunday editions of eastern North Carolina newspapers included articles on Bertha, the hurricane history of the Carolinas, and hurricane Hazel, 1954. A central theme woven through the articles was that extensive coastal development in recent years has dramatically raised the potential for financial loses and casualties from a strong-category storm like Hazel.

Fran came ashore the evening of September 6; the eye made landfall just west of Wilmington and subsequently the storm followed a north-northwest track through the Raleigh-Durham area and into central Virginia. Damage to piers, boats, and homes in the Topsail Island area was extensive; many structures that survived Bertha succumbed to Fran. Revised setback locations (see the *special feature* E-Lines and E-Zones, p. 250) based on the post-storm berm line prohibit any rebuilding on most pre-storm, oceanfront properties and many properties originally in compliance with setback regulations before the storm are now too close to the beach to allow rebuilding. In any case, owners of homes built after 1985 are not eligible for federal disaster assistance.

Over the last two decades, hurricanes and other storm events have become big news. Television networks such as CNN and The Weather Channel give continuous updates on storms and place reporters at strategic sites along the projected paths. Reporters brave driving rains and powerful winds to give their commentaries; their lives are in danger from rising storm surges, downed powerlines, and wind-carried projectiles moving at hurricane speeds. Public safety officials, the mayor, and the TV reporters are the last ones to evacuate! Post-storm coverage at the national level fades quickly, but damage assessment, relief efforts, and longer-term coastal management issues continue to attract statewide and local media coverage well after a storm.

Hurricane Camille was probably the last U. S. hurricane in which misguided self-confidence and excessive testosterone levels overruled common sense advice to head for safer, inland locations. The public has gotten the message; these storms are dangerous; evacuation orders are heeded! Modern tracking and forecasting methods are accurate enough to convince the public of the impending danger and provide governmental agencies and public safety personnel with ample lead times in which to implement evacuation orders and other safety precautions. Coastal residents now ride out the storm well inland with friends or relatives or in an easily assessable public emergency shelter. Gone are the days of the "hurricane party"! During the height of Fran, a forlorn North Topsail Beach resident phoned emergency officials in Wilmington requesting assistance and, convinced of impending doom for himself, his family, and his dog, he gave his final will and testament. Of course by that time, 100 mile per hour winds, downed powerlines, and flooded streets and bridges insured that no help was forthcoming. The family and their dog survived the storm, but in post-storm television interviews, the survivors acknowledged the foolishness of their decision and exhorted us all to heed future evacuation orders.

Bertha, Fran, and more recently Josephine (early October, 1996) have done much to publicize long-term, coastal management issues and policies in North Carolina and neighboring states. Based on statewide media coverage, general support in North Carolina for the current, "soft" coastal defense policies has increased since the storms and support for hardened structures to protect vulnerable private homes and property has been considerably eroded. Politicians of both major parties find it harder and harder to defend what the public visualizes as foolish behavior and speculative greed on the part of coastal developers and landowners, while reasoned voices supporting pragmatic, science-based, coastal policies have essentially been vindicated. Out-of-state or distant, non-local ownership also has worked against hardening of coastal policies. Why should hardworking, local taxpayers bailed out absentee owners of lavish beachfront homes, especially when increasing privatization has severely restricted access to popular public beaches in recent years?

Shell Island Resort Condominium

The Shell Island "case" attracted, steady, statewide media coverage during the summer of 1996 and hurricanes Bertha and Fran just served to heighten the controversy. Seeds for the problem were sown in 1985 when a barrier island resort condominium complex, originally valued at $22 million, was built at Wrightsville Beach near Wilmington, NC. At that time, the complex was about one-half mile from Mason Inlet; subsequently the inlet has migrated steadily toward the resort, and as of October, 1996, the inlet was less than 200 feet away.

In view of the obvious threat, condo owners and developers petitioned the State's Coastal Resources Commission for a variance from existing coastal regulations. They want to build a seawall of sand-filled pipes and giant plastic bags to stop erosion and "stabilize" the inlet before the buildings are undermined and rendered worthless. Media coverage focused on the Commission's 6 to 5 nay vote in late July. An editorial "Close call on the seawalls", carried in the State's two leading newspapers, The Charlotte Observer and the Raleigh News & Observer (8/4/96), praised the Commission's decision and exhorted the Governor to continue with his policy of appointing well-educated, environmentally-responsible individuals to the Commission. In addition, the editorial writer specifically cautioned the Governor against using his appointment powers to change the composition of the Commission to insure a favorable decision for the condo owners. Such a course of action, the writer noted, would make a shambles of the State's coastal management policy. Subsequently in late September, the Commission reaffirmed its nay vote, again by a one-vote margin (6 to 5). The owners are threatening to take the issue to court, but as noted by commission members and coastal experts, the plaintiffs can hardly argue that the problem was "unforeseen".

In August, two feature articles on coastal management issues were carried in the Raleigh News & Observer. One (Worries about Shell Island Resort reach high tide; 8/11/96, Todd Richissin) focused specifically on the Shell Island situation and the other (Beachfront property: going .... going ....; 8/12/96, David Kirkpatrick, reprinted from The Wall Street Journal) is a more general article on coastal management and land use in the Carolinas and other eastern seaboard states. The lead photo in the Wall Street Journal article featured Orrin Pilkey, a highly respected faculty member at Duke University, an expert on Carolina coastal processes, and a highly visible public advocate for sensible, natural-science based, coastal management policies. Professor Pilkey is shown on the beach at Pawley's Island, South Carolina, where very expensive, beachfront lots, homes, and developments are being threatened by erosion. Growing pressures for sidestepping state-imposed limits on building and beach hardening activities have already led to numerous legal snarls and lawsuits; one (Lucas versus the State of South Carolina, 1992) went all the way to the U. S. Supreme Court. In his Wall Street Journal article, Kirkpatrick argues that inflated property values on private beaches, bought and developed by highly affluent, largely out-of-state owners, are being subsidized by Uncle Sam while "blue collar, public beaches" are left "to Nature". In other words, taxpayer-financed beach projects are designed to benefit wealthy property owners and home owners. This is quite a startling conclusion, given the fact that the Journal is not exactly a vigorous proponent of environmental regulation nor an enemy of private property rights!

The Richissin article frames the Shell Island situation as a battle between affluent, out-of-state condo owners, real estate speculators, and developers versus second- and third-generation, blue collar, beer-drinking locals who have grown up with the public sections of Wrightsville Beach and have come to love them. A hardened structure to keep the resort complex from tipping into Mason Inlet could interfere with longshore sand transport and jeopardize stability of the public beaches. Why should the State cave in to wealth and affluence and rescue those who have knowingly made unwise, ignorant investment decisions? Would anyone be interested if this was essentially a "blue collar, public beach" with a few modest cottages or beach bum shacks? How many of the affluent condo owners are dead set against "against big gov'ment" and welfare system handouts? Are the wealthy and affluent guaranteed financial success, even in the face of their own ignorance and foolish judgment?

The Shell Island case is an important lesson for earth science teachers and professionals whose knowledge, expertise, and recommendations are often brushed aside and ignored in the public arena. Compelling social and political arguments can help tip the scales toward public policies based on scientific expertise and sound management and away from those based solely on doctrinaire obsessions with "real estate profits" and the "sanctity of private property rights".

This latter issue was being debated (November, 1996) in the North Carolina Joint Legislative Commission on Governmental Operations. Beach erosion during Hurricane Fran placed many privately-owned, beachfront lots partly or entirely below the "new" maximum high tide line. According to state regulations, the State owns all coastal lands below the high tide level, thus in many cases, private homes are completely or in part on state-owned lands. Obviously, this is a thorny legal issue with no satisfactory solution. It does, however, highlight the common-sense approach of leaving fragile, oceanfront lands relatively undeveloped and under public management.

Supplemental References

Davis, R. A. Jr., 1996, Coasts: Prentice Hall, Upper Saddle River, NJ, 320 p.

Davis, R. E. and Dolan, R., 1993, Nor'easters: American Scientist, v. 81, p. 428-439

Hinrichsen, D., 1996, Pushing the limits: Humanity and the world's coasts; a status report: The Amicus Journal, v. 18, p. 16-20

Korgen, B. J., 1995, Seiches: American Scientist, v. 83, no. 4 (July-August), p. 330-341

Pilkey, O. H. Jr., Neal, W. J., and Pilkey, O. H. Sr., 1978, From Currituck to Calabash: Living with North Carolina's barrier islands: North Carolina Science and Technology Research Center, Research Triangle Park, NC, 228 p.

Salvesen, D., 1996, Sand castles: On Topsail Island, homeowners discover that building on barrier islands is risky business: The Amicus Journal, v. 18, no. 4, p. 28-31

Environmental Awareness Inventory
Part 3: Human Interaction with the Environment

1. A widely-read, highly-influential book concerning the detrimental effects of pesticides and other toxic manmade chemicals in the environment was published in 1962. What was the title and who was the author? 1 pt
a. Stillness at Dawn; Robert Woodson   b. Ecology Lost; Margaret Mitchell
c. Toxic Chemicals and Sterility; Alicia Silverstone   d. Silent Spring; Rachel Carson

2. Which **one** is now the leading candidate for disposal of domestic, high-level, nuclear waste from weapons-producing facilities and reprocessed fuel pellets from nuclear electrical generating plants? 1 pt
a. Canisters stored in Greek-style mausoleums manned round-the-clock by armed guards; Arlington, VA
b. underground chambers excavated in volcanic tuff; Yucca Mountain, NV
c. a huge excavation left when the Bingham Canyon open-pit copper mine is abandoned; near Salt Lake City, UT
d. burial in seafloor sediment; deepest part of the Alaskan trench
e. underground openings left from mining salt, Gulf Coast, southern Louisiana

3. What is leachate? 1 pt
a. a potent aphrodisiac prepared from the dried skin of southeast Asian leeches
b. an aqueous residue rich in dissolved potassium, magnesium, and iodine left over after sodium chloride has been precipitated from seawater
c. solute-rich water films and droplets in pores and openings of the vadose zone
d. an unpleasant, potentially dangerous pollutant formed where groundwater interacts with the buried contents of a waste dump or sanitary landfill

4. Which was the largest, maritime, tanker oil spill as of 1996? 1 pt
a. Mega Borg; Gulf of Mexico off Galveston, Texas; 1990
b. Exxon Valdez; Prince William Sound, Alaska; 1989
c. Amoco Cadiz; Brittany Coast of France; 1978
d. B. P. Von Hindenburg; southern New Jersey Atlantic Coast; 1937

5. Which formerly idyllic, middle-class suburban neighborhood went from the "American Dream" to "your worst nightmare" when property owners learned that their homes were built on lands once used as a chemical waste dump? 1 pt
a. Cottonwood Acres, Grand Junction, Colorado; 1970
b. River Front Estates; Camden, New Jersey; 1988
c. Love Canal, near Niagara Falls, New York; 1976
d. the Haight-Asbury District, San Francisco, California, 1969

6. Which was an important result of the Comprehensive Environmental Response Compensation Liability Act, passed into law by Congress in 1980? 1 pt
a. the Superfund was authorized and thus Superfund sites were identified
b. required that specific industries report total, annual emissions of toxic materials
c. specified the preferred design and materials for constructing sanitary landfills
d. legislated that dissolved fluorine be added to municipal water supplies to lower the incidence of dental caries in children

7. Which two chemical compounds are largely responsible for "cultural eutrophication"? **one answer, 1 pt**
a. nitrate, phosphate   b. ozone, sulfur dioxide   c. gold, platinum   d. DDT, PCBs

8. A dietary deficiency in which micronutrient is known to be associated with goiter and cretinism?   a. chlorine   b. iron   c. calcium   d. iodine

9. Which health problem unknowingly afflicted many citizens of Imperial Rome, especially those of the upper, wealthy, ruling class?
a. giardia, an intestinal parasite in drinking water
b. syphilis, a sexually-transmitted infection
c. lead poisoning from cooking utensils and wine storage containers
d. dioxin poisoning related to red mineral pigments used for clothing dyes

10. What is the Reading Prong? 1 pt
a. a protruding device to make books easier to handle
b. an area in the U. K., England, noted for anomalously high-levels of dissolved lead and zinc in soils and groundwater
c. a well-defined, geologic area in eastern Pennsylvania and western New Jersey noted for elevated levels of radon in soil gases
d. the portion of West Virginia projecting northward along the Ohio River between Pennsylvania and Ohio; noted for soils deficient in iodine

11. Which response concerning "hard water" and "soft water" is correct?
a. calcium, magnesium, and bicarbonate are the dominant dissolved chemical species in hard water; hard water is statistically associated with lower rates of heart disease than soft water
b. sodium and sulfate are the dominant dissolved constituents in hard water; soft water is statistically associated with lower rates of heart disease than hard water

12. Which is still the dominant legal statute concerning surface water usage in the eastern United States? 1 pt   a. Darcy Doctrine
b. Public Trust Doctrine   c. Safe Yield Doctrine   d. Riparian Doctrine

Answers

1. d	2. b	3. d	4. b	5. c	6. a
7. a	8. d	9. c	10. c	11. a	12. d

Performance Evaluation; Scoring

Each correct answer is worth one point; add up your total. Now be honest, and subtract one point for each lucky guess!

Score	**Awareness Ranking; advice**	INTERNET Address
9-12	**nerd;** keep up the good work, don't let up	http://envnerd.com
6-9	**blue collar achiever;** earn an A with hard work	http://envblcol.com
3-6	**slowpoke;** don't look back, the clueless are gaining	http://envslpoke.com
0-3	**clueless;** extra hard work makes up for a slow start	http://envcluless.com

Water: Process, Supply, and Use • Chapter 10; Home Study Assignment

Questions 1-8; Refer to Figure 10.1 and Tables 10.1 and 10.2, p. 259-260.

1. Which component of the global hydrologic cycle has the larger mass?
a. precipitation falling on land   b. precipitation falling on the oceans

2. Which component of the global hydrologic cycle has the smaller mass?
a. evaporation from land to the atmosphere   b. runoff from the land to the oceans

3. Which of Earth's water reservoirs contains the largest amount of water?
a. rivers and streams   b. Antarctic ice sheet   c. lakes   d. the atmosphere

4. Which one contains the larger volume of freshwater?
a. the world's rivers and streams   b. the world's groundwater systems

5. Which best characterizes the residence times for water in the oceans?
a. hundreds of years   b. tens of years   c. thousands of years   d. less than ten years

6. What percentage of Earth's water is not in the oceans and glacial ice?
a. between 5% & 10 %   b. less than 1 %   c. between 3 % & 4 %   d. about 2.5 %

7. Which continent receives the highest yearly average precipitation?
a. South America   b. North America   c. Asia   d. Antarctica

8. On an annual basis, which continent loses the least water to the atmosphere by evaporation?   a. South America   b. North America   c. Asia   d. Antarctica

9. Which end of the dipolar water molecule has a net positive electrical charge?
a. end with the two hydrogen atoms   b. end with the single oxygen atom

10. What is the specific heat of water and how does this value compare to specific heats of other low molecular weight liquids?
a. 0.5 calories/gram; substantially lower than for other liquids
b. 1.0 calories/gram; substantially higher than for other liquids

11. Which refers to the only temperature and pressure at which liquid water, water vapor, and ice are in equilibrium?
a. evapo point   b. triple point   c. tri-state point   d. reflux point

12. Which pair of terms are exact synonyms?
a. watershed, drainage basin   b. water cycle, watershed
c. drainage net, drainage basin   d. triple point, specific heat

13. In general, which area of a drainage basin would show a higher drainage density?
a. the area with slower surface infiltration and higher runoff
b. the area with faster surface infiltration and lower runoff

14. Other climatic, topographic, and hydrologic factors being equal, which drainage basin would exhibit a higher runoff associated with a given storm?
a. one with deep, sandy soils overlying highly fractured bedrock
b. one with deep, clay-rich soils overlying unfractured, horizontal shale strata

15. Two watersheds have comparable topography, vegetation cover, and equal areas. Which major stream would be more susceptible to flash flooding?
a. one in a nearly circular watershed   b. one in a long, narrow watershed

16. Given two otherwise equivalent drainage basins, which would show the more rapid rise in discharge of the major stream following a specific storm event?
a. a basin with lower relief   b. a basin with higher relief

Questions 17-22; Agree (A) or Disagree (D)? Briefly explain the basis for any disagreements.

17. ( ) In general, stream sediment loads will increase in a drainage basin altered through clear-cutting or wildfires, especially in the next few years following removal of the vegetation.

18. ( ) Freshly-burned, steep hillslopes are highly susceptible to debris slides and mudflows, especially if the first rainfall event following the burning is unusually heavy and intense.

19. ( ) Increased sediment loads in mountain watersheds usually result in more pronounced pool-and-riffle stream channels and improved habitat for fish and other aquatic organisms.

20. ( ) During days and weeks between precipitation events, discharge of a perennial stream is derived mainly from shallow throughflow in the vadose zone.

21. ( ) The discharge of an ephemeral stream generally increases in the downstream direction because water from the groundwater system is continuously flowing into the channel.

22. ( ) Infiltration to the shallow throughflow zone and deeper groundwater zone is maximized when precipitation rates greatly exceed soil infiltration rates.

23. Which common, rock-forming mineral or mineral group is most readily dissolved by groundwater?   a. calcite   b. quartz   c. feldspars   d. clay minerals

24. Which bedrock is most readily dissolved by water in the vadose zone and groundwater system?   a. granite   b. limestone   c. shale   d. andesite

25. Which one is not characteristic of groundwater?
a. constant temperature year round   b. supply is normally independent of droughts
c. rarely contains suspended sediment   d. rarely contains dissolved constituents

26. Which stream is said to be perennial and effluent?
a. the channel bottom is above the local water table year round
b. the local water table is above the channel bottom year round

27. Which **one** is the most accurate definition of the water table?
a. a volume of rock saturated with water
b. a boundary between unsaturated rock below and saturated rock above
c. an underground mass of partly saturated rock
d. a boundary between saturated rock below and unsaturated rock above

28. Which soil or rock material would make the best aquifer?
a. one with high porosity and hydraulic conductivity
b. one with a low permeability and high potability
c. one with a high potability and high porability
d. one with a low porosity and hydraulic conductivity

29. Which **two** properties measure the ease (or difficulty) of groundwater transmission through a saturated, porous material? **two answers**
a. potability   b. porous resistivity   c. hydraulic conductivity   d. permeability

30. A perched water table develops under what conditions?
a. a horizontal aquiclude above the regional water table lies below an aquifer
b. an aquifer above the regional water table is overlain by a horizontal aquitard
c. an aquifer below the regional water table is underlain by a horizontal aquiclude
d. an aquitard below the regional water table lies above a horizontal aquifer

31. What response best describes an artesian well?
a. water rises above the top of the aquifer without any pumping
b. pressurized groundwater rises from a deep, unsaturated aquifer
c. the well produces warm, fairly saline water recharged by an affluent stream
d. the well is drilled horizontally into a perched water table aquifer

32. Which **two** statements are true for artesian aquifers? **two answers**
a. The water table in the recharge area is at a higher elevation than the top of the aquifer in the subsurface.
b. Upward flow from a permeable aquiclude is prevented by a confining aquifer.
c. The pressure at any point in the aquifer exceeds the weight of the water column between the point and the top of the aquifer.
d. Artesian aquifers commonly feed stream-side springs and seeps in deep canyons and stream valleys.

33. Excessive groundwater withdrawals can cause which **two** results? **two answers**
a. the water table drops or declines in elevation
b. an influent stream becomes an effluent stream
c. the land surface subsides   d. porosity in the aquifer increases as water is removed

34. Which aquifer material would have the highest groundwater velocities and be least effective in removing unwanted pollutants from the water? **one answer**
a. a medium-grained, unconsolidated sand with a dusting of clay minerals
b. a well-cemented, unfractured sandstone   c. a cavernous limestone
d. an unconsolidated, clay-rich, muddy silt deposit

35. What response correctly states the behavior of some hazardous substances that might accidentally enter the groundwater system? **one answer**
a. Gasoline and kerosene would float on the water table; ethyl alcohol would dissolve and disperse in the groundwater.
b. Gasoline and kerosene would float on the water table, but most pesticides break down chemically when they reach the water table.
c. Sulfuric and nitric acids would sink to the bottom of the aquifer; kerosene would accumulate as a layer just below the water table.

36. Which response best describes an aquifer?
a. the porous, permeable, saturated, cone of depression in an aquiclude
b. a layer or stratum in which groundwater flows downward to the water table
c. a saturated, porous, and permeable layer or stratum
d. an unsaturated, influent-flow bed or stratum below a spring

37. Which **two** statements about aquifers on barrier islands and atolls are correct?
a. wells drilled below sea level will produce only saline water
b. the water table must be 40 feet above sea level to keep the salty water in the aquifer below sea level
c. any salty water in the aquifer will rise if the water table is lowered by pumping
d. pumping freshwater from one, large capacity well is more likely to cause salty water to rise in the aquifer than pumping from several, widely spaced, smaller capacity wells

38. Which **one** statement concerning the vadose zone is true?
a. lies above the water table   b. the pore spaces are saturated with water
c. is a well-oxygenated, shallow aquifer   d. lies below the capillary fringe zone

39. For unconfined aquifers, what hydrologic factor is approximated by the slope of the water table?
a. porosity head   b. hydraulic conductance   c. effluent incline   d. hydraulic gradient

40. Which **one** describes an influent stream?
a. groundwater flows into a perennial stream from a shallow aquifer
b. water from an ephemeral stream infiltrates downward to the groundwater system

41. Fresh groundwater along a coastline overlies deeper, salty groundwater. If the water table elevation drops by one foot, how far will the interface between the freshwater and salty water rise?   a. 10 feet   b. 30 meters   c. 20 meters   d. 40 feet

42. Which aquifer material would have the largest capacity to naturally remove sewage pollutants?   a. fractured granite   b. well-sorted, coarse gravel
c. slightly clayey sand   d. limestone with solution channels and caverns

43. What basic force pushes groundwater from pore to pore below the water table?
a. inverted, saturation impulse   b. permeability slope factor
c. pressure gradient or hydraulic gradient   d. affluent-flow factor

44. Which best describes the configuration of an unconfined water table around a pumping well?   a. upside-down siphon   b. withdrawn and depressed
c. cone of depression   d. inverted cone head

45. Which **one** statement concerning artesian wells is not true?
a. the well penetrates an aquifer overlain by an aquiclude
b. the well penetrates an aquifer underlain by an impermeable bed
c. the aquifer is generally inclined, and it is saturated to an elevation above the point where the well penetrates the aquifer
d. when a well penetrates the aquifer, the water rises to the bottom of the aquitard above the aquifer

46. Which **one** statement is inappropriate for an area subsiding because of excessive pumping of groundwater? **one answer**
a. the aquifer material compacts when dewatered
b. the water table is substantially lowered   c. the aquifer is well-consolidated bedrock
d. effluent streams may change to influent streams

47. Which term denotes the volume of voids or open space in a rock or unconsolidated material?
a. permeability   b. space yield   c. porosity   d. saturation index

48. Which material has a very high porosity and very low hydraulic conductivity?
a. uncompacted water-saturated clay   b. unfractured granite
c. shale bedrock   d. coarse, uncemented gravel

49. Which are more easily polluted by surface discharges and dumping?
a. groundwater systems associated with gravel-filled valleys of influent streams
b. groundwater systems associated with gravel-filled valleys of effluent streams

50. What is the basic relationship describing groundwater flow?
a. porous media netscape field modeling   b. Marcy's Model
c. the hydroflux equation   d. Darcy's Law

51. Which formulation best estimates the average velocity of groundwater flow between two points in a shallow, channellike, alluvial aquifer?
a. hydraulic conductivity (K) multiplied (*) by the hydraulic gradient (l); (K*l)
b. hydraulic conductivity (K) multiplied (*) by the hydraulic gradient (l) divided by (/) the aquifer porosity (n); (K*l/n)

52. Which **two** statements concerning the overall water budget for the lower 48 states of the U. S. are correct? What is incorrect about the other two statements? Refer to Figure 10.12, p. 272.
a. On a daily average basis, streamflow to the Pacific Ocean exceeds groundwater discharge along the Pacific Coast.
b. On a daily average basis, streamflow of the lower Colorado River into Mexico exceeds the discharge of streams flowing northward into Canada.
c. On a daily average basis, net evaporation from reservoirs exceeds the quantity of water delivered by streamflow to Mexico.
d. On a daily average basis, evapotranspiration equals about ten percent (10 %) of precipitation.

53. What are the three most severely overdrafted groundwater systems in the lower 48 states of the U. S.? **one answer**
a. central Florida; San Joaquin Valley, CA; Ogallala aquifer, High Plains region
b. San Joaquin Valley, CA; Ogallala aquifer, High Plains region; valleys of central and southern Arizona
c. San Joaquin Valley, CA; Ohio River Valley, Ohio, Kentucky, and Indiana; valleys of central and southern Arizona
d. western North and South Dakota and eastern Montana; Ogallala aquifer, High Plains region; valleys of central and southern Arizona

54. Which **one** is the severely overdrawn, High Plains, regional aquifer?
a. Comanche   b. Lakota   c. Cheyenne   d. Ogallala

55. Which **one** represents an offstream water use?
a. irrigation   b. hydroelectric power generation

56. Which **two** statements concerning trends in water use in the U. S. are correct? See Figure 10.17, p. 278.
a. During the years between 1950 and 1990, groundwater use grew consistently and has exceeded offstream surface water use since 1975.
b. During the years since 1950, instream surface water use has greatly exceeded combined groundwater and offstream surface water uses.
c. Instream water use in 1990 amounted to approximately 3,200 million cubic meters per day.
d. Instream use of surface water grew steadily from 1950 to 1975-80 and has declined very slightly since then.

57. Which **two** statements concerning trends in water use in the U. S. are correct? See Figure 10.18, p. 278.
a. Since 1950, more water has been used for irrigation than for any other single use.
b. Industrial water use has been rapidly increasing since 1950.
c. Water used in association with thermoelectric plants increased fivefold between 1950 and 1980; since 1980, this use has shown a slight decline.
d. Public water use has remained steady or slightly declined since 1950; in 1985, industrial uses topped public use for the first time ever.

58. Which river has, since 1922, given rise to a litany of litigation known as the Law of the River?   a. Tennessee   b. Colorado   c. Mississippi   d. Hudson

Questions 59-64; Matching: The Colorado River Basin; see p. 281.
a. Flaming Gorge   b. Grand Canyon   c. Hoover   d. Powell   e. Salt   f. San Juan

59. ( ) spectacular canyon and national park in northern Arizona

60. ( ) major upper basin tributary; drainage basin includes parts of northwestern New Mexico, southwestern Colorado, southeastern Utah, and northeastern Arizona

61. ( ) scenic canyon segment and national recreational area; only large dam and reservoir on the Green, the major upper basin tributary of the Colorado River

62. ( ) lake impounded behind Glen Canyon dam; named for a famous nineteenth century geologist, explorer, and pubic servant

63. ( ) major lower basin tributary; supplies surface water to Phoenix and the Valley of the Sun, AZ

64. ( ) dam impounding Lake Mead near Las Vegas, NV

Questions 65-69; Do you agree (A) or disagree (D) with the following statements? State reasons for any disagreements. See p. 282 and Table 10.7.

65. ( ) Excessively high dissolved salt contents in Lower Colorado River water have forced the U. S. to construct a desalination plant in order to insure adequate quality of water delivered to Mexican users.

66. ( ) Yearly evaporation, mostly from reservoirs, exceeds the amount of Colorado River water allocated to the state of Nevada.

67. ( ) The Salton Sea in the Imperial Valley of southern Arizona, was formed following major floods on the lower Salt River early in the twentieth century.

68. ( ) California is allocated more Colorado River water than any other state in the compact and more than Mexico.

69. ( ) Actual annual deliveries of Colorado River water to compact members are generally higher than apportioned allotments because the 1922 reference river discharge was substantially lower than the average yearly discharge.

70. Which **two** statements concerning the changes in the Colorado River and its riparian habitat in the Grand Canyon are reasonably correct and factual? What is wrong with the two incorrect statements?
a. Since Glen Canyon Dam was closed in 1963, the natural spring floods have been eliminated and sediments loads have been significantly reduced.
b. Very coarse, bouldery, flash-flood debris dams deposited at mouths of tributaries resist erosion because the maximum discharges now are substantially lower than those formerly associated with the natural spring floods.
c. Sand bars and sandy beach areas along the river are now larger and more numerous than before the dam was built. These deposits had formerly been scoured out and carried on downstream during the annual spring floods; now they are left behind to gradually accrete additional sand.
d. Native new world plant species such as mesquite, catclaw acacia, salt cedar, and tamarisk now grow down to the water's edge; native river fishes, such as rainbow trout, have prospered in the cooler, sediment-free water.

71. What is schistosomiasis?
a. a dangerous fever carried by mosquitoes indigenous to swampy, wetlands along the Persian Gulf coast of Saudi Arabia
b. a South American, freshwater mussel species whose habitat will be seriously degraded and diminished by construction of the proposed Tucurui dam in Brazil
c. a dangerous, debilitating snail-born parasitic infection reaching epidemic proportions in irrigated lands downstream from the Aswan High Dam, Egypt
d. a plant species native to the dry lands of the Middle East; has spread widely in the southwestern U. S., mainly into formerly well-watered riparian habitats deprived of surface water by dams, irrigation diversions, and overdrafting of groundwater

72. Which **one** is not characteristic of dam-impounded reservoir pools in tropical areas?   a. excessive growth of water hyacinth and other aquatic plants
b. possibly acidic lake waters   c. poorly oxygenated or anaerobic bottom waters
d. seasonal cooling and overturn of the lake waters

73. Which **two** may represent reasonable and environmentally sound solutions to suppress the rapid growth of water hyacinth and other aquatic plants in dam-impounded, tropical reservoirs such as the proposed Tucurui dam in Brazil?
a. Import manatees and other aquatic mammals that eat aquatic vegetation.
b. Pour on the herbicides including Agent Orange left over from the Vietnam War.
c. Clear-cut the drainage basin above the dam to reduce water transpiration and raise base streamflow into the reservoir.
d. Teach native peoples, displaced from their traditional homes and livelihoods by the rising waters of the reservoir, to harvest the aquatic plants for soil conditioners, food for livestock, and fertilizer.

74. Which taxonomy order would be most indicative of wetlands soils? See Table 3.1, p. 66.   a. ultisols   b. oxisols   c. mollisols   d. histosols

75. On the provided map of the U. S., accurately plot and label the following features or locations.

a. Grand Canyon, AZ   b. Mississippi River   c. Ohio River
d. Chesapeake Bay   e. Missouri River   f. Hudson River
g. Glen Canyon Dam; Lake Powell   h. Salton Sea, Imperial Valley, CA
i. Owens Valley and Owens River   j. Columbia and Snake Rivers

76. On the provided world map, accurately plot and label the following countries, features, or locations.

a. Indus River   b. Yangtze River   c. Egypt; Nile delta
d. Amazon River   e. Ganges River   f. Nepal; Arun River
          g. Lesotho; upper Orange River Basin, southern Africa
              h. Tucurui damsite, Tocantins River, Brazil

Environmental Geology Quiz • Chapter 10

1. Which component of the global hydrologic cycle has the larger mass?
a. precipitation falling on the oceans   b. precipitation falling on land

2. Which of Earth's water reservoirs contains the largest amount of water?
a. the atmosphere   b. lakes   c. Antarctic ice sheet   d. rivers and streams

3. Which best characterizes the residence times for water in the oceans?
a. thousands of years   b. tens of years   c. less than ten years   d. hundreds of years

4. Which end of the dipolar water molecule has a net positive electrical charge?
a. end with the two hydrogen atoms   b. end with the single oxygen atom

5. What is the specific heat of water and how does this value compare to specific heats of other low molecular weight liquids?
a. 1.0 calories/gram; substantially higher than for other liquids
b. 0.5 calories/gram; substantially lower than for other liquids

6. Which pair of terms are exact synonyms?
a. triple point, specific heat   b. water cycle, watershed
c. drainage net, drainage basin   d. watershed, drainage basin

7. Other climatic, topographic, and hydrologic factors being equal, which drainage basin would exhibit a higher runoff associated with a given storm?
a. one with deep, clay-rich soils overlying unfractured, horizontal shale strata
b. one with deep, sandy soils overlying highly fractured bedrock

8. Two watersheds have comparable topography, vegetation cover, and equal areas. Which major stream would be most susceptible to flash flooding?
a. one in a nearly circular watershed   b. one in a long, narrow watershed

9. Which bedrock is most readily dissolved by water in the vadose zone and groundwater system?   a. limestone   b. granite   c. shale   d. andesite

10. Which stream is said to be perennial and effluent?
a. the local water table is above the channel bottom year round
b. the channel bottom is above the local water table year round

11. Which two properties measure the ease (or difficulty) of groundwater transmission through a saturated, porous material? two answers
a. permeability   b. pore resistivity   c. hydraulic conductivity   d. potability

12. A perched water table develops under what conditions?
a. an aquitard below the regional water table lies above a horizontal aquifer
b. an aquifer above the regional water table is overlain by a horizontal aquitard
c. an aquifer below the regional water table is underlain by a horizontal aquiclude
d. a horizontal aquiclude above the regional water table lies below an aquifer

13. Which one statement is true for artesian aquifers? one answer
a. The water table in the recharge area is at a higher elevation than the top of the aquifer in the subsurface.
b. Artesian aquifers commonly feed stream-side springs in deep canyons and valleys.

14. For unconfined aquifers, what hydrologic factor is approximated by the slope of the water table?
a. effluent incline   b. porosity head   c. hydraulic conductance   d. hydraulic gradient

15. Excessive groundwater withdrawals can cause which two results? **two answers**
a. an influent stream becomes an effluent stream   b. water table elevation declines
c. porosity in the aquifer increases as water is removed   d. the land surface subsides

16. What response correctly states the behavior of some hazardous substances that might accidentally enter the groundwater system? **one answer**
a. Sulfuric and nitric acids would sink to the bottom of the aquifer; kerosene would accumulate as a layer just below the water table.
b. Gasoline and kerosene would float on the water table; ethyl alcohol would dissolve and disperse in the groundwater.
c. Gasoline and kerosene would float on the water table, but most pesticides break down chemically when they reach the water table.

17. Which **one** statement about aquifers on barrier islands and atolls is correct?
a. pumping freshwater from one, large capacity well is more likely to cause salty water to rise in the aquifer than pumping from several, widely spaced, smaller capacity wells
b. wells drilled below sea level will produce only saline water
c. the interface between the freshwater and salty water in the aquifer will rise about 10 feet for each one foot lowering of the water table elevation

18. Which aquifer material would have the largest capacity to naturally remove sewage pollutants?   a. slightly clayey sand   b. well-sorted, coarse gravel
c. fractured granite   d. limestone with solution channels and caverns

19. What basic force pushes groundwater from pore to pore below the water table?
a. pressure or hydraulic gradient   b. affluent-flow factor   c. porous slope impulse

20. Which **one** statement is inappropriate for an area subsiding because of excessive pumping of groundwater? **one answer**
a. aquifer compacts when dewatered   b. aquifer is well-consolidated bedrock
c. water table declines   d. effluent streams may change to influent streams

21. What is the basic relationship describing groundwater flow?
a. Netscape modeling   b. Marcy's Model   c. the hydroflux equation   d. Darcy's Law

22. Which are more easily polluted by surface discharges and dumping?
a. groundwater systems associated with gravel-filled valleys of influent streams
b. groundwater systems associated with gravel-filled valleys of effluent streams

23. Which **one** denotes the volume of voids in a rock or unconsolidated material?
a. permeability   b. saturation index   c. porosity   d. space yield

24. Which formulation best estimates the average velocity of groundwater flow between two points in a shallow, channellike, alluvial aquifer?
a. hydraulic conductivity (K) multiplied (*) by the hydraulic gradient (l); (K*l)
b. hydraulic conductivity (K) multiplied (*) by the hydraulic gradient (l) divided by (/) the aquifer porosity (n); (K*l/n)

25. What are the three most severely overdrafted groundwater systems in the lower 48 states of the U. S.? **one answer**
a. San Joaquin Valley, CA; Ogallala aquifer, High Plains region; valleys of central and southern Arizona
b. San Joaquin Valley, CA; Ohio River Valley, Ohio, Kentucky, and Indiana; valleys of central and southern Arizona
c. central Florida; San Joaquin Valley, CA; Ogallala aquifer, High Plains region

26. Which **one** represents an offstream water use?
a. irrigation   b. hydroelectric power generation

27. Which river has, since 1922, given rise to a litany of litigation known as the Law of the River?   a. Tennessee   b. Colorado   c. Mississippi   d. Hudson

Questions 28-30; Matching: The Colorado River Basin
a. Flaming Gorge   b. Powell   c. Hoover

28. ( ) lake impounded behind Glen Canyon dam; named for a famous nineteenth century geologist, explorer, and pubic servant

29. ( ) dam impounding Lake Mead near Las Vegas, NV

30. ( ) scenic canyon segment and national recreational area; only large dam and reservoir on the Green, the major upper basin tributary of the Colorado River

Questions 31-35; Agree (A) or Disagree (D)? Briefly explain the basis for any disagreements.

31. ( ) Freshly-burned, steep hillslopes are highly susceptible to debris slides and mudflows, especially if the first rainfall event following the burning is unusually heavy and intense.

32. ( ) Increased sediment loads in mountain watersheds usually result in more pronounced pool-and-riffle stream channels and improved habitat for fish and other aquatic organisms.

33. ( ) Infiltration to the shallow throughflow zone and deeper groundwater zone is maximized when precipitation rates greatly exceed soil infiltration rates.

34. (   ) The Salton Sea in the Imperial Valley of southern Arizona, was formed following major floods on the lower Salt River early in the twentieth century.

35. (   ) Actual annual deliveries of Colorado River water to compact members are generally higher than apportioned allotments because the 1922 reference river discharge was substantially lower than the average yearly discharge.

36. Which statement concerning the changes in the Colorado River and its riparian habitat in the Grand Canyon is reasonably correct and factual?
a. Very coarse, bouldery, flash-flood debris deposited at mouths of tributaries resist erosion because the maximum discharges now are substantially lower than those associated with the natural spring floods before Glen Canyon dam was built.
b. Sand bars and beaches along the river are now larger and more numerous than before the dam was built, and native new world plant species such as mesquite, catclaw acacia, salt cedar, and tamarisk now grow down to the water's edge.

37. What is schistosomiasis?
a. a South American, freshwater mussel species whose habitat will be seriously degraded and diminished by construction of the proposed Tucurui dam in Brazil
b. a dangerous, debilitating snail-born parasitic infection reaching epidemic proportions in irrigated lands downstream from the Aswan High Dam, Egypt
c. a dangerous fever carried by mosquitoes indigenous to swampy, wetlands along the Persian Gulf coast of Saudi Arabia

38. Which one is not characteristic of dam-impounded reservoirs in tropical areas?
a. excessive growth of water hyacinth and other aquatic plants
b. possibly acidic lake waters   c. poorly oxygenated or anaerobic bottom waters
d. seasonal lake overturn

39. Which would be most indicative of wetlands soils?
a. ultisols   b. oxisols   c. mollisols   d. histosols

40. On the map/maps provided, plot and label the following features or locations.
a. Grand Canyon, AZ   b. Chesapeake Bay   c. Hudson River
d. Owens Valley and Owens River, CA   e. Salton Sea, Imperial Valley, CA
        f. Lesotho; upper Orange River Basin, southern Africa
            g. Tucurui damsite, Tocantins River, Brazil

Answers; Home Study Assignment
Water: Process, Supply, and Use • Chapter 10

1. b	2. b	3. b	4. b	5. c
6. b	7. a	8. d	9. a	10. b
11. b	12. a	13. a	14. b	15. b
16. b	17. A	18. A	19. D*	20. D*
21. D*	22. D*	23. a	24. b	25. d
26. b	27. d	28. a	29. c, d	30. a
31. a	32. a, c	33. a, c	34. c	35. a
36. c	37. c, d	38. a	39. d	40. b
41. d	42. c	43. c	44. c	45. d
46. c	47. c	48. a	49. a	50. d
51. b	52. a, c*	53. b	54. d	55. a
56. b, d	57. a, c	58. b	59. b	60. f
61. a	62. d	63. e	64. c	65. A
66. A	67. D*	68. A	69. D*	70. a, b*
71. c	72. d	73. a, d	74. d	

*

19. Pool-and-riffle stream channel morphology is degraded or destroyed by increased sediment loads. Loss of the pools and riffles generally causes habitat for fish and other aquatic organism to deteriorate, not improve.

20. During times between precipitation events, perennial stream discharge is derived from an influx of groundwater; vadose zone throughflow is limited to a few days following heavy rainfall events.

21. Discharge generally decreases downstream in ephemeral streams because some water is lost by evapotranspiration and some by infiltrating downward to the groundwater system. Tributaries are also likely to be ephemeral, so little or no additional water is flowing into the stream from tributaries.

22. When precipitation rates exceed infiltration rates, runoff is vastly increased since the water "piles up" on the land surface faster than it can infiltrate into the soil. Infiltration would be maximized when precipitation rates are equal to or slightly less than infiltration rates. In this case, rainfall infiltrating the shallow through-flow zone and would have a chance to sink deeper into the groundwater system.

52. response b; The Colorado River is about the only stream of any consequence that flows southward into Mexico; the Red River of the North that forms the boundary between Minnesota and North Dakota is a major stream flowing northward into Canada. Many north-flowing tributaries to the St. Lawrence River also flow northward across the Canadian border. Thus the statement should read that streamflow to Canada exceeds stream flow to Mexico.

response d; On a daily basis, evapotranspiration equals about 66 % or two-thirds of precipitation, not 10 % as stated.

67. The Salton Sea is in southern California, not Arizona. It was formed early in the twentieth century during severe flooding on the Lower Colorado River, not on the Salt River as stated.

69. The assumed discharge used in the 1922 agreements was higher than the average measured for the years since that time. Thus in general, water deliveries to most states are somewhat less than allotted amounts.

70. response c; The statement is completely backwards. Sand bars and sandy beach areas had all but disappeared from the river since the Glen Canyon Dam was closed, preventing the normal sediment supply and annual spring flood discharges from reaching the Canyon. An "artificial flood" in the spring of 1996, produced by large, controlled releases of water from the dam, had the hoped for result of replenishing many of the sandy, river-edge areas eroded away over the years since the dam had been closed.

response d; The highly altered riparian habitat in the Canyon now features mostly non-native, old-world imports such as salt cedar and tamarisk. Non-native fish including trout have prospered in the cold, sediment-free water, but native Colorado River fishes, adapted to the natural river conditions, have been severely stressed since the dam was closed.

Answers: Environmental Geology Quiz • Chapter 10

1. a	2. c	3. a	4. a	5. a
6. d	7. a	8. b	9. a	10. a
11. a, c	12. d	13. a	14. d	15. b, d
16. b	17. a	18. a	19. a	20. b
21. d	22. a	23. c	24. b	25. a
26. a	27. b	28. b	29. c	30. a
31. A	32. D*	33. D*	34. D*	35. D*
36. a	37. b	38. d	39. d	

*

32. Pool-and-riffle stream channel morphology is degraded or destroyed by increased sediment loads. Loss of the pools and riffles generally causes habitat for fish and other aquatic organism to deteriorate, not improve.

33. When precipitation rates exceed infiltration rates, runoff is vastly increased since the water "piles up" on the land surface faster than it can infiltrate into the soil. Infiltration would be maximized when precipitation rates are equal to or slightly less than infiltration rates. In this case, most of the rainfall would infiltrate into the shallow throughflow zone and would have a chance to sink deeper into the groundwater system.

34. The Salton Sea is in southern California, not Arizona. It was formed early in the twentieth century during severe flooding on the Lower Colorado River, not on the Salt River as stated.

35. The assumed discharge used in the 1922 agreements was higher than the average measured for the years since that time. Thus in general, water deliveries to most states are somewhat less than allotted amounts.

Commentary/Media Watch

Environmental Concerns: Do They Stop at the State Line?

Legal fights concerning water rights, dams, riparian habitat, and long-distance water transfers are not limited to the "thirsty" American West. For ten years the State of North Carolina, allied with environmental groups, has been fighting a massive project to divert water (60 million gallons/day) from Lake Gaston on the Roanoke River in North Carolina, to Virginia Beach, VA. The proposed diversion amounts to an interbasin transfer and obviously, any diverted water would be permanently removed from the river. The state and its allies have argued that the project is unnecessary, that other viable, less expensive and less environmentally damaging solutions are available, and that water quality and fish and wildlife habitat in the lower Roanoke basin would be permanently damaged if the project were completed (see lower Roanoke River, Environmental Stewardship section, Commentary/Media Watch, Chapter 18). The project includes a very expensive, 75-mile-long pipeline, already under construction despite the fact that final permits are being vigorously contested in court. As of 12/15/96, forty other states had submitted briefs supporting North Carolina's position contesting the legality of the project. For the past ten years, Virginia Beach officials have justified the Lake Gaston-Roanoke River pipeline project as the only possible solution to their projected future water needs. An Associated Press article, 6/17/96, reported on a new study that showed how alternative sources of water were available in southeastern Virginia and how modest "improvements" in the Norfolk city water system over the last few years have significantly reduced projected water needs in the area to the point that the Lake Gaston pipeline is unnecessary. This conclusion seriously undermines the city's apparent desperate need for Roanoke River water; after all, they do seem to have survived quite well for the last ten years without it.

Smaller, less-distant rivers in southeastern Virginia, such as the Blackwater, Nottoway, and Meherrin, could be further exploited as alternative sources of water, but Virginia Beach officials cited growing public opposition and potentially serious environmental damage as reasons for rejecting this idea. North Carolina editorial writers (for example, 6/20/96; "A telling response; Concern apparently ends at the state line"; The Daily Reflector, Greenville, NC) and opponents of the Roanoke River diversion project gleefully pounced on these remarks, noting that Virginia Beach officials had finally acknowledged legitimate environmental concerns in southeastern Virginia similar to the ones they had ignored for years among residents of the lower Roanoke River basin just across the state line in North Carolina.

More on the Lower Roanoke River

While flood-control measures are based on frequency estimates of a given flood stage, water diversion projects are highly sensitive to estimates of the frequency and discharges of low- or minimum-stage flows. Obviously, removal of a fixed volume of water from a river or stream has far more impact when the discharge is low than when it is high. Thus for prudent water-supply planning, we need good estimates of minimum discharges and of lengths of time over which such conditions might be expected. These same considerations apply to rivers with "managed" discharges. Below Lake Gaston and the Roanoke Rapids dam, discharges on the Roanoke are managed by releases from the dams. These in turn, depend on electrical power demand and to a lessor extent on flood control. Water is stored during periods of high river discharge and released during low river stages, thus insuring that power can be generated during the hot, occasionally dry summer months when usage of electricity is high. Discharge management was largely responsible for a major fish kill on the lower Roanoke during the summer of 1995. For some time the river was high enough that many of the floodplain swamps, forested areas, and wetlands were inundated. Then dam managers significantly reduced releases to increase storage, and of course, downstream river levels dropped and water temporarily stored in the floodplain flowed back into the river. Unfortunately, the floodplain waters were rich in organic constituents and very low in dissolved oxygen. With addition of the floodplain water, dissolved oxygen levels in the river dropped to the point that thousands of fish suffocated including low-oxygen-tolerant species such as catfish. In hindsight, more gradual reductions in dam release rates could have been avoided this unfortunate event.

Intrastate Interbasin Water Squabbles

Future disputes over intrastate interbasin water transfers in many areas, including North Carolina, seem inevitable. An Associated Press article, 6/20/96, reported that the rapidly growing cities of Cary and Apex were requesting more water from Lake Jordan, a large water supply-flood control reservoir in the upper Cape Fear River watershed near Raleigh. However treated wastewater from the two cities would be discharged into the upper Neuse River near Raleigh; thus the withdrawal/discharge process would constitute interbasin water transfer. Cary and Apex insist that the additional Lake Jordan water is needed to sustain their explosive growth. Municipalities and other water uses farther downstream on the Cape Fear are concerned that reduced river flows will jeopardize their prospects for future growth and they are strongly opposed to the transfers. Unfortunately, the dispute is being narrowly framed in either/or terms; additional water means continued growth; no additional water means limits on growth; less water means growth may not even be possible! Before investing in more new, expensive public works projects, why not try wasting less water and better conserving what we do use? Perhaps then some "desperately needed" new water supply projects could be put off for a decade or two or even delayed indefinitely.

However, even water conservation has its "dark side". An Associated Press article, 8/4/96, noted problems, such as more frequent flushing and more pipe blockages, associated with water-saving, low-volume toilets dictated under the 1994 National Energy Policy Act. These problems have led to growing, nationwide black-market sales and installations of the outlawed, older, larger-volume commodes.

Large Dams and Large-Scale Water Diversion Projects in Developing Countries

In addition to capital expenditures, water made available for irrigation and other uses, and the value of electricity to be generated, environmental impacts of massive dams and water-diversion projects are now carefully considered in cost-benefits analyses. In the United States, environmental impact statements (EISs) are mandated by law for such projects. In developing countries, huge, "visionary" hydrologic development projects have great appeal as a way to rapidly improve living conditions, raise living standards, and contribute to national pride. The Aswan High Dam on the Nile in Egypt taught us all some valuable lessons about the painful tradeoffs involved is such projects. The dam was financed, first by the United States and later by the Soviet Union. Egypt and the dam were pawns in the Cold War. The completed dam paid off as planned; electricity was produced, new land were irrigated, water was made available for year-round agriculture, and new clean water supplies were provided.

A few consequences of the dam were ignored and others may have come as surprises. Evaporative water losses are very large; sediments are trapped behind the dam so prime farmland in the delta is eroding away; year-round farming demands chemical fertilizers, pesticides, and herbicides, all of which cost money and have unwanted negative environmental consequences; reduced freshwater flows through the delta resulted in formerly valuable coastal-estuarine fisheries being wiped out; and the many new, permanently wet canals and waterways allowed the snail host for the schistosomiasis flatworm to greatly expand its range, thus spreading the disease to new areas. Other adverse social impacts and environmental impacts such as excessive growth of aquatic plants, water hyacinth for example (p. 286), low dissolved oxygen levels, and poor water quality in general have troubled large dam projects elsewhere in subtropical/tropical parts of Africa. Vertical convective circulation, a major factor in oxygenating temperate-zone lakes and reservoirs, is severely restricted in areas with warm climatic conditions year-round.

The World Bank has for many years been a main financing source for large dam and water diversion projects in developing countries. However, the bank has come under increasing scrutiny and pressure to adopt more environmentally-responsible lending practices, especially in regard to massive projects with potentially enormous, largely unknown, environmental consequences. Environmental and social impacts are now accorded serious consideration in the loan application process, although this has not always been the case. Thus national and international environmental groups routinely publicize the potential environmental and social consequences of projects requesting World Bank financing. In some cases, the sponsoring government may be so deeply committed to the project that domestic opposition and adverse environmental consequences are purposefully suppressed and/or ignored.

In 1996, the bank rejected a loan request to finance a huge dam, reservoir, and hydroelectric project on the Arun River in Nepal. This project had been actively opposed by leading international environmental organizations and local Nepalese groups. An enormous, $8 billion dam and water diversion project on the upper Orange River in Lesotho, southern Africa, is currently funded and under construction (see Anonymous, 1996). The most controversial of these massive-scale projects is the Three Gorges Dam in the upper Yangtze River basin of western China (Magagnini, 1994; Qing, 1996; Walker and others, 1996). In 1994, persuasive environmental, social, and economic arguments helped persuade the bank to "pull the plug" on the project. However, the Chinese Government turned to private American financial sources for loans to continue with construction. This source of funding has evidently also evaporated. In late 1996, the U. S. Export-Import Bank announced that it will not lend to U. S. financial institutions seeking to provide funding to the Chinese Government for completion of the controversial project. Despite these setbacks, the Chinese Government has vowed to complete the project by 2009 with or without foreign financing.

Environmental and social impacts strongly influenced the Export-Import Bank's decision. The dam would probably destroy the only remaining habitat for the baiji or Chinese river dolphin; evidently only about 300 of the animals are known to still exist, most of them in the relatively unpolluted reaches of the Yangtze downstream from the dam. In 1986, the Center for Marine Conservation successfully lobbied to have the baiji added to the endangered species list to insure that any U. S. involvement in projects impacting on baiji habitat would be subject to provisions of the Endangered Species Act. Being an American bank, the Export-Import Banks felt obligated to follow U. S. laws in regard to this project, even though it's outside of the United States. The dam will have minimal value in controlling severe, middle and lower basin flooding such as occurred in 1954 and 1996, reputable international environmental organizations are strongly opposed to the project, and there is surprisingly strong and outspoken domestic opposition. In August, 1996, American news media reported that a prominent Chinese geophysicist had predicted earthquakes up to magnitude 6 could be triggered when the reservoir is filled.

Early North American Casualties of Large Water Diversion Projects

By the late 1930s, Owens Valley, California and the Colorado River delta lands in Mexico were dried up vestiges of their former selves, ecological victims of large-scale water diversion projects. The City of Los Angeles bought water rights all up and down the valley and diverted streams draining eastward from the Sierra Nevada into the Los Angeles aqueduct. The environmental troubles of Mono Lake (see *case history* Mono Lake and the Public Trust Doctrine, p, 314) also derive from diverting Sierran streams into the aqueduct. When Hoover and other dams on the lower Colorado River were completed and large volumes of water were diverted for irrigation, freshwater flows into the delta were effectively stopped. Although Owens Valley diehards fought for years to save their streams, marshes, and pasture lands, the ecological tragedy of the Colorado delta went largely unnoticed. Aldo Leopold's essay, "The Green Lagoons", in A Sand County Almanac (see supplemental references to Chapter 1), is a poignant reminder of the richly diverse aquatic and terrestrial vegetation and wildlife once hosted by today's barren, mud-cracked, salt-encrusted Colorado River delta lands.

Supplemental References

Anonymous, 1996, Notes on World Bank participation in the Arun River dam, Nepal and the $8 billion dollar, upper Orange River basin dam and water diversion, Lesotho: EDF Letter, v. XXVII, no. 3, May, 1996

Dennison, M. S and Berry, J. F., 1993, Wetlands; Guide to science, law, and technology: Noyes Publications, Park Ridge, NJ 07656, 439 p.

Fetter, C. W., 1994, Applied hydrogeology, third edition: Prentice Hall, Upper Saddle River, NJ, 691 p.

Freeze, A. R. and Cherry, J. A., 1979, Groundwater: Prentice Hall, Upper Saddle River, NJ, 604 p.

Hinrichsen, D., 1995, Waterworld; A hundred years of plumbing, plantations, and politics in the Everglades: The Amicus Journal, v. 17, no. 2, p. 23-27

Levin, T., 1996, Immersed in the Everglades: Sierra, v. 81, May/June, p. 56-63 & 86-87

Magagnini, S., 1994, Review of Yangtze! Yangtze! and On Leaving Bai Di Cheng (see Qing, Dai, 1996, reference listed below): The Amicus Journal, v. 16, no. 3, p. 44

Mairson, C., 1994, The Everglades: Dying for help: National Geographic Magazine, v. 185, no. 4, April, p. 2-35

McDonnell, J. J and six other authors, 1996, New method developed for studying flow on hillslopes: EOS, v. 77, no. 47, p. 465 & 472

Mitchell, J. G., 1992, Our disappearing wetlands: National Geographic Magazine, v. 182, no. 4, October, p. 3-45

O'Connell, K. A., 1996, Gore unveils Everglades Plan: in Park News, National Parks, v. 70, no. 5-6, p. 13-14

Pollack, S., 1996, Holding the world at bay: Two men, one goal; Saving the Chesapeake: Sierra, v. 81, May/June, p. 50

Qing, Dai, 1996, Yangtze! Yangtze!; Debate over the Three Gorges Project, 2nd edition, Earthscan Publications Ltd., 295 p.

Schwarzbach, D. A., 1995, Promised land. But what about the water? Heated politics over Israel's most precious commodity: The Amicus Journal, v. 17, no. 2, p. 35-39

Smith, P., 1996, Heart of the Hudson; National Geographic Magazine, v. 189, no. 3, March, p. 72-95

Stevens, W. K., 1994, Severe ancient droughts: A warning to California: in Themes of the Times; The Changing Earth, Fall 1996 edition; distributed by Prentice Hall, Upper Saddle River, NJ, p. 11

Walker, R., Mallot, R. L., Shipley, R., and Lin, F. K., 1996, On Leaving Bai Di Cheng; The cultures of China's Yangtze Gorges: NC Press Ltd., 263 p.

Warne, A. G. and Stanley, D. J., 1993, Archaeology to refine Holocene subsidence rates along the Nile delta margin, Egypt; Geology, v. 21, p. 715-718

Water Pollution and Treatment • Chapter 11; Home Study Assignment

1. In general which hydrologic systems have longer residence times for pollutants?
a. rivers and streams   b. groundwater aquifers

Questions 2-4; Matching: See Figure 11.2, p. 293.

a. recovery zone   b. pollution zone   c. active decomposition zone

2. ( ) downstream BOD increases rapidly and dissolved oxygen decreases rapidly

3. ( ) downstream BOD decreases gradually and dissolved oxygen gradually increases to normal levels

4. ( ) BOD reaches a maximum; dissolved oxygen reaches a minimum

5. Which one is routinely used to assess microbial pollution in waters?
a. biological oxygen demand   b. oxygen-depleting microbial capacity
c. fecal coliform bacteria counts   d. anaerobic methane production

6. Which pathogen in the city's water supply stunned Milwaukee, WI, 1993?
a. cholera   b. typhoid   c. ebola   d. cryptosporidium   e. giardia

7. Which two dissolved constituents are most responsible for cultural eutrophication?   a. nitrates   b. sulfates   c. carbonates   d. phosphates

8. Runoff from which lands would be expected to deliver the highest concentrations of soluble nitrogen and phosphorous compounds to streams and rivers?   a. forested   b. urban   c. agricultural   d. recreational beaches

Questions 9-17; Matching: Localities and Environmental Issues

a. Love Canal   b. Detroit   c. Prince William Sound
d. Hudson   e. northern Persian Gulf   f. Cuyahoga
g. lower Pechora basin, Russia   h. Gulf of Mexico, off Galveston
   i. West Virginia, western Pennsylvania, and eastern Ohio

9. ( ) stream flowing though Cleveland, OH, so polluted with petroleum and other industrial wastes that its surface ignited and burned in 1969; has since been cleaned up to become an urban asset

10. ( ) a lovely, scenic river tragically contaminated with polychlorinated biphenyls in the 1960s and 70s; concentrations in water, bottom muds, and aquatic organisms have gradually decreased since 1980

11. ( ) site of massive crude oil spill from the supertanker Mega Borg, 1990

12. ( ) largest and most ecologically disastrous crude oil spill in North America; the supertanker Exxon Valdez grounded, ripping a hole in her hull

13. ( ) wakeup call concerning past chemical waste disposal practices and environmental health

14. ( ) deliberate crude oil spills by Iraqis during Operation Desert Storm, 1991

15. ( ) surface waters degraded by acidic waters draining from abandoned coal mining areas

16. ( ) surface waters and land areas polluted by crude oil leaking from ruptured pipelines
17. ( ) waterway in a heavily urbanized and industrialized area; connects Lakes Huron and Erie; substantial improvements in water quality are evident since the dismal, highly polluted conditions of the 1950s and 60s

Questions 18-20; Matching: Classes of Pollutants and Specific Substances

a. radioactive materials   b. heavy metals   c. hazardous chemicals
18. ( ) trichloroethylene   19. ( ) plutonium   20. ( ) lead, zinc

21. Which **two** are characterized as nonpoint sources of surface water pollution?
a. storm-water runoff from urban streets
b. effluent discharged from a municipal sewage treatment plant
c. agricultural runoff containing herbicides and pesticides
d. highly acidic mine waters draining from an old tunnel or adit

22. In general, a moving groundwater mass contaminated by one or more toxic and or hazardous chemicals is called a what?   a. cone   b. ribbon   c. plume   d. rung

23. Which **one**, a trace element necessary for life in low concentrations and toxic in high concentrations, is locally concentrated in drainage waters from some irrigated fields?   a. chlorodane   b. selenium   c. mercury   d. hydrogen sulfide

24. The water table in an unconfined, sandy, coastal aquifer is three feet above sea level. How will the freshwater-salt water interface at depth in the aquifer change in response to a reduction of the water table to two feet above sea level?
a. rise one foot   b. rise to an elevation of 78 feet below sea level
c. rise to an elevation of 40 feet below sea level   d. sink 40 feet

25. Which dissolved groundwater constituent, if present in higher than normal levels, is most likely have an anthropogenic origin?
a. iron   b. calcium   c. hydrogen sulfide   d. nitrate

26. Which pair of contaminants would most likely be found along a heavily used older road or highway in New York or Michigan? **one answer**
a. excessive DDT in the soils: a gasoline plume floating on the shallow water table
b. excessive lead in the soils; excessive dissolved chlorides in the near-surface groundwater
c. excessive selenium in the soils; excessive dissolved phosphates in the shallow groundwater
d. excessive zinc in the soils; a plume of lindane in the shallow, water table aquifer

Questions 27-30; True (T)/False (F): See Figure 11.12 and *case history* The Threatened Groundwater of Long Island, p. 303-305.

27. ( ) In the Long Island, NY, groundwater system, the Raritan clay aquitard has inhibited the excess dissolved nitrate plume from penetrating downward into the Lloyd aquifer.

28. ( ) The Magothy aquifer of the Long Island, NY, groundwater system is mainly recharged with water moving downward from the unconfined, surficial aquifer comprised of Pleistocene glacial deposits.

29. ( ) Nassau and Suffolk Counties, New York, are in the easternmost portion of Long Island, close to New York City.

30. ( ) Much of the water in the Magothy aquifer, Long Island, NY, groundwater system is confined and under some artesian pressure.

31. Which **two** interpretations or conclusions are logical and reasonable? See Figure 11.14, p. 308.
a. Lead and arsenic levels in U. S. surface waters evidently declined steeply from 1976 to 1978, then declined more gradually from 1978 to 1986.
b. DDT and PCBs in U. S. surface waters both evidently declined during the years between 1970 and 1986.
c. Mercury, one of the heavy metals in the aquatic environment, is of little concern because its concentration in fish tissue has remained relatively constant in the ten years between 1976 and 1986.
d. Banning the use of DDT and PCBs and switching from leaded to unleaded gasoline have not resulted in any noticeable improvements in surface water quality in the U. S.

32. Which **one** statement concerning domestic wells and/or septic tank sewage disposal systems is unreasonable and/or seriously flawed?
a. A moderately permeable vadose zone soil and a relatively deep water table enhance natural purification processes before the wastewater from the drain field reaches the local water table.
b. An impermeable, compacted clay in the drain field vadose zone is advantageous because under those conditions, the wastewater will seldom if ever reach the water table before being completely purified.
c. A small mound in the water table will typically form beneath the wastewater drain field and a cone of depression will form around a pumping well.
d. A well and septic tank drain field should be sited such that the water table elevation in the well during pumping is higher than the water table elevation beneath the drain field.

33. Which **two** can cause a normally well-functioning septic tank wastewater treatment system to temporarily fail or malfunction? **two answers**
a. Infiltration from a heavy rain raises the water table to the elevation of the drain field exit pipes.
b. The wastewater drain field is developed in a hardpan soil with an extremely low hydraulic conductivity.
c. The wastewater drain field is developed below the water table of a shallow, unconfined aquifer.
d. A large quantity of a strong, anti-bacterial, chemical agent is accidentally released into the septic tank system.

34. Which **two** chemicals are commonly used to disinfect effluent wastewater from secondary treatment plants?   a. chlorine   b. iodine   c. ammonia   d. ozone

Questions 35-37; Matching: Wastewater Recycling Terms

a. renovation   b. reuse   c. return

35. ( ) disinfected wastewater from a secondary treatment plant is applied to the land surface; water is further purified as it percolates through the vadose zone to the water table

36. ( ) cleaned, purified recycled wastewater is pumped from the groundwater system

37. ( ) disinfected wastewater from a secondary treatment plant is applied to agricultural or forest crops

38. Which two could be serious, negative consequences of using sludge from municipal sewage treatment plants for soil reclamation?
a. soil porosity and fertility could be significantly degraded
b. heavy metals could become concentrated in the soils
c. dissolved nitrates could build up to unacceptable levels in the shallow groundwater system
d. sand and grit from the sludge could cause excessive sediment deposition in local surface streams

Questions 39-41; Matching: Wastewater Recycling Programs

a. Arcata, CA   b. Muskegon, MI   c. Clayton County, GA

39. ( ) Young salmon are raised in water from wastewater treatment ponds.

40. ( ) Spray irrigation and vadose zone percolation provide effective, post-secondary, wastewater treatment.

41. ( ) Secondarily-treated wastewater promotes forest growth and recycling of purified, former wastewater into the local, domestic water supply.

42. Which one is a potentially low-cost method, involving growing plants, that could effectively substitute for conventional, secondary and post-secondary wastewater treatment processes?
a. nutrient-film treatment   b. intensive anaerobic digestion
c. sludge activation composting   d. biogenetic nutrient decomposition

43. Which one is a potentially recoverable, clean-burning fuel formed during anaerobic digestion and decomposition of organic matter in wastewater and sludge from municipal sewage treatment plants and animal feed lots?
a. carbon dioxide; CO_2 b. ammonia; NH_3 c. ozone; O_3 d. methane; CH_4

44. What is a reasonable definition for *safe yield* of an aquifer?
a. the quantity of water withdrawn from an aquifer each year required to support all water needs of private land owners above the aquifer
b. an unnecessarily small amount of groundwater decreed by Big Gov'ment regulators conspiring to bankrupt small businesses and wonderful entrepreneurs dependent on abundant, low-cost supplies of clean groundwater
c. the total amount of water that can be withdrawn from an aquifer each year in compliance with the Absolute Ownership Doctrine
d. an amount of water equal to or less than that recharged to the aquifer each year

Questions 45-52; Matching: Combine the following words in such a way as to match the surface water and/or groundwater legal principle with its description. Responses must be in their correct order; some responses may be used more than once. For example, the Reasonable Use Doctrine would be identified as (l) (e) (i).

a. Prior b. Trust c. Correlative d. American e. Use
f. Rights g. Ownership h. Public i. Doctrine j. Riparian
k. Appropriation l. Reasonable m. Absolute n. Rule

45. ( ) ( ) dominant surface water legal principle in the eastern U. S.; asserts that land owners adjacent to lakes and streams have a right to use the water but do not actually have direct ownership of the water

46. ( ) ( ) ( ) asserts that the first user has an established right to the water; that right can be passed on through inheritance or sale to others; the prevalent surface water legal principle in the water-deficient western states

47. ( ) ( ) ( ) same as the American Rule

48. ( ) ( ) property owners' groundwater withdrawal rights are limited or restricted to what are judged as reasonable and beneficial uses of that water; commonly applied in the western United States

49. ( ) ( ) ( ) affirms that states have an obligation and responsibility to consider ecosystem protection as being in the legitimate public interest

50. ( ) ( ) ( ) basically the same as the English Rule; states that a property owner can withdraw as much groundwater as is necessary, even though the aquifer and negative impacts from overdrafts may extend far beyond the individual's property boundaries

51. ( ) ( ) ( ) peripheral land owners and groundwater users in a groundwater basin with one or more common aquifers are extended some protection from excessive withdrawals or other damaging activities of a single landowner in the same basin

52. ( ) ( ) ( ) basis for the 1983 California Supreme Court ruling on the Mono Lake water diversion case

53. Which statement concerning water diversion from the Mono Lake, CA, drainage basin is not correct?
a. Formerly safe breeding grounds for California gulls and other waterfowl were devastated by predators.
b. Alkali-rich dust derived from dried-out parts of the lake bottom has polluted the atmosphere during windstorms.
c. The lake's salinity essentially doubled between 1941 and 1981.
d. Citing the Public Appropriations Doctrine, the federal courts reaffirmed that the City of Los Angeles had exclusive rights to the surface waters of Mono Basin.

54. Which legislative act established the Superfund and Superfund Sites?
a. Refuse Act, 1899 b. Water Quality Act, 1987
c. Fish and Wildlife Coordination Act, 1958
d. Comprehensive Environmental Response, Compensation, and Liability Act, 1980

55. Which legislative act established control of nonpoint pollution sources as national policy?

a. National Environmental Policy Act, 1969   b. Water Quality Act, 1987
c. Federal Water and Pollution Control Act, 1956
d. Fish and Wildlife Coordination Act, 1958

56. On the U. S. map provided, accurately plot and label the following locations.

a. Mono Lake, CA	b. Long Island, NY	c. Lake Erie
d. Hudson River	e. Niagara Falls, NY	f. Lake Huron
g. Atlantic City, NJ	h. Detroit River	i. Cleveland, OH
j. Lake Michigan	k. Milwaukee, WI	l. Arcata, CA

Environmental Geology Quiz • Chapter 11

1. In general which hydrologic systems have longer residence times for pollutants?
a. rivers and streams   b. groundwater aquifers

2. Assume that a point source mass of some biodegradable pollutant is introduced to a stream or river. Which **one** describes the recovery zone?
   a. downstream BOD increases rapidly and dissolved O2 decreases rapidly
   b. downstream BOD decreases gradually and dissolved O2 gradually increases to normal levels
   c. BOD reaches a maximum; dissolved O2 reaches a minimum

3. Which **one** is routinely used to assess microbial pollution in waters?
a. biological oxygen demand   b. oxygen-depleting microbial capacity
c. fecal coliform bacteria counts   d. anaerobic methane production

4. Which pathogen in the city's water supply stunned Milwaukee, WI, 1993?
a. giardia   b. typhoid   c. ebola   d. cryptosporidium

5. Which **two** dissolved constituents are most responsible for cultural eutrophication?   a. nitrates   b. sulfates   c. phosphates   d. carbonates

6. Runoff from which lands would be expected to deliver the highest concentrations of soluble nitrogen and phosphorous compounds to streams and rivers?   a. recreational beaches   b. urban   c. agricultural   d. forested

Questions 7-12; Matching: Localities and Environmental Issues
a. Hudson   b. Love Canal   c. Detroit   d. Cuyahoga
e. West Virginia, western Pennsylvania, and eastern Ohio
f. lower Pechora River basin, Russia

7. ( ) stream flowing though Cleveland, OH, so polluted with petroleum and other industrial wastes that its surface ignited and burned in 1969; has since been cleaned up to become an urban asset

8. ( ) a lovely, scenic river tragically contaminated with polychlorinated biphenyls in the 1960s and 70s; concentrations in water, bottom muds, and aquatic organisms have gradually decreased since 1980

9. ( ) national wakeup call concerning past chemical waste disposal practices and environmental health

10. ( ) surface waters degraded by acidic waters draining from abandoned coal mining areas

11. ( ) surface waters and lands polluted by crude oil leaking from ruptured pipelines

12. ( ) waterway in a heavily urbanized and industrialized area; connects Lakes Huron and Erie; substantial improvements in water quality are evident since the dismal, highly polluted conditions of the 1950s and 60s

13. In general, a moving groundwater mass contaminated by one or more toxic and or hazardous chemicals is called a what?   a. rung   b. plume   c. ribbon   d. cone

14. Which **two** are characterized as nonpoint sources of surface water pollution?
a. highly acidic mine waters draining from an old tunnel or adit
b. effluent discharged from a municipal sewage treatment plant
c. agricultural runoff containing herbicides and pesticides
d. storm-water runoff from urban streets

15. Which dissolved groundwater constituent, if present in higher than normal levels, is most likely have an anthropogenic origin?
a. calcium   b. nitrate   c. hydrogen sulfide   d. iron

16. Which pair of contaminants would most likely be found along a heavily used older road or highway in New York or Michigan? **one answer**
a. excessive zinc in soils; a plume of lindane in the shallow, water table aquifer
b. excessive DDT in soils: a gasoline plume floating on the shallow water table
c. excessive lead in soils; excessive dissolved chlorides in near-surface groundwater
d. excessive selenium in soils; excessive dissolved iron in the shallow groundwater

Questions 17-18; True (T) or False (F)

17. ( ) The Magothy aquifer of the Long Island, NY, groundwater system is mainly recharged with water moving downward from the unconfined, surficial aquifer comprised of Pleistocene glacial deposits.

18. ( ) Much of the water in the Magothy aquifer, Long Island, NY, groundwater system is confined and under some artesian pressure.

19. Which **one** interpretation or conclusion is incorrect or badly misleading?
a. Mercury, one of the heavy metals in the aquatic environment, is of little concern because its concentration in fish tissue has remained relatively constant in the ten years between 1976 and 1986.
b. DDT and PCBs in U. S. surface waters both evidently declined during the years between 1970 and 1986.
c. Lead and arsenic levels in U. S. surface waters evidently declined steeply from 1976 to 1978, then declined more gradually from 1978 to 1986.

20. Which **one** statement concerning domestic wells and septic tank sewage disposal systems is unreasonable and/or seriously flawed?
a. An impermeable, compacted clay in the drain field vadose zone is advantageous because under those conditions, the wastewater will seldom if ever reach the water table before being completely purified.
b. A moderately permeable vadose zone soil and a relatively deep water table enhance natural purification processes before the wastewater from the drain field reaches the local water table.
c. A well and septic tank drain field should be sited such that even when the well is being pumped, the water table elevation at the well is higher than the water table elevation beneath the drain field.

21. Which **two** chemicals are commonly used to disinfect effluent wastewater from secondary treatment plants?   a. chlorine   b. iodine   c. ammonia   d. ozone

22. Which **one** can cause a normally well-functioning septic tank wastewater treatment system to temporarily fail or malfunction?
a. A large quantity of a strong, anti-bacterial, chemical agent is accidentally released into the septic tank system.
b. The wastewater drain field is developed in a hardpan soil with an extremely low hydraulic conductivity.
c. The wastewater drain field is developed below the water table of a shallow, unconfined aquifer.

23. Which **one** could be a serious, negative consequence of using sludge from municipal sewage treatment plants for soil reclamation?
a. soil porosity and fertility could be significantly degraded
b. sand and grit from the sludge could cause excessive sediment deposition in local surface streams
c. heavy metals could become concentrated in the soils

Questions 24-25; Matching: Wastewater Recycling Programs
a. Clayton County, GA   b. Arcata, CA

24. ( ) Young salmon are raised in cleaned water from wastewater treatment ponds.

25. ( ) Secondarily-treated wastewater is recycled to promotes forest growth and recharge shallow aquifers in the watershed area devoted to the local, domestic water supply.

26. Which **one** is a potentially low-cost method, involving growing plants, that could effectively substitute for conventional, secondary and post-secondary wastewater treatment processes?
a. nutrient-film treatment   b. biogenetic nutrient decomposition
c. sludge activation composting   d. intensive anaerobic digestion

27. Which **one** is a potentially recoverable, clean-burning fuel formed during anaerobic digestion of organic matter in wastewater and sludge from municipal sewage treatment plants and large animal raising facilities?
a. ammonia; NH_3 b. carbon dioxide; CO_2 c. ozone; O_3 d. methane; CH_4

28. What **one** is an accurate, reasonable definition for *safe yield* of an aquifer?
a. the quantity of water withdrawn from an aquifer each year required to support all water needs of private land owners above the aquifer
b. an amount of water equal to or less than that recharged to the aquifer each year
c. the total amount of water that can be withdrawn from an aquifer each year in compliance with the Absolute Ownership Doctrine

29. Which **one** statement concerning the environmental consequences of water diversion from the Mono Lake, CA, drainage basin is not correct?
a. As the lake level fell, formerly safe breeding grounds for California gulls and other waterfowl were devastated by predators.
b. Alkali-rich dust derived from dried-out parts of the lake bottom has contributed to air pollution during windstorms.
c. Citing the Riparian Doctrine, the federal courts reaffirmed that the City of Los Angeles had exclusive rights to the surface waters of Mono Basin, California.

30. Which legislative act established the Superfund and Superfund Sites?
a. Water Quality Act, 1987   b. Fish and Wildlife Coordination Act, 1958
c. Comprehensive Environmental Response, Compensation, and Liability Act, 1980

31. Which legislative act established control of nonpoint pollution sources as national policy?   a. National Environmental Policy Act, 1969
b. Water Quality Act, 1987   c. Fish and Wildlife Coordination Act, 1958

32. Which **one** is the dominant surface water legal principle in the eastern U. S.? It asserts that land owners adjacent to lakes and streams have a right to use the water but do not actually have direct ownership of the water.
a. Riparian Doctrine   b. Reasonable use Doctrine   c. Prior Appropriation Doctrine

33. Which **one** asserts that the first user has an established right to the water and that right can be passed on to others through inheritance or sale? It is the prevalent surface water legal principle in the water-deficient western states.
a. American Rule   b. Prior Appropriation Doctrine   c. Reasonable Use Doctrine

34. Which **one** affirms that states have an obligation and legal responsibility to consider ecosystem protection as being in the legitimate public interest?
a. Public Trust Doctrine   b. Absolute Ownership Doctrine   c. Public Rights Rule

35. Which **one** states that a property owner can withdraw as much groundwater from beneath his property as necessary, even though the aquifer and the negative impacts from overdrafts may extend far beyond the individual's property boundaries?
a. Reasonable Use Doctrine   b. Absolute Ownership Doctrine   c. American Rule

36. On the U. S. map provided, accurately plot and label the following locations.

a. Mono Lake, CA	b. Long Island, NY	c. Lake Erie
d. Atlantic City, NJ	e. Milwaukee, WI	f. Arcata, CA
g. Lake Huron	h. Detroit River	i. Cleveland, OH

Answers; Home Study Assignment
Water Pollution and Treatment • Chapter 11

1. b	2. b	3. a	4. c	5. c
6. d	7. a, d	8. c	9. f	10. d
11. h	12. c	13. a	14. e	15. i
16. g	17. b	18. c	19. a	20. b
21. a, c	22. c	23. b	24. b	25. d
26. b	27. T	28. T	29. F	30. T
31. a, b	32. b	33. a, d	34. a, d	35. a
36. b	37. c	38. b, c	39. a	40. b
41. c	42. a	43. d	44. d	45. j i
46. a k i	47. l e i	48. d n	49. h b i	50. m g i
51. c f i	52. h b i	53. d	54. d	55. b

Answers: Environmental Geology Quiz • Chapter 11

1. b	2. b	3. c	4. d	5. a, c
6. c	7. d	8. a	9. b	10. e
11. f	12. c	13. b	14. c, d	15. b
16. c	17. T	18. T	19. a	20. a
21. a, d	22. a	23. c	24. b	25. a
26. a	27. d	28. b	29. c	30. c
31. b	32. a	33. b	34. a	35. b

Commentary/Media Watch

The Pigeon River Controversy

Tennessee's leading environmental official, as quoted in local newspapers, 11/20/96, described the Pigeon River as "black" where it enters Tennessee from western North Carolina. Blame for the discoloration rests with upstream, wastewater discharges from Champion International's paper mill at Canton, NC. The comment was made in the context of an ongoing dispute involving the corporation, the EPA, and the states of North Carolina and Tennessee. In 1994, an EPA report cited the Champion mill as the third-largest point source polluter in North Carolina. Investments ($330 million) in pollution abatement and modernized industrial processes at the plant have resulted in substantial reductions in air pollutants and wastewater discharges, and in recent years, water quality in the Pigeon River has improved and fish populations have increased. However, Tennessee still rates Champion's efforts as "not good enough". A variance from North Carolina's standard wastewater permitting policies allows Champion to continue discharging the same quantity of discolored wastewater for the next five years. The EPA stated that it will approve this variance only if Tennessee does not raise objections. Clearly, Tennessee does object. A follow-up Associated Press article, 12/1/96, reported that Tennessee's governor and environmental commissioner were to formally announce the state's objections at a news conference held at the river's edge on December 3. The Dead Pigeon River Council, a community-based environmental organization in eastern Tennessee, is leading the opposition to continued permitting of the wastewater discharges. During the annual pre-Christmas parade in Newport, TN, its members prominently displayed placards reading Save the Pigeon River and distributed black armbands to remind parade watchers of the river's once sparkling clean waters.

This dispute has a long and eventful history (see Bartlett, 1995, supplemental references) and its outcome is still in doubt. There has been progress however. Dioxin discharges from the Canton plant were eliminated many years ago, but the highly stable and biopersistent chemical is still present in sediments downstream from the plant and in Douglas Lake, east of Knoxville, TN.

Safe Drinking Water and Water Treatment

Safe drinking water supplies, municipal water treatment plants, and wastewater treatment plants are frequent newsmakers, locally and nationally. In this writer's community, the water-supply system and other utilities are city-owned. In July, 1996, local media featured the dedication of the city's new $14 million wastewater treatment facility. Ultraviolet light is being utilized to kill off any bacteria and viruses that survive the primary and secondary treatment steps. Chlorination of wastewaters is avoided. Chlorine is known to combine with natural or anthropogenic organic molecules in surface waters and with charcoal in water purification/filtration systems to form toxic and carcinogenic chloro-organic compounds. The plant has enough excess capacity to allow for continued population and economic growth and state-of-the-art, nitrate and phosphate removal processes result in effluent with lower levels of these nutrients than the state's standards require. With its new facility in operation, the city is well-positioned to profit from any trading in water quality credits and debits that may develop among interbasin wastewater dischargers in the future.

Chloramines are used for final purification of the city's drinking water. Chloramines have enough oxidizing power to eliminate infectious bacteria and viruses from the water without contributing a "chlorine" taste and odor. However for a week or so each year, the city switches back to the conventional hypochlorite source for chlorine. Residents immediately notice the chlorine taste and odor and often call to complain or ask "why the change". The reasoning is simple. Even in the presence of chloramines, local pockets of microorganisms can develop and survive in a few "dead ends" and low-throughflow sections of the maize of mains and pipes that constitutes the water delivery system. Thus a once-a-year dose of the more strongly oxidizing form of chlorine is employed to protect against any such possibility.

A newspaper article, 7/11/96, reported on the growing threat that live microorganisms pose to the nation's water supply. After years of focusing on potential dangers from chemicals and toxins etc. in water supplies, bacteria and other microorganisms are showing growing resistance to chlorine and are making a strong comeback. For example, giardia is spreading over most of the country and cryptosporidium, the protistan that stunned Milwaukee in 1993 (p. 294) is evidently immune to many forms of chlorination. Thus municipal drinking waters and water supply systems need to be carefully monitored and checked!

Local papers also featured a lead article, 6/19/96, noting that the city of Farmville, NC and its largest employer, Collins & Aikman, were engaged in negotiations concerning the company's metal-rich wastewaters. Such wastewaters are far more expensive to clean than conventional municipal sewage and may require specialized equipment and/or treatment processes that would not normally be incorporated into small-community wastewater treatment plants. The city and company are both seeking an agreement that holds down costs and satisfies state water quality standards.

Pesticides in North Carolina's Groundwater

An Associated Press article (8/4/96) carried in local and statewide newspapers reported that pesticides or pesticide residues were detected in 17 of 100 test wells located in all regions of North Carolina. Follow-up tests showed that water from two wells in the eastern part of the state contained pesticides at levels high enough to threaten public health. Four other wells had high levels of pesticides, and DDT derivatives were found in six, attesting to the longevity (biopersistence) of that compound and its derivatives. DDT has not been used since 1972. Other contaminants identified include alachlor, atrazine, cyanazine, metolachlor, simazine, and compounds of the triazine family. State officials noted that for many years, the agricultural community in general and specifically pesticide regulators and industry representatives were saying that we're using small amounts and they'll never reach the groundwater. These were wishful, highly self-serving notions! The test results support numerous critics who have maintained all along that pesticide contamination of the state's groundwaters has been consistently underestimated.

Releases of Untreated Sewage During Hurricanes Bertha and Fran

An Associated Press article (7/26/96) carried in local and area newspapers reported that during hurricane Bertha, falling trees broke numerous power lines and interrupted electrical service to many municipal sewage treatment plants, causing untreated sewage to be released into the Neuse River and other streams in southeastern North Carolina. Similar power outages and raw sewage releases were triggered by Hurricane Fran. These discharges have seriously polluted parts of the state's southern coastal waters and estuaries, resulting in dangerously high coliform bacteria counts. Backup electricity generators are needed to prevent future repetitions of this problem. This same story was also carried on local television. Video images featured signs posted in affected areas announcing that these waters were temporarily closed to fishing, shellfish harvesting, and recreation.

These shutdowns, of course, occurred during heavy storms when the untreated wastewater was released into streams and rivers already swollen to flood-stage levels. Thus dilution worked to reduce the impact of the releases. Consider the worst possible case of the "shutdown scenario". Due to extra heavy demand or an accidental equipment failure, the region-wide electrical transmission grid collapses after three weeks of the hottest, driest, weather on record when surface water discharges are at record or near-record lows. Then releases of untreated sewage would have a devastating effect on dissolved oxygen levels and overall water quality. This writer remembers his favorite, childhood. trout-fishing brook in eastern Pennsylvania. The lower mile or so was polluted by the town's raw sewage. For most of the year however, the polluted reach of the stream was quite "healthy" with abundant trout and other aquatic organisms. Only in the late summer and early fall, when the creek's discharge dropped to its lowest levels for the year, did the added BOD of the raw sewage lower dissolved oxygen levels to the point that the trout migrated upstream into the clean part of the creek.

Declines in Fish Stocks and Fish Habitat; Outdoor Sportsmen Respond

In North Carolina and other states with large commercial and/or recreational fishing enterprises, more and more, sports-section editorialists are taking strong positions in favor of environmental cleanup and protection. After all, these profitable and popular activities, with their attendant large sales of supporting equipment and supplies, are seriously jeopardized by deteriorating water quality in freshwater, estuarine, and coastal environments. For example, in his editorial of 8/11/96 that appeared in the Sports Section of the Raleigh (NC) News & Observer, writer Bob Simpson recounts a story of how years ago, fishermen off southeast Alaska could expect to routinely land large halibut with only a hand line. Now such catches are a rarity. The editorial then moves on to discuss falling fish stocks, decreasing average sizes, environmental concerns over fish habitat and reproduction especially in estuaries, and worldwide pollution and overfishing in general. Simpson finishes his editorial by noting that all things are intertwined, a familiar notion to Gaia followers (see the *special feature* "The Gaia Hypothesis", p. 17), and that each of use will have to be more concerned with "good ecological health" if our mouth-watering daily seafood specials and pleasurable recreational fishing experiences are to survive.

Excessive Nutrient Loads in Rivers and Coastal Waters of North Carolina

Over the past two decades, once rare algal blooms and fish kills in North Carolina's lower river basins and estuaries have became all too common, especially in the late summer and early fall when river discharges are at their seasonal lows. Decaying and decomposing algae deplete dissolved oxygen in the water, causing widespread fish mortality. The algal blooms are fueled by anthropogenic phosphate and nitrate contributed from agricultural runoff, large-scale animal-raising operations, airborne sources, storm water runoff, and municipal wastewaters. The nationwide phase out of phosphate-based detergents gave the rivers a short reprieve, but the temporary improvements in water quality have slowly been overwhelmed by increased loads of nitrates and phosphates from other sources. The Neuse River has been the focus for much of the discussion and debate over water quality issues.

Early in 1995, the Raleigh News & Observer, one of the state's leading newspapers, ran a series of articles under the umbrella title "Boss Hog; North Carolina's pork revolution"(in Supplemental References, see Warrick and Stith, 1995). These highly-visible, well-documented, and well-illustrated articles blew the whistle on public indifference and complacency concerning the recent, massive growth in corporate hog-producing facilities, mostly in the eastern part of the state. Featured topics include leaking "hog lagoons" and their detrimental effects on surface water and groundwater quality, lack of public access to basic information concerning the hog industry, lack of effective state regulatory policies combined with understaffed regulatory agencies, the powerful, invasive, financial and political clout of the hog-producers, and the way too cozy relationships between big pork's power brokers and the state's leading politicians entrusted with "serving the public interests". This series won the John B. Oakes Award for Distinguished Environmental Journalism.

During ten days of heavy rains in June of that same year, the dire inferences of these articles seemed to come true. Local flooding and overflows and structural failures of hog-waste lagoons resulted in large volumes of nutrient-rich, organic-rich wastewaters being released into rivers and streams. The results were devastating. In addition to massive fish kills, some of the state's prime shellfish harvesting and recreational coastal waters were closed for the remainder of the summer due to high levels of coliform bacteria. Tough-minded environmental journalism and a largely preventable ecological disaster combined to focus the public spotlight on excessive nutrient loadings and other water quality issues.

An article written for the News & Observer, 2/3/96, by a representative of a leading national environmental organization outlined the necessary steps to decrease nitrogen loads and improve the poor water quality now evident in the Neuse River and estuary. The article recommended tough new controls on agriculture and large animal-raising facilities. About a month later, the Neuse's water quality troubles were forcibly brought to the public's attention in a series of four extensive articles in the News & Observer entitled "Sold down the river" (see Leavenworth and Warrick, 1996; Supplemental References). Industry, consumers, municipalities, and agriculture all came in for a share of the blame, but agriculture and the rapidly expanding animal-raising industry came under the heaviest scrutiny. Similar concerns over surface water quality issues in North Carolina were also reported in the New York Times (3/2/96). Studies of Chesapeake Bay cited largely unregulated, nonpoint sources as major contributors of excess nutrients, particularly nitrates. These sources include agricultural runoff and airborne nitrates derived from nitrogen oxides produced in fossil fuel combustion and from ammonia released to the atmosphere from decomposing animal wastes. For each nitrate unit prevented from entering the bay, an equivalent amount from a new or expanded older source takes its place. Overall reductions are difficult to achieve and maintain. Under the current conditions of rapidly-expanding hog populations and locally explosive urban growth, even a major, comprehensive cleanup of the Neuse may just be enough to "break even".

During the spring and summer months, pressures mounted for new rules and legislative action to address the water quality issues. A 50-foot vegetated buffer along streams of the Neuse basin was one of many proposals put forth. Such buffers have been widely shown to significantly lower nitrate/phosphate loads in agricultural runoff entering drainage ditches and streams. Agricultural interests angrily opposed the buffer requirement, arguing that it would be too expensive, impossible to implement, and would put the family farmer out of business just to save a few fish. Of course, most of the actual opposition came from large, well-capitalized, corporate farms and animal-raising operations. Being classified as agricultural enterprises, they are largely unregulated and have so far escaped from tight, water quality permitting requirements imposed on point source discharges. Thus at the same time that municipalities and industry are investing in upgraded wastewater treatment facilities to meet tougher standards, agricultural interests continue to ignore their significant role in contributing nonpoint source nutrients to rivers and streams.

On June 11, a major fish kill was reported on a small, upper-basin tributary to the Black River, itself a major, lower-basin tributary to the Cape Fear River. The kill was traced to an accidental release of slurried, waste potato and sweet potato cattle feed from a facility owned by North Carolina's junior United States senator. Dissolved oxygen in the creek was quickly depleted and an estimated 6000 fish were killed. This was the third major fish kill of the year in the Cape Fear basin. Follow-up stories reported that a careless worker had left the slurry tank valve open and that the damaging consequences of the spill could have been averted had farm management followed recommendations mandated in a 1995 state inspection and closed off an old drainage pipe that conveyed the waste directly to the creek. A retention basin placed between the slurry tanks and the creek would also have averted all or most of the damage. The senator was eventually assessed a civil penalty and clean-up/restocking fees totaling about $48,000. Washington critics dubbed the whole affair "yam scam and "yam gate", but by publicly accepting responsibility for the spill, the senator seemed to largely defuse the issue.

Additional summer, 1996, news reports on pending state legislative actions and issues affecting surface water quality are summarized below.

6/13/96   The Associated Press reported that federal funds for research on hog manure were granted to Iowa State and North Carolina State Universities. Research and development costs associated with large-scale animal-raising facilities and technical advice on waste disposal and environmental problems are by-and-large paid for by the taxpayer and provided free of charge to hog producers by state and federal agencies.

6/13/96   The Associated Press reports that a North Carolina Senate panel unveils tougher rules on animal wastes.

6/20/96   The Associated Press reports that the state Senate approved much tougher regulations for large-scale, hog-raising facilities. The new rules will protect surface water quality and reduce the nuisance potential, mainly odor, of these facilities. The Senate bill incorporates many of the suggestions made by the governor's "Blue Ribbon" Committee.

6/21/96   The Associated Press reports that the state House went along with the toughened Senate bill to regulate large factory-farm hog producers. These regulations are designed to lessen nutrient runoff into surface waters and to prevent waste lagoons from leaking or failing.

6/22/96   Newspapers reported that a hog-raising enterprise featured in statewide television commercials sponsored by the pork producers has a troubled environmental record, including numerous violations and fines for illegal discharges of untreated wastewaters dating back to 1982. The commercial portrays the facility as a "family farm" with a clean environmental record that contributes generously to the state's economy. This writer viewed the commercial only once. The negative publicity evidently caused it to be quickly yanked.

6/23/96   The Associated Press reported on continuing environmental and social impacts of last June's (1995) collapse of a large hog waste lagoon that sent a massive dose of lethal, polluted water into the lower New River near Jacksonville, NC. The offending corporation was later assessed a fine of $100,000 for the spill. Area television newscasts featured a ceremony on the Jacksonville waterfront commemorating the first anniversary of the spill.

7/27/96   The Raleigh News & Observer noted that drawn out negotiations on state spending between the House and Senate have tilted in favor of strong support for comprehensive programs to solve some of the state's severe, water quality issues. Hog producers, municipalities, and agriculture may receive substantial funding to reduce nutrient levels in runoff and wastewaters discharged to the state's rivers and streams.

8/2/96   The Associated Press reported that a major food processing corporation had filed an extensive analysis indicating minimal environmental impacts from its proposed slaughter of an additional 40,000 hogs/week at its Bladen County meat-packing plant.

8/4/96   The Associated Press reported that a spill of untreated wastewater from a small manufacturing plant was responsible for killing more than 50 fish along a 100-yard reach of the Swannanoa River near Asheville, NC. It is significant that even relatively small, local fish kills such as this one now attract media interest.

8/11/96   The Associated Press reported that municipal and industrial water users in the upper Yadkin-Pee Dee River basin, North Carolina, are joining together to develop the Yadkin River Basin Association in an effort to head off the deterioration in water quality and fish kills that have plagued the Neuse River and its estuary in recent years. Membership would be limited to entities with permits to discharge >= one million gallons of treated wastewater per day. The idea is to cut costs by funding joint, comprehensive monitoring of treated effluent and river water. Farming and animal raising were not mentioned, despite being substantial contributors of excess nitrate derived from direct runoff and oxidation of ammonia released from urine and manure.

8/13/96   The News & Observer reported that sometime during the weekend, wastewaters broke through the bottom of a hog lagoon into an old, buried drainage pipe, releasing a million gallons of lagoon waste into a small tributary of the Trent/Neuse Rivers near New Bern, NC. This is the second major lagoon spill this summer (1996). These same owners have had a similar-type lagoon leak at one of their other properties. Many years ago, original wetland areas in eastern North Carolina were drained by networks of underground tile drains that discharged into ditches and small creeks. Unaware or unconcerned corporate land owners have sited hog waste lagoons directly above these old buried drainage pipes. If the lagoon bottom fails, the underground drains carry the wastewaters directly into drainage ditches and streams.

This spill continued for two days after state officials were first notified and farm officials were yet to respond. No fish kills were reported, but the nearest parts of the Neuse River estuary were closed to shellfish harvesting!

Pfiesteria Make Its Debut in North Carolina's Coastal Waters

Pfiesteria is the "mystery" dinoflagellate responsible for ulcerous infections that begun showing up in fish and crabs of the Neuse estuary and other coastal waters over the last two decades. These symptoms were rare at first but became much more widespread in recent years. The organism bores into the flesh of its victims, including Man. In addition to ulcers, researchers and river workers have reported numerous other symptoms from contact with the organism. Some are similar to those of amnesic shellfish poisoning, known to be caused by one or more of the highly dangerous, "red tide" toxins. Pfiesteria researchers at Duke and North Carolina State Universities and affected river workers were interviewed for CNN's Science and Technology Week (November 10) and on CNN's evenings news the following day. Neither the organism nor its effects were recognized prior to the influx of excess nutrients. Thus the present-day, predatory form might only develop under environmental conditions characterized by excess nutrients. Such conditions would have occurred very rarely in the past, but over the past decade, they have become more-or-less established as standard, year-round environmental conditions.

Water Quality "Blues" on the Danube

The Danube, once one of Europe's most lovely, scenic rivers, is an historic, international river. The upper basin includes parts of southern Germany, Switzerland, northern Italy, Liechtenstein, western Austria, and the Czech Republic. The central portion of the river winds through northern Austria and Vienna, forms a short segment of the border between Hungary and the Slovak Republic, then flows south through Hungary and its capital city Budapest and the Serbian Republic and its capital city of Belgrade. In Serbia, the river again turns to the east, forming most of the border between Bulgaria and Romania. About a hundred kilometers from the Black Sea, the river bends sharply north for about 150 km, then again turns to the east forming a short segment of Moldavian-Romanian border. The river flows into the Black Sea through an extensive series of distributaries and wetlands that comprise the Danube delta, most of which is in Romania. The river that inspired the most famous of the Strauss waltzes survived fairly well through the wars and political turmoil of first half of the twentieth century, but, four decades of rigid, state-dominated Soviet Block industrial policy brought serious pollution and water quality deterioration to the middle and lower basin. The clean waters, diverse riparian habitats, and once-productive fisheries in the river and delta wetlands have largely disappeared and two massive water diversion and hydroelectric projects, the Gabcikovo Dam in the Slovak Republic and the Nagymaros Dam project in Hungary, further threaten the river's ecological health.

The Danube would make an excellent international class project in water quality, environmental attitudes, detrimental effects of unregulated industrial pollution, and future pollution abatement, The current degraded environmental condition of the middle and lower Danube basin is in sharp contrast to the overall improvements in water quality that have been achieved in western Europe and the United States over those same forty years. Environmental degradation and its negative impacts on health and ecological well-being will rank among the worst legacies of hardline Communist rule in eastern Europe.

Supplemental References

Anonymous, 1996, US-Mexico cooperation benefits border environment: EDF Letter, v. XXVII, no. 4, July, p. 1 & 5

Bartlett, R. A., 1995, Troubled waters; Champion International and the Pigeon River controversy: The University of Tennessee Press, Knoxville, TN, 376 p.

Edwards, M., 1994, Soviet pollution: National Geographic Magazine, v. 186, no. 2, p. 70-99

Hinrichsen, D., 1994, Putting the blue back in the Danube: The Amicus Journal, v. 16, no. 3, p. 41-43

Hinrichsen, D., 1989, Blue Danube: International Notebook Section, The Amicus Journal, v. 11, no. 1, p. 4-8.

Leavenworth, S. and Warrick, J., 1996, Sold down the river: A special four-part series (March) of the Raleigh, NC, News & Observer
These four major articles focus on the science, economics, and politics of water quality and ecological deterioration of the Neuse River in North Carolina.

Mitchell, J. G., 1996, Our polluted runoff; National Geographic Magazine, v. 189, no. 2, February, p. 106-125

Pringle, C., Vellidis, G., Heliotis, F., Randacu, D., and Cristofor, S., 1993, Environmental problems of the Danube Delta: American Scientist, v. 81, p. 350-361

Stith, P. and Warrick, J., 1996, Boss Hog: North Carolina's pork revolution: The Amicus Journal, v. 18, no. 1, p. 36-42

Stumm, W. and Morgan, J. J., 1996, Aquatic chemistry, third edition: John Wiley & Sons, New York, NY, 1022 p.

Warrick, J. and Stith, P., 1995, Boss Hog; North Carolina's pork revolution: A special edition (March) of the Raleigh, NC, News & Observer

Waste Management • Chapter 12; Home Study Assignment

1. Which **one** does not reflect the prevalent, present-day viewpoint regarding waste materials and how to handle them?
a. dilute and disperse   b. resources out of place
c. concentrate and contain   d. integrated waste management

2. Which **two** are not significant components of the integrated waste management concept?   a. composting   b. dispersal   c. recycling   d. dilution

3. An old gravel pit is used for a sanitary landfill; after the landfill is closed, the land is then rezoned for parklands. Which response describes this process?
a. optional resource redeployment   b. sequential land use concept
c. multiple use management principle   d. municipal zoning analysis

4. Which **one** response does not apply to the Fresh Kills Landfill, NY? See p. 322.
a. located in Manhattan a few blocks south of the Wall Street financial district
b. projected to rise higher than the Statue of Liberty in New York harbor
c. will probably be filled by the year 2000, despite its enormous planned capacity
d. receives about 3 million tons of solid waste from New York City each year

5. Which potentially recoverable fuel gas is routinely generated in capped, sanitary landfills?   a. carbon dioxide   b. radon   c. ammonia   d. methane

6. Which **one** could be a temporary, negative impact of a newly-implemented, highly successful, municipal, partial recycling program?
a. new supplies of waste paper glut local markets, and waste paper prices plunge
b. composting significantly reduces recycling targets set for aluminum and glass
c. reduced volumes of waste materials are destined for the local sanitary landfill
d. the useful lifetime of the existing sanitary landfill is extended for a few years

7. Which **two** statements concerning integrated waste management are reasonably correct and accurate? The other two are seriously flawed in some way.
a. Ash disposal and air-quality issues are important environmental concerns associated with incineration.
b. Over the next ten years, partial recycling programs can probably not achieve more than a 10 percent reduction in volume of urban waste needing disposal.
c. The city of Seattle, WA, has implemented a very successful waste management strategy mainly utilizing incineration and ocean disposal.
d. Composting may be a cost-effective method of treating organic wastes.

8. Which **two** sets of soil and bedrock conditions would make the better landfill sites? **two answers**
a. a large sinkhole; limestone bedrock; unconfined water table, average elevation a few feet below the bottom of the sinkhole
b. thick zone of clay rich soil and partially weathered shale; shale bedrock
c. river floodplain site; deep, sandy soil; water table a few feet below the land surface
d. very thick residual, soil horizons A through C overlying massive, relatively unfractured granite bedrock; water table is normally well below the soil-bedrock boundary

9. In general, federal or state groundwater monitoring requirements for landfills can be expected to extend how far beyond the closure date for the facility?
a. no more than ten years   b. six months or less   c. one year   d. thirty years

10. The capping material added to seal the surface of a filled, sanitary landfill should exhibit which two characteristics? Review Chapter 3. **two answers**
a. high hydraulic conductivity   b. low erodibility
c. high montmorillonite content   d. strong cohesion

11. In which **one** would laterally migrating methane be most likely to first accumulate just outside a capped, sealed, sanitary landfill?
a. a vadose zone monitoring well   b. leaves and stems of deep-rooted plants
c. open fractures in the bedrock   d. a groundwater monitoring well

12. Which material is normally used as a low-cost, pollutant-free liner and cap for sanitary landfills?
a. size-sorted crushed shale   b. hardened asphalt and sand
c. highly expansive soil   d. compacted clay

13. Which category of environmental impacts is most prevalent at U. S. Superfund sites? **one answer**   a. air quality problems   b. groundwater contamination
c. vegetative impacts   d. surface water pollution

14. Which is a safe, legal, and relatively inexpensive way for individuals and corporations to avoid liability for cleanup and remediation costs associated with discovery of hazardous wastes on a recently-purchased property?
a. Enlist the aid of your congressman and strong-arm the EPA or appropriate state agency.
b. Have an environmental audit done on the property before the sale is finalized.
c. Sue the prior owners, real estate agency, lending agency, and any other organizations or individuals involved in the sale of the property.

15. Which **two** statements concerning land application of hazardous wastes are reasonable and defensible? The other two are seriously flawed in some way.
a. Wastes rich in heavy metals can be safely and effectively detoxified by land application and mixing with the surface soil layer.
b. Soil microorganisms can be very effective in gradually decomposing or rendering harmless many kinds of organic wastes and petroleum products.
c. Soil microorganisms, such as bacteria and fungi, are typically most abundant in the topmost, organic-rich soil layers.
d. Saline brines produced in a water desalination plant can cheaply and safely be disposed of by land application.

16. For which **two** hazardous wastes would deep-well disposal generally be preferable to other possible disposal methods?
a. oil field or industrial waste brines
b. extremely toxic, liquid organic waste with a very long biopersistence
c. liquid, high-level, radioactive wastes
d. liquid wastes produced in cattle and hog feeding facilities

17. Which is the best set of subsurface geologic and hydrologic conditions for deep-well disposal of hazardous liquid wastes? **one answer**
a. fractured limestone or sandstone aquifer; top of aquifer is 900 meters below the surface; aquifer contains saline water and is overlain by thick shales and interbedded thin sandstone layers
b. thick, unfractured shale beds at a depth of 140 feet; shales have low hydraulic conductivity and the top of the shale beds is 90 feet below the surface; uncemented, coarse-grained sands and gravel beds overlie the shale
c. highly fractured granite bedrock at a depth of 900 meters; same highly fractured granite extends upward to the base of the weathered granite and surface soil zone
d. highly fractured, cavernous limestone at a depth of 30 feet; the limestone is overlain by a continuous, 20 foot-thick shale layer and a clay-rich soil at the surface; the regional water table is unconfined and 40 feet below the surface

18. Which is, by far, the most abundant, naturally-occurring uranium isotope in rocks and minerals? a. U-236 b. U-234 c. U-235 d. U-238

19. Which **two** isotopes are used as the fuel in nuclear fission reactors?
a. U-235 b. Th-232 c. U-238 d. Pu-239

    Questions 20-25; Matching: U. S. Locations and Environmental Issues

a. Seattle, WA	b. Denver, CO	c. Oak Ridge, TN
d. Love Canal, NY	e. Elizabeth, NJ	f. Yucca Mountain, NV

20. ( ) has ambitious plans for recycling and waste volume reduction
21. ( ) dump site for very dangerous, undocumented chemical wastes including phosgene, also known as mustard gas, used in WW I and the Iran-Iraq War
22. ( ) previously buried, chemical wastes turned suburbanite homeowners' dreams into nightmares
23. ( ) probable future burial site for America's high-level radioactive waste
24. ( ) in the 1960s, earthquakes were triggered by deep-well injection of liquid, toxic wastes; later, sequential land use involved converting gravel pits into landfills and landfills into a sports stadium complex
25. ( ) manmade, radioactive isotopes accumulated in the local fish and waterfowl and also escaped into the shallow groundwater system

26. Which **one** statement concerning the proposed Yucca Mountain, Nevada, high-level nuclear waste repository site is incorrect or seriously flawed?
a. The area is quite dry; the water table is about 500 meters below the land surface.
b. The welded tuff bedrock extends far below the depth of the local water table.
c. Highly absorptive zeolite minerals in the welded tuff below the proposed repository site would retard movements of dissolved constituents in downward-percolating, vadose-zone water.
d. Vertical to steeply-inclined fractures and inactive faults present no problems since the wastes will be stored as solids, not as gases or liquids.

27. Which **two** statements concerning nuclear weapons production and nuclear fission for electric power in the U. S. are basically correct? The other two are incorrect, seriously flawed, or highly controversial.
a. In general, small quantities of high-level, radioactive wastes from weapons production facilities pose a lessor, environmental problem than storing spent fuel rods removed from reactors used for electric power generation.
b. Small quantities of the chemically unreactive gaseous fission product krypton-85, with a half-life of about 10 years, are released directly to the atmosphere from nuclear power plants.
c. Low-level radioactive waste needs to be isolated for about 500 years to insure safe radiation levels; high-level wastes, particularly those rich in Pu-239, need to be isolated for about 250,000 years.
d. Nuclear weapons production facilities, such as Rocky Flats near Denver, CO, and Hanford, WA, have had exemplary environmental and safety records.

28. Which process would probably be the safest and most prudent way to dispose of instantly lethal or very dangerous, manmade chemicals such as chemical weapons agents, dioxin, and polychlorinated biphenyls, PCBs?
a. ocean dumping beyond the continental shelf   b. high-temperature incineration
c. store them in a surface impoundment   d. use land application methods

29. On the U. S. map provided, accurately plot and label the following locations.

a. Seattle, WA	b. Denver, CO	c. Oak Ridge, TN
d. Love Canal, NY	e. Elizabeth, NJ	f. Yucca Mountain, NV

Environmental Geology Quiz • Chapter 12

1. Which **one** notion does not reflect the prevalent, present-day viewpoint regarding waste materials and how to handle them?
a. integrated waste management   b. dilute and disperse
c. resources out of place   d. concentrate and contain

2. Which **two** are not significant components of the integrated waste management concept?   a. dilution   b. recycling   c. dispersal   d. composting

3. An old gravel pit is used for a sanitary landfill; after the landfill is closed, the land is then rezoned for parklands. Which response describes this process?
a. multiple use management principle   b. municipal zoning analysis
c. optional resource redeployment   d. sequential land use concept

4. Which potentially recoverable fuel gas is routinely generated in capped, sanitary landfills?   a. carbon dioxide   b. methane   c. radon   d. ammonia

5. Which **one** could be a temporary, negative impact of a newly-implemented, highly successful, municipal, partial recycling program?
a. new supplies of waste paper glut local markets, and waste paper prices plunge
b. reduced volumes of waste materials are destined for the local sanitary landfill
c. the useful lifetime of the existing sanitary landfill is extended for a few years

6. Which **one** statement concerning integrated waste management is reasonably correct and accurate? The other two are seriously flawed in some way.
a. The city of Seattle, WA, has implemented a very successful waste management strategy mainly utilizing incineration and ocean disposal.
b. Ash disposal and air-quality issues are important environmental concerns associated with incineration.
c. Over the next ten years, partial recycling programs can probably not achieve more than a 10 percent reduction in volume of urban waste needing disposal.

7. Which **one** set of soil and bedrock conditions would make the best landfill site?
a. a large sinkhole; limestone bedrock; unconfined water table with an average elevation a few feet below the bottom of the sinkhole
b. river floodplain site; deep, sandy soil; water table a few feet below the land surface
c. very thick residual, soil overlying massive, relatively unfractured granite bedrock; water table is normally well below the soil-bedrock boundary

8. In general, federal or state groundwater monitoring requirements for landfills can be expected to extend how long after the closure date for the facility?
a. six months or less   b. no more than ten years   c. thirty years   d. one year

9. The capping material used to seal a filled, sanitary landfill should exhibit which **two** characteristics?   a. high hydraulic conductivity   b. low erodibility
c. high montmorillonite content   d. strong cohesion

10. Which **one** category of environmental impacts is most prevalent at U. S. Superfund sites?   a. vegetative impacts   b. surface water pollution
c. air quality problems   d. groundwater contamination

11. Where just outside a capped, sealed, sanitary landfill would laterally migrating methane be most likely to first accumulate?
a. in a vadose zone monitoring well   b. in leaves and stems of deep-rooted plants
c. in open fractures in the bedrock   d. in a groundwater monitoring well

12. Which material is normally used as a low-cost, pollutant-free liner and cap for sanitary landfills?   a. highly expansive soil   b. compacted clay
c. pea sized crushed granite   d. hardened asphalt and sand

13. Which is a prudent and relatively inexpensive way to avoid cleanup and remediation liabilities associated with hazardous wastes discovered on a recently-purchased property?
a. Strong-arm the EPA or appropriate state agency and pray to your chosen God.
b. Insist on an environmental audit before purchasing any property.
c. Sue the prior owners, real estate agency, lending agency, and any other organizations or individuals involved in the sale of the property.

14. Which **one** statement concerning land application of hazardous wastes is reasonable and defensible? The other two are seriously flawed in some way.
a. Soil microorganisms can be very effective in gradually decomposing or rendering harmless many kinds of organic wastes and petroleum products.
b. Wastes rich in heavy metals can be safely and effectively detoxified by land application and mixing with the surface soil layer.
c. Brines generated in seawater desalination plants can cheaply and safely be disposed of by land application.

15. For which **two** hazardous wastes would deep-well disposal generally be preferable to other disposal methods?
a. oil field or industrial waste brines   b. liquid, high-level, radioactive wastes
c. extremely toxic, liquid organic waste with a very long biopersistence
d. liquid wastes produced in cattle and hog feeding facilities

16. Which is the best set of subsurface geologic and hydrologic conditions for deep-well disposal of hazardous liquid wastes? **one answer**
a. fractured limestone or sandstone aquifer; top of aquifer is 900 meters below the surface; aquifer contains saline water and is overlain by thick shales and interbedded thin sandstone layers
b. highly fractured granite bedrock at a depth of 900 meters; same highly fractured granite extends upward to the base of the weathered granite and surface soil zone
c. thick, unfractured shale beds at a depth of 140 feet; shales have low hydraulic conductivity and the top of the shale beds is 90 feet below the surface; uncemented, coarse-grained sands and gravel beds overlie the shale

17. Which **two** isotopes are used as the fuel in nuclear fission reactors? **two answers**
a. U-235   b. Pu-239   c. Th-232   d. U-238

18. Which process would probably be the safest and most prudent way to dispose of lethal or dangerous manmade chemicals such as nerve gas, dioxin, and PCBs?
a. use land application methods   b. store them in a surface impoundment
c. high-temperature incineration   d. ocean dumping beyond the continental shelf

Questions 19-22; Matching: U. S. Locations and Environmental Issues

a. Oak Ridge   b. Elizabeth   c. Denver   d. Yucca Mountain

19. ( ) NJ; dump site for very dangerous, undocumented chemical wastes including phosgene, also known as mustard gas, used in WW I and the Iran-Iraq War
20. ( ) NV; probable future burial site for America's high-level radioactive waste
21. ( ) CO; in the 1960s, earthquakes were triggered by deep-well injection of liquid, toxic wastes; later, sequential land use involved converting gravel pits into landfills and landfills into a sports stadium complex
22. ( ) TN; manmade, radioactive isotopes accumulated in the local fish and waterfowl and also escaped into the shallow groundwater system

23. Which **one** statement concerning nuclear weapons production and nuclear fission for electric power in the U. S. is basically correct? The other two are incorrect, seriously flawed, or highly controversial.

a. In general, small quantities of high-level, radioactive wastes from weapons production facilities pose a lessor, environmental problem than storing spent fuel rods removed from reactors used for electric power generation.
b. Low-level radioactive waste needs to be isolated for about 500 years to insure safe radiation levels; high-level wastes, particularly those rich in Pu-239, need to be isolated for about 250,000 years.
c. Nuclear weapons production facilities, such as Rocky Flats near Denver, CO, and Hanford, WA, have had exemplary environmental and safety records.

24. On the U. S. map provided, accurately plot and label the following locations.

a. Seattle, WA   b. Denver, CO   c. Oak Ridge, TN
d. Love Canal, NY   e. Elizabeth, NJ   f. Yucca Mountain, NV

Answers; Home Study Assignment
Waste Management • Chapter 12

1. a	2. b, d	3. b	4. a	5. d
6. a	7. a, d	8. b, d	9. d	10. b, d
11. a	12. d	13. b	14. b	15. b, c
16. a, b	17. a	18. d	19. a, d	20. a
21. e	22. d	23. f	24. b	25. c
26. d	27. b, c	28. b		

Answers: Environmental Geology Quiz • Chapter 12

1. b	2. a, c	3. d	4. b	5. a
6. b	7. c	8. c	9. b, d	10. d
11. a	12. b	13. b	14. a	15. a, c
16. a	17. a, b	18. c	19. b	20. d
21. c	22. a	23. b		

Commentary/Media Watch

Your Local Sanitary Landfill

Sanitary landfills and problems associated with municipal waste disposal can be expected to generate some level of media interest in most areas. This writer's local media occasionally report on city and county waste disposal problems and on the plight of the Pitt County landfill, officially closed as of January, 1995. Leachate has shown up in shallow wells screened below the bottom of the landfill and a few, small plumes extend laterally to the edges of the landfill. In addition, mixed, highly toxic pesticides and herbicides from a burned out agricultural chemical plant and storage facility were buried in the landfill. These are now cause for concern and the State has mandated certain, long-lasting and potentially costly monitoring/remediation requirements that must eventually be satisfied. Unfortunately, this problem has been largely taken over by out-of-town attorneys more interested in finding "deep pockets" to "help share the costs" than in taking appropriate steps to delineate the groundwater problem and limit any environmental damages. At the present time, solid refuse from the city is trucked to a distant, private landfill at considerable expense. Eventually, combustible wastes will be burned at a new facility, now under construction, to produce process heat and electricity for a large industrial plant near Kinston, NC.

Dioxin Disaster Averted

A short, terse, newspaper article, 7/31/96, reported on a successful, very "delicate" marine salvage operation to avert potentially severe environmental damage in the Gulf of St. Lawrence north of Prince Edward Island, eastern Canada. In 1970, the barge Irving Whale, loaded with heating oil and waste oil laced with PCBs in separate compartments, sank in the Gulf of St. Lawrence. After twenty-six years on the bottom, the sunken barge was carefully raised to the surface and refloated without causing any structural damage. Much of the non-PCB-bearing oil had long-since leaked away, but the original load of PCB-laced oil was recovered.

This was Canada's most complex and costly salvage operation to date. Recent advances in marine salvage equipment and methods made the operation feasible whereas previously, it had been judged as "too risky".

Recycling News from "Earth Matters"

The television program Earth Matters, show each Sunday on CNN, often includes material of interest to environmental geology. For example, the program for August 4, 1996, included the following reports directly relevant to Chapter 12.

A small, innovative New Jersey company has devised and patented a manufacturing process to recycle roof shingles into an inexpensive and effective, asphalt road-patching product. Contractors and roofing companies sidestep landfill fees for old roofing shingles, the product requires no virgin asphalt, and the manufacturing and pothole-patching processes don't produce the smoke particulates and hydrocarbon vapors usually associated with asphalt preparation and highway repair work. The company is now supplying local markets in New Jersey, but has plans to expand nationwide in the near future.

Another small, innovative firm in Seattle, WA, collects used/obsolete computer disks from computer industry giants such as Microsoft and from smaller companies in the area. Some are salvaged, reconditioned, and sold as new disks; others are scraped and converted into useful, low-cost products. The company's president estimates that for each ton of "raw disks" collected, 1990 pounds are recycled back into productive uses and only 10 pounds of solid waste are carried to a landfill.

The same program also featured an ecologically "complete" solar greenhouse and vegetable garden on Cape Cod, MA. The greenhouse employs solar heating and uses solar panels for generating electricity. Chickens and rabbits, raised for eggs and fur respectively, use up excess oxygen produced by photosynthesis, resupply carbon dioxide to spur vigorous plant growth, provide "natural" fertilizers year-round, and supply "body heat" during the winter months. Lady bugs are cultured to control aphids. The greenhouse's organically-grown, high quality, fresh vegetables and salad fixings are snapped up daily by upscale restaurants in the Boston area and on Martha's Vineyard.

Fresh Kills Update (see *case history*, p. 322)

According to an Associated Press article, 12/1/96, the gigantic Fresh Kills landfill on Staten Island has been the final destination for New York City's residential trash (garbage) since the late 1930s when the city stopped dumping its refuse in the ocean. Staten Island borough officials and residents are looking forward to the year 2001 when the smelly eyesore will finally be closed. Residents of neighboring states such as Pennsylvania are worried that commercial waste disposal companies may open new sites in their states to accommodate the Big Apple's discarded refuse and unwanted leftovers. From time to time, abandoned underground coal mines in eastern Pennsylvania have been suggested as possible super-size landfill sites. Acidic groundwater generated in many of these mines eventually reaches the surface and pollutes rivers and streams. Acidic mine waters filtering through landfill refuse would produce a "super leachate" that would be far more environmentally damaging than acidic mine drainage alone. For many years now, the City has been sending ("exporting") its commercial trash to out-of-state landfills and incinerators. One possible solution would be for the City to help finance new plants to burn trash and generate electricity. These new facilities would provide plenty of additional carbon dioxide for global warming, but fitted with the latest anti-pollution technology, emissions of other gaseous air pollutants and particulates could be held to a minimum. Thus the City could lower its waste disposal costs and generate some of its own electricity while at the same time reducing its enormous volume of residential trash to a much smaller volume of ash which, of course, also requires permitted-only disposal.

Repository for Low-Level Radioactive Wastes, Wake County, North Carolina
    News of the proposed repository for low-level radioactive waste in Wake County, North Carolina, has many characteristics in common with a slow-moving, emotion-filled soap opera. But unfortunately low-level nuclear waste is sorely lacking in sexiness and glamour! The proposed repository has its roots in an organization known as the Southeast (States) Compact Commission, an organization funded by seven southeastern states entrusted with planning for prudent disposal of low-level radioactive wastes produced in its member states. For many years, compact states, including North Carolina, shipped their waste to the federal facility at Barnwell, South Carolina. Looking forward to the time when the Barnwell repository would have to be closed, the Commission voted to locate a successor in North Carolina and agreed to provide adequate funds for site exploration and development on a year to year basis. An extensive search for an appropriate site was completed, and the project's major contractor proceeded with site evaluation and geotechnical studies. As of 6/18/96, the Commission had already spent $90 million on the project and has yet to see an operating repository materialize. Meanwhile low-level waste is being temporarily stored at nuclear generating plants and other facilities. An uninformed public, widespread opposition, and less-than-forthright political leadership virtually guarantee additional court challenges and continued "real slow" progress.
    The following "blow-by-blow" account of news reports, June to December, 1996, provides some insight into the complications and complexities involved in this project. Political concerns are at least as important as the scientific and technological issues. Opponents and supporters both cite the same geologic evidence to defend their positions, and demagoguery is easily disguised as legitimate environmental concern.

             Media Reports; The Wake County, NC, Repository
6/7/96   Local and statewide papers carried brief articles reporting on the brewing controversy over continued public funding for the repository.
6/18/96   An Associated Press article reported that the Southeast Compact Commission was nearing a vote on a request from the State of North Carolina for an additional $15 million to continue site evaluation studies. Commissioners were clearly "not pleased" with progress to date, and noted that $90 million had already been appropriated for the project

6/19/96 An Associated Press article reported that the Southeast Compact Commission had voted to reject North Carolina's request for additional funds. The Commission was especially critical of the State's slow progress in moving toward site approval and licensing, and was leery of the State's efforts to effectively defuse growing local and statewide opposition to the repository project.

7/19/96 The Associated Press reported that with no additional funding available, Chem-Nuclear Systems Inc., the project's major contractor, had suspended further operations and was laying off most of its employees working on the project. This is a major blow to a project that is already three years behind schedule.

8/13/96 A highly respected staff writer on environmental issues for the Raleigh (NC) News & Observer reported that local opposition to the proposed waste repository site in southern Wake County has been growing since funds for the project were cut off in August. In an article entitled Activists Question Meeting on Dump, it was noted that representatives from local citizens groups and neighboring counties were not informed of nor invited to a meeting concerning the repository attended by state compact commission members, industry representatives, and state regulatory personnel. Opposition groups protested vigorously and argued that the meeting may, in fact, have violated the state's open meeting requirement. According to the participants, this meeting's purpose was to discuss ways to get additional funding and to make some visible progress on the project. Two major points are evident from this flap: 1) the State needs to move forward with the project, and 2) ignoring the concerned public is certainly not the way to gain widespread support for the project.

16/11/96 The News & Observer, in its headline story for the Metro Section, reported that the North Carolina Low-Level Radioactive Waste Authority had conditionally decided to provide $26 million to fund the repository project over the next four years. A new prime contractor, Harding Lawson Associates, was hired to carry on with site studies and evaluation. Harding Lawson was cited as having successfully helped the company U. S. Ecology win a license for a similar waste repository in California. Authority members seemed generally pleased with this outcome, but they were concerned about pay rates in the contract for clerical help, $55/hr, and for consultations with Harding Lawson's vice president, $185/hr. The Harding Lawson plan evidently includes additional geotechnical site studies despite the fact that further progress on the repository project now seems to hinge much more on political actions than on technological factors. Private consulting geologists engaged by Wake and Chatham Counties to follow the project described the new expenditures as a waste of money, and state regulatory officials voiced concerns over bedrock fracture that could allow contaminated groundwater to spread into surrounding areas should the repository ever leak.

Geologically, the proposed southern Wake County site is about the best one can expect to find in North Carolina. The Coastal Plain section of the state includes abundant wetlands and lands subjected to temporary inundation; the surficial, unconfined water table is very shallow in most areas. Much of the region is underlain by a prolific limestone aquifer that provides water to numerous municipal and commercial users. The shallow water tables and important aquifers should be enough to eliminate any potential coastal plain sites from consideration. The eastern and central Piedmont areas locally have fairly thick surficial soil and decomposed rock (saprolite) layers overlying bedrock. These weathered zones constitute fairly permeable, local shallow aquifers in many areas, and open fractures in the near-surface bedrock provide additional sources of groundwater in many areas. Potential repository sites in this region would be superior to those in the coastal plain, but groundwater conditions are still far from ideal for a waste repository site. Conditions are fairly similar in the western, mountainous part of the state. Here again, highly permeable, shallow aquifers in surficial soil and weathered-rock zones and open, shallow fractures in bedrock would provide unfavorable groundwater conditions for the repository. That, basically leaves only areas of Triassic bedrock such as the proposed southern Wake County site.

As in other eastern states such as Pennsylvania, Connecticut, and Virginia, the Triassic rocks are remnants of deeply eroded, sediment-filled, fault-bounded basins that formed as Pangaea was beginning to fragment and the Atlantic Ocean was just beginning to open. Extensive geological investigations in this and other parts of the state done in site evaluations for nuclear generating plants showed that most faults in areas underlain by Triassic and older bedrock were fairly well-sealed, had low permeabilities, and had not moved during Cenozoic time. Conglomerates and conglomeratic sandstones are found close to the basin-bounding faults, but much finer-grained sandstones and mudstones and a few coals occupy the axial portions of the basins. For the most part, the Triassic strata have shallow to subhorizontal dips and most, especially the finer-grained sandstones and mudstones, have relatively low permeabilities. The Triassic strata are not prolific groundwater sources and being softer and less indurated than the Piedmont crystalline rocks, they have much lower fracture permeabilities. Thus based solely on geologic and hydrogeologic considerations, the proposed Wake County site was well chosen.

Supplemental References

Devine, R., 1996, The little things that run the world: SIERRA, v. 81, no. 4, July/August, p. 32-37 & 62-63

Frosch, R. A., 1994, Industrial ecology: Minimizing the impact of industrial waste: Physics Today, v. 47, p. 63-68

Grove, N., 1994, Recycling: National Geographic Magazine, v. 186, no. 1, July, p. 92-116

Hasan, S. E., 1996, Geology and hazardous waste management: Prentice Hall, Upper Saddle River, NJ, 450 p.

Hinrichsen, D., 1996, Stratospheric maintenance; Fixing the ozone hole is a work in progress: The Amicus Journal, v. 18, no. 3, p. 35-38

The author notes how the Montreal Protocol of 1987 spurred Texas Instruments to radically change its manufacturing processes, resulting in lowered production costs, higher quality products, elimination of CFCs, and elimination of 3000 pounds of hazardous lead-bearing waste per year.

Page, G. W., 1997, Contaminated sites and environmental cleanup; International approaches to prevention, remediation, and reuse: Academic Press, Inc., Orlando, FL and London, U. K., 195 p.

Testa, S., 1994, Geological aspects of hazardous waste management: Lewis Publishers, Boca Raton, FL, 560 p.

The Geologic Aspects of Environmental Health • Chapter 13
Home Study Assignment

Questions 1-6; Matching: Human Environmental Health Conditions

a. excessive lead intake   b. goiter   c. schistosomiasis
d. Burkitt's tumor   e. "Mad Hatter" syndrome   f. lung cancer

1. ( ) caused by mercury poisoning; prevalent among nineteenth century felt industry workers
2. ( ) risks greatly magnified by smoking and environmental radon exposure
3. ( ) caused by iodine deficiency
4. ( ) snail-borne fever common along the Nile River, Egypt
5. ( ) lymph system disorder; prevalent in parts of Africa characterized by specific climatic and topographic conditions
6. ( ) known to occur among modern-day moonshine drinkers; was evidently endemic in higher social classes of ancient Rome

7. Which **one** element is not known to be essential for life in any way?
a. selenium   b. iron   c. calcium   d. titanium

8. Which element is most abundant in rocks of Earth's crust and in the human body?   a. carbon; C   b. potassium; K   c. nitrogen; N   d. oxygen; O

9. Which **two** are bulk elements with respect to human life?
a. arsenic   b. calcium   c. phosphorous   d. copper

10. Which is an essential trace element with respect to human life?
a. copper   b. nitrogen   c. bismuth   d. cadmium

11. Which **two** are characterized as age elements with respect to human life?
a. lead   b. arsenic   c. potassium   d. magnesium

12. Which chemical symbols for the two elements are both correct? **one answer**
a. selenium, S; calcium, Ca   b. potassium, K; mercury, Hg
c. sulfur, S; beryllium, B   d. phosphorous, P; copper, Co

13. Which toxic chemical element is very dangerous to living organisms because of its biochemical similarity to zinc, an essential trace element?
a. cadmium   b. lead   c. iodine   d. fluorine

14. A residual soil shows very low calcium and anomalously high chromium contents? What is the most probable, parental bedrock for this soil?
a. ultramafic rocks   b. limestones   c. sandstones   d. granitic igneous rocks

15. Among the common sedimentary rocks, which typically have the highest concentrations of many different trace elements?
a. those lithified from mud sediments and clays
b. those lithified from sand sediments   c. those composed mainly of calcite

16. Which element is not an important component of bones and teeth?
a. fluorine   b. phosphorous   c. calcium   d. lithium

17. In general, which kind of bedrock contains the highest concentrations of copper and zinc?
a. black shales   b. granites   c. sandstones   d. limestones

18. Which toxic element exhibits natural, bacterial recycling and dangerous biomagnification in aquatic environments?
a. potassium   b. lead   c. mercury   d. cobalt

19. Which **two** human afflictions are closely linked to dietary iodine deficiency?
a. thyroid troubles   b. cretinism   c. infant death syndrome   d. osteoporosis

20. Which trace element, ingested in small amounts, gives protection against dental caries and osteoporosis?   a. chlorine   b. iodine   c. bromine   d. fluorine

21. Which chemical conditions enhance dissolution and bioavailability of metallic elements, such as copper and zinc, and inhibit dissolution and bioavailability of selenium?   a. alkaline soils and waters   b. acidic soils and waters

Questions 22-26; Matching: Environmental Health Conditions and Locations

a. surface waters with dissolved SO_4/CO_3 less than 0.6

b. excessive dissolved selenium in agricultural wastewater; parts of the San Joaquin Valley, California

c. multiple health disorders in cattle; the clay pits area, central Missouri

d. iodine deficient soils and drinking water; includes Wisconsin, Michigan, and the northern plains states

e. Zintsu River Basin, Japan

22. (   ) birth defects and growth deformities in amphibians and aquatic birds

23. (   ) itaiitai disease; very painful bone disorder related to excessive dietary cadmium

24. (   ) probable excess metal concentrations in plants, soils, and surface waters, including the very poisonous element beryllium

25. (   ) the so-called goiter belt

26. (   ) show a spatial correlation with areas of low death rate from stroke, Japan

27. Which **two** statements concerning hard and soft water are correct and/or reasonably accurate? The other two are erroneous and/or seriously flawed.
a. Ca and Mg bicarbonates are the major dissolved ions in hard water
b. hard waters are usually slightly acidic and dissolve heavy metals, such as lead and zinc, more readily than do soft waters
c. environmental health studies in selected areas suggest that soft water may convey some protection against heart disease and stroke
d. groundwaters from limestone bedrock areas are usually hard; in water softening systems, dissolved calcium and magnesium are exchanged for dissolved sodium

28. Which airborne asbestos fibers are known to be very dangerous to human health, especially to the lungs?
a. white; the chrysotile variety   b. blue; the crocidolite amphibole variety

29. What is an optimum content, in parts per million, of dissolved fluoride in the water supply to protect children against dental caries? a. 0.1 b. 1 c. 10 d. 100

30. Which two statements concerning iodine are reasonably accurate?
a. iodine in soils is highly soluble and is readily incorporated into vegetation
b. marine algae, eaten in Japan and other countries, is a rich dietary source of iodine
c. in highly leached soils, iodine tends to be concentrated in the B horizon
d. residual soils associated with granitic bedrock would be expected to have higher iodine contents than residual soils associated with marine shale bedrock

31. Which environmental health correlation is evidently not significant? The other three are supported by credible studies? **one answer**
a. iodine deficiency; higher than normal rates of breast cancer
b. saline dust and soils; higher than normal rates of esophageal cancer
c. smoking and exposure to radon; high rates of lung cancer
d. exposure to airborne chrysotile asbestos fibers; high incidence of lung cancer

Questions 32-39; Matching: Radioactivity
a. gamma radiation   b. beta radiation   c. alpha radiation

32. ( ) travels farther and penetrates more deeply than the other two
33. ( ) essentially composed of highly energetic electrons emitted during nuclear conversion of neutrons into protons
34. ( ) after emission, atomic number of nucleus drops by two, atomic mass decreases by four atomic mass units
35. ( ) very energetic, very short wavelength radiation emitted from all nuclei undergoing radioactive decay
36. ( ) no change in atomic number of nucleus after emission
37. ( ) composed of highly energetic helium nuclei
38. ( ) after emission, atomic number increases by one, same atomic mass number
39. ( ) radioactive decay method of radon-222

Questions 40-43; Matching: Half-Lives of Radionuclides
a. carbon-14   b. uranium-238   c. radon-222   d. radium-226
40. ( ) 1,600 yrs   41. ( ) 3.8 days   42. ( ) 4.5 billion yrs   43. ( ) 5570 yrs

44. Some radon-222 is stored in an airtight container. How much of the original amount of radon will still be present at the end of a five-day period?
a. more than 50 %   b. less than 25 %   c. exactly 12.5 %   d. between 25 and 50 %

45. A radiation detector counts 200 nuclear disintegrations per second in a liter of air. What is the radiation activity of the sample?
a. 200 curies/liter   b. 200 becquerels/liter   c. 200 picocuries/liter

46. Which two are correct characterizations of radiation dose measurements?
a. absorbed radiation; the rad; grays   b. equivalent radiation; the rem; sieverts
c. absorbed radiation; the rem; grays   d. equivalent radiation; the rad; sieverts

47. Which site would probably have the higher background radiation level?
a. limestone bedrock outcrop; 5000 feet above sea level
b. granite bedrock outcrop; 5000 feet above sea level

48. Which site would probably have the higher background radiation level?
a. limestone bedrock outcrop; sea level
b. limestone bedrock outcrop; 5000 feet elevation

49. What is a reasonable average value for the dose of natural, background radiation received by Americans each year?
a. 1.5 mSv   b. 0.15 mSv   c. 15 mSv   d. 0.015 mSv

50. Which **one** correctly characterizes the Reading Prong?
a. area in the United Kingdom characterized by elevated dissolved heavy metal contents in groundwaters and higher than average risks of heart disease
b. area of phosphate-rich sedimentary strata in central Florida
c. area of glacial deposits in northern Ohio noted for soft groundwaters and high rates of heart disease
d. area of igneous and metamorphic bedrock in eastern Pennsylvania and New Jersey note for elevated radon levels in homes

51. You plan to build a home with a basement and water supply well in a rural setting known to be associated with elevated, interior, radon levels. Which **two**, relatively low-cost steps could be taken to reduce interior radon activities to 2 pCi/l or less?
a. install a well-ventilated, outdoor, water supply tank
b. seal the foundation or basement excavation with a plastic liner and compacted clay layer
c. install four vadose zone wells around the home and pump out the soil gases and air on a regular basis
d. install a well in the bedrock to a depth of one hundred feet below the original water table; pump the well at a high-enough rate to dewater the shallow fractures in bedrock beneath the home

52. Which **one** set of subsurface and climatic conditions would probably be associated with the highest upward migration rates for Rn-222?
a. atmospheric pressure is below average after four consecutive days of above average atmospheric pressure; very shallow water table; deep clayey soil; horizontal, unfractured shale bedrock
b. atmospheric pressure is above average after four consecutive days of below average atmospheric pressure; shallow sandy soil; granite bedrock with closely-spaced, vertical fractures; deep water table
c. atmospheric pressure is above average after four consecutive days of below average atmospheric pressure; very shallow water table; deep clayey soil; horizontal, unfractured shale bedrock
d. atmospheric pressure is below average after four consecutive days of above average atmospheric pressure; shallow sandy soil; granite bedrock with closely-spaced, vertical fractures; deep water table

53. Which **one** step can lower a homeowner's risks of dying in a home fire or from lung cancer?
a. reduce interior radon levels from 4 to 2 pCi/l
b. prohibit smoking in the home and on the property
c. make sure the tap water contains between 1 and 2 ppm F
d. make sure all food is prepared with iodized salt

54. Which **two** statements concerning the *case history* Heart Disease in Georgia, p. 363, are essentially correct? The other two contain factual errors or serious flaws in interpretation.
a. Coastal plain soils have higher and more diverse, trace metal contents than soils derived from Paleozoic and older rocks of the Piedmont and Blue Ridge terranes.
b. In areas with substantial topographic relief, death rates from heart disease in Georgia were generally lower than in low-relief areas with highly leached soils derived from Cenozoic sedimentary strata.
c. Higher heart disease mortality rates in the coastal plain counties are in part due to low-level toxicity from soluble zirconium in the sandy soils.
d. Soils in central and north Georgia contained higher levels of chromium, manganese, and copper than soils in the coastal plain counties. In small amounts, these three trace elements are essential for human life.

55. On the U. S. map provided, accurately plot and label the following locations.

a. Ohio   b. Reading Prong   c. central Missouri
d. coastal plain of Georgia   e. sandhills region, Nebraska

56. On the world map provided, accurately plot and label the following locations.

a. Rome, Italy   b. Caspian Sea coast, Iran   c. Sweden

Environmental Geology Quiz • Chapter 13

Questions 1-4; Matching: Human Environmental Health Conditions
a. Keshan disease   b. excessive dietary lead   c. "Mad Hatter" syndrome   d. goiter

1. ( ) caused by mercury poisoning; prevalent among nineteenth century felt industry workers
2. ( ) often fatal condition in parts of China caused by dietary selenium deficiency
3. ( ) caused by iodine deficiency
4. ( ) known to occur among modern-day moonshine drinkers; was evidently endemic in higher social classes of ancient Rome

5. Which **one** element is not known to be essential for life in any way?
a. titanium   b. iron   c. calcium   d. selenium

6. Which **two** are bulk elements with respect to human life?
a. calcium   b. copper   c. arsenic   d. phosphorous

7. Which **two** gradually accumulate in the body with age?
a. lead   b. magnesium   c. potassium   d. arsenic

8. Which chemical symbols for the two elements are both correct? **one answer**
a. phosphorous, P; copper, Co   b. selenium, S; calcium, Ca
c. sulfur, S; beryllium, B   d. potassium, K; mercury, Hg

9. Which toxic chemical element is very dangerous to living organisms because of its biochemical similarity to zinc, an essential trace element?
a. fluorine   b. lead   c. iodine   d. cadmium

10. A residual soil shows very low calcium and anomalously high chromium contents? What is the most probable, parental bedrock for this soil?
a. granitic igneous rocks   b. ultramafic rocks   c. limestones   d. sandstones

11. Among the common sediments and sedimentary rocks, which typically have the highest concentrations of many different trace elements?
a. sands, sandstone   b. muds, clays, mudstone, shale   c. those composed of calcite

12. Which toxic element exhibits natural, bacterial recycling and biomagnification in aquatic environments?   a. cobalt   b. mercury   c. lead   d. potassium

13. Which **two** human afflictions are closely linked to dietary iodine deficiency?
a. thyroid troubles   b. osteoporosis   c. infant death syndrome   d. cretinism

14. Which airborne asbestos fibers are known to be very dangerous to human health, especially to the lungs?
a. white; the chrysotile variety   b. blue; the crocidolite amphibole variety

15. What is an optimum content, in parts per million, of dissolved fluoride in the water supply to protect children against dental caries?   a. 0.1   b. 1   c. 10   d. 100

16. Which chemical conditions enhance dissolution and bioavailability of metallic elements, such as copper and zinc, and inhibit dissolution and bioavailability of selenium?   a. alkaline soils and waters   b. acidic soils and waters

Questions 17-20; Matching: Locations and Environmental Health Conditions
a. surface waters with dissolved SO4/CO3 less than 0.6   b. Zintsu River Basin, Japan
c. agricultural wastewater in parts of the San Joaquin Valley, California
d. multiple health disorders in cattle; the clay pits area, central Missouri

17. ( ) birth defects and growth deformities in amphibians and aquatic birds due to excessive selenium intake
18. ( ) itaiitai disease; painful bone disorder related to excessive dietary cadmium
19. ( ) probable excess metal concentrations, including the very poisonous element beryllium, in plants, soils, and surface waters
20. ( ) show a spatial correlation with areas of low death rate from stroke, Japan

21. Which one statement concerning iodine is inaccurate and flawed?
a. residual soils associated with granitic bedrock would be expected to have higher iodine contents than residual soils associated with marine shale bedrock
b. algae from the sea, eaten in Japan and elsewhere, is a rich dietary source of iodine
c. iodine in soils is highly soluble and is readily incorporated into vegetation

22. Which one statement concerning hard and soft water is correct and/or reasonably accurate? The other two are erroneous and/or seriously flawed.
a. hard waters are usually slightly acidic and dissolve heavy metals, such as lead and zinc, more readily than do soft waters
b. environmental health studies in selected areas suggest that soft water may convey some protection against heart disease and stroke
c. groundwaters from limestone bedrock areas are usually hard; in water softening systems, dissolved calcium and magnesium are exchanged for dissolved sodium

Questions 23-25; Matching: Radioactivity
a. alpha radiation   b. beta radiation   c. gamma radiation

23. ( ) highly energetic, short wavelength radiation; travels farther and penetrates more deeply than the other two
24. ( ) essentially comprised of highly energetic electrons emitted during nuclear conversion of neutrons into protons
25. ( ) after emission, atomic number of nucleus drops by two, atomic mass decreases by four atomic mass units; radioactive decay method of radon-222

Questions 26-28; Matching: Half-Lives of Radionuclides
a. radon-222   b. carbon-14   c. radium-226
26. ( ) 1,600 years   27. ( ) 3.8 days   28. ( ) 5570 years

29. Some radon-222 is stored in an airtight container. How much of the original amount of radon will still be present at the end of a five-day period?
a. more than 50 %   b. less than 25 %   c. exactly 12.5 %   d. between 25 and 50 %

30. Which two are correct characterizations of radiation dose measurements?
a. absorbed radiation; the rad; grays   b. equivalent radiation; the rem; sieverts
c. absorbed radiation; the rem; grays   d. equivalent radiation; the rad; sieverts

31. Which site would probably have the higher background radiation level?
a. limestone bedrock outcrop; 1000 feet above sea level
b. granite bedrock outcrop; 6000 feet above sea level

32. What is a reasonable average value for the annual, background dose of radiation received by Americans?   a. 1.5 mSv   b. 0.15 mSv   c. 15 mSv   d. 0.015 mSv

33. Which **one** correctly characterizes the Reading Prong?
a. area in the United Kingdom characterized by elevated dissolved heavy metal contents in groundwaters and higher than average risks of heart disease
b. area of glacial deposits in northern Ohio noted for soft groundwaters and high rates of heart disease
c. area of igneous and metamorphic bedrock in eastern Pennsylvania and New Jersey note for elevated radon levels in homes

34. You plan to build a home with a basement and water supply well in an area known for elevated, interior, radon levels. Which **two**, relatively low-cost steps could be taken to reduce interior radon activities to 2 pCi/l or less?
a. install four vadose zone wells around the home and pump out the soil gases and air on a regular basis
b. seal the foundation excavation with a compacted clay layer and plastic liner
c. install a well-ventilated, outdoor, water supply tank

35. Which set of subsurface and climatic conditions would probably be associated with the highest upward migration rates for Rn-222?
a. atmospheric pressure is above average for four consecutive days; very shallow water table; deep clayey soil; horizontal, unfractured shale bedrock
b. atmospheric pressure is below average for four consecutive days; shallow sandy soil; granite bedrock with closely-spaced, vertical fractures; deep water table

36. Which **one** step can lower risks of dying in a home fire or from lung cancer?
a. reduce interior radon levels from 4 to 2 pCi/l
b. prohibit smoking in the home and on the property
c. use water with 1 to 2 ppm F and iodized salt in all food preparations

37. Which statement concerning soil studies and heart disease in Georgia is essentially correct? The other two contain factual errors or serious flaws.
a. Coastal plain soils have higher and more diverse, trace metal contents than soils derived from Paleozoic and older rocks of the Piedmont and Blue Ridge terranes.
b. In areas with substantial topographic relief, death rates from heart disease in Georgia were generally higher than in low-relief areas with highly leached soils derived from Cenozoic sedimentary strata.
c. Soils in central and north Georgia contained higher levels of the essential trace elements chromium, manganese, and copper than soils in the coastal plain.

38. On the map/maps provided, accurately plot and label the following locations.
a. Ohio
b. Reading Prong, PA & NJ
c. central Missouri
d. coastal plain of Georgia
e. sandhills region, Nebraska
f. Rome, Italy
g. Caspian Sea coast, Iran

Answers; Home Study Assignment
The Geologic Aspects of Environmental Health • Chapter 13

1. e	2. f	3. b	4. c	5. d
6. a	7. d	8. d	9. b, c	10. a
11. a, b	12. b	13. a	14. a	15. a
16. d	17. a	18. c	19. a, b	20. d
21. b	22. b	23. e	24. c	25. d
26. a	27. a, d	28. b	29. b	30. a, b
31. d	32. a	33. b	34. c	35. a
36. a	37. c	38. b	39. c	40. d
41. c	42. b	43. a	44. d	45. b
46. a, b	47. b	48. b	49. a	50. d
51. a, b	52. d	53. b	54. b, d	

Answers: Environmental Geology Quiz • Chapter 13

1. c	2. a	3. d	4. b	5. a
6. a	7. a	8. d	9. d	10. b
11. b	12. b	13. a, d	14. b	15. b
16. b	17. c	18. b	19. d	20. a
21. a	22. c	23. c	24. b	25. a
26. c	27. a	28. b	29. d	30. a, b
31. b	32. a	33. c	34. b, c	35. b
36. b	37. b			

Commentary/Media Watch

Micronutrients and Health

Inorganic chemical compounds, naturally derived for the most part from bedrock and soils, play extremely diverse and complex roles in ecological health and in the well-being of individual organisms. Unfortunately, health professionals, nutritionists, and health foods advocates consistently refer to these chemical elements and compounds as "minerals"; just consider the commonly used phrase "vitamins and minerals". Inorganic nutrient is a much better descriptive term and avoids the obviously, to earth scientists at least, incorrect use of the term mineral. Given the strong, current interest in physical fitness and health, we should assume that our students will be more familiar with "mineral nutrients" than with inorganic nutrients! Some, such as selenium, are essential micronutrients for humans and most other animals; too little is debilitating or fatal, the right amount is essential for good health and successful reproduction, but too much is detrimental, toxic, and ultimately fatal (Figure 13.3, p. 354).

The essential micronutrients characteristically play important regulating and catalytic roles in day-to-day cell metabolism. Most general, well-known, daily nutritional supplements include selenium (Se) bound into yeast, copper (Cu) as an oxide or gluconate, cobalt (Co) as cyanocobalamin, vitamin B-12, chromium (Cr) also bound in yeast, zinc (Zn) as an oxide or gluconate, and iodine (I) as potassium iodide. Magnesium supplements have recently been shown to protect against low-birth-weight infants; as a group, these babies are commonly beset with severe, life-threatening and costly health problems. Fluorine functions to toughen mammalian bones and tooth enamel by promoting the precipitation of fluorapatite in place of hydroxyapatite. Other elements, such as lead and mercury, are not essential for life and are toxic in excessive amounts. Cadmium's well-known toxicity derives from its close geochemical similarity to zinc; vital cell functions regulated by zinc-bearing enzymes are detoured or shut down completely when cadmium atoms substitute for zinc atoms in the enzyme. Still other inorganic constituents, such as strontium and barium, are evidently neither essential nor particularly toxic at higher-than-normal levels.

Trace element and micronutrient abundances exhibit wide natural variations in soils, surficial deposits, and bedrock. Compared to average background abundances, some areas are notably enriched in one or more trace elements; others are strongly depleted. Thus natural variations can result in both the "too little" and "too much" condition. Man-caused anomalies are exclusively of the "too much" type, being related mainly to mining, smelting, and metal-processing industries, widely used "tainted" commodities such as leaded gasoline, and past uses of toxic elements such as arsenic and mercury in industrial processes, insecticides, and fungicides. Soils along older highways will all exhibit elevated lead contents, courtesy of lead-based gasoline additives.

In general, muds and mud-derived rocks (shales, mudstones, argillites, slates, schists, etc.) carry adequate abundances of most essential micronutrients and non-essential trace elements as well. Cations and anions alike are readily adsorbed on the surfaces of fine-grained clay mineral particles and organic matter; they remain attached during transportation, deposition, burial, and all but the most intense and prolonged forms of weathering. In fact, some black, organic-rich shales are well-known for their unusually high abundances of uranium, zinc, selenium, and other trace elements. Thus soils derived from muds or mudrocks typically contain average or above average, background abundances of most trace elements. Rocks and soils subject to severe geochemical differentiation typically show strong enrichment in specific trace elements and strong depletion in others. For example, granites and rhyolites may be strongly enriched in molybdenum, tin, beryllium, and rubidium and granitic pegmatites may contain unusually high contents of beryllium, boron, and lithium. Clean, quartz sandstones normally have very low abundances of trace elements. Those that are present, such as zirconium, titanium, and boron, are tightly bound up in tough, insoluble, residual minerals such as zircon, rutile, and tourmaline and are not readily available (not bioavailable) to the soil microflora, soil microfauna, and plants.

Like shales and mudrocks, the intermediate silica-content igneous rocks, andesite and diorite, generally have well balanced trace element abundances while more mafic basalts and gabbros may show elevated contents of elements such as cobalt, nickel, selenium, and chromium, and depletion in zinc and molybdenum. The most severe trace element excesses and deficiencies show up in soils derived from olivine/Mg-silicate-dominated bedrock such as dunite and serpentinite. Such soils may contain near-toxic levels of chromium and exhibit severe deficiencies in calcium, potassium, and essential micronutrients such as selenium, molybdenum, and zinc. In such areas, floral assemblages may be severely stunted or fail to thrive.

Bioavailability and Geochemical Mobility

Geochemical availability, bioavailability, is often the most important single factor controlling ecological effects, damaging or otherwise, associated with environmental deficiencies and excesses in trace elements. Assuming that the trace elements in question are not tightly bound in insoluble mineral grains or isolated as inclusions in such grains, soil moisture pH is the major player in bioavailability; enhanced uptake by algae and plants insure that elevated, local, trace element abundances will be passed through to organisms higher up the food chain. Most metal element cations such as Cu, Co, Zn, and Ni have enhanced chemical mobility and bioavailability under acidic soil conditions; these ions are rendered relatively immobile and unavailable by alkaline soil conditions. Thus plants, such as corn, growing in metal-rich acidic soils may show damaging effects from excessive metal uptake, but those grown in limed or naturally alkaline soils with similar bulk metal element abundances will be healthy and show no harmful effects. Important micronutrients such as copper and zinc may have low bioavailability in organic-rich soils, even under acidic conditions because the metal cations are quickly and tightly bonded to the organic matter. Despite repeated applications of Cu-bearing fertilizers, poor copper availability is an ongoing concern of large-scale farming enterprises based on organic-rich, peaty soils in easternmost North Carolina.

Selenium, on the other hand, is mobilized and made bioavailable in alkaline soils and aquatic environments. Under typical, terrestrial oxidizing conditions, selenium as the selenate 2- anion (SeO_4) precipitates as relatively insoluble compounds with iron and other elements under acidic conditions. The selenate and molybdenate anions are much more soluble in alkaline waters and soils than in acidic environments. In general, higher solubilities mean enhanced bioavailability and lower solubilities mean lowered bioavailability; but other factors such as soil organic matter may play important roles.

Health Effects of Selenium

Selenium is highly instructive in illustrating the changing notions, diverse effects, and complexities of trace elements and health. Today it is recognized as essential for life, is widely available in dietary supplements, has been used to reduce toxic effects caused by excess cadmium and arsenic, and combined in organic form, has generated interest as an antioxidant and possible anticancer agent. As recently as the 1950s, selenium was generally regarded as a poison and possible carcinogen as well. This negative reputation derived mainly from animal afflictions such as "alkali disease" and the "blind staggers", known since the historic travels of Marco Polo to be caused by ingestion of certain, selenium-concentrating plants growing in specific areas of central Asia. These afflictions and their cause were subsequently rediscovered on the American northern plains in the nineteenth century. The plants in question are selenium accumulators; when grown in selenium-bearing, alkaline soils, they accumulate large-enough doses of organically-bound, readily available selenium compounds to sicken and kill grazing animals such as horses, mules, camels, and cattle. On the northern plains, the seleniferous soils derive mainly from Cretaceous-age marine shales. Seasonally dry to semiarid climatic conditions result in alkaline soil conditions; thus during relatively short periods of abundant soil moisture in late spring and summer, selenium, as selenate, is highly mobile in the wetted soils and is readily taken up by accumulator plants. People living largely "off the land" in these same areas have only slightly elevated selenium intakes and exhibit relatively minor discomfort and irritations or no symptoms at all. However, locally high rates of birth deformities may signal the dangers of too much selenium.

Selenium mobilized from irrigated, agricultural soils and parental, Tertiary, marine mudstones in the San Joaquin Valley, California, was identified as the main culprit in the ecological tragedy that struck the Kesterson National Wildlife Refuge (p. 357; p. 521-524).

Keshan's Syndrome/Selenium Deficiency

Equally interesting examples of the "too little selenium" scenario are also well documented. Unexplained, sudden, often fatal, cardiac and pulmonary malfunction were known to strike seemingly robust, healthy inhabitants, many of them young people, of specific regions in China. The condition, named Keshan's disease after a city in northern China, was endemic in inland parts of northern and southwestern China. Authorities suspected a soil-related geochemical cause; unusually low selenium abundances in local soils, foodstuffs, and inhabitant's blood implicated selenium dietary deficiency as the prime suspect. In the late 1970s, clinical trials involving dietary selenium supplements achieved excellent results. General use of selenium supplements in the affected areas has essentially eliminated the condition today.

The most interesting geological question concerning Keshan's disease is why was the condition so site specific? The disease was endemic in towns and villages of one valley and absent from towns and villages in adjacent valleys. The answer probably lies in local cultural and hydrologic conditions. All food is produced locally; little or none is brought in from "the outside". Alluvial soils of river floodplains are replenished by silt and clay from distant sources and almost certainly provide adequate selenium to grains and other locally-produced foodstuffs. However, much of the Keshan's area coincides with karst areas typified by small, highly localized hydrologic basins, subsurface drainage conditions, and sparse surface streams. Soils in these areas are entirely residual and subject to constant leaching by slightly alkaline soil moisture and groundwater. Without input of selenium-bearing silt and clay from distant sources, these soils eventually will become strongly depleted in selenium, and locally grown food products may not contain enough selenium to prevent the disease.

Parts of New Zealand's South Island devoted mainly to sheep raising are underlain by dunite, an olivine-rich rock notoriously low in selenium. However, white muscle disease and other pathological effects of selenium deficiency in sheep were insignificant until the ranchers changed the source of sulfur used as a dietary supplement for their sheep. For many years, native sulfur from volcanic areas or from sedimentary deposits in salt-dome caps was used. As the economics of the sulfur industry changed, much less expensive sulfur became available as a byproduct of oil refining and natural gas purification. Pathological effects of selenium deficiency appeared soon after ranchers began using sulfur from these new sources.

In retrospect, the causes and solution for the problem were fairly simple. Sulfur and selenium have strong geochemical similarities; a few are noted in the following table. As a result, native sulfur and sulfur-bearing minerals almost

reduced forms	oxidized forms
hydrogen selenide, H_2Se	selenate ion, SeO_4^{2-}
hydrogen sulfide, H_2S	sulfate ion, SO_4^{2-}
selenide ion, Se^{2-}; sulfide ion, S^{2-}	selenite ion, SeO_3^{2-}, sulfite ion, SO_3^{2-}

always contain small abundances of selenium. Thus the sulfur from natural deposits fortuitously provided enough selenium to protect the sheep from selenium deficiency. The highly purified "manufactured sulfur" contained too little selenium to protect the sheep and selenium deficiency symptoms soon appeared. Additional dietary selenium supplements solved the problem. Ranchers and their families living in the low-selenium areas exhibit relatively low-selenium levels in blood, but have not shown symptoms of selenium deficiency.

Mercury in the Environment

Mercury is a particularly dangerous environmental contaminant. Excessive mercury intake causes severe nervous system and brain impairment. Symptoms of mercury toxicity have been known among mercury miners and workers in Europe since at least the Middle Ages. As late as the nineteenth century, mercury was used in the manufacture of felt. The amusing Alice In Wonderland character, the Mad Hatter, was fashioned after the not-so-amusing, jerky physical movements and wacky behavior of hat industry workers suffering from mercury toxicity.

Most metallic elements are gradually removed from terrestrial, aquatic, and marine environments by burial, particularly in mud sediments. However bacterial action in aquatic environments converts relatively insoluble and immobile ionic forms of mercury into methylated mercury compounds. These are highly soluble and readily available in the food chain. Thus one of the early environmental mercury scares involved Lake Winnipeg and other large lakes in Manitoba Province, Canada. Organic mercury compounds, used as fungicide on seed-grain wheat, made their way from the Saskatchewan and Assiniboine river systems to the large lakes of southern Manitoba. Eventually, excessive mercury levels showed up in fish and waterfowl from the lakes. Environmental mercury levels have slowly declined since mercury-based fungicides were discontinued.

Although most, if not all, significant commercial and industrial point sources of mercury contamination have been eliminated or drastically curtailed, we still see occasional reports of excessive mercury levels in some marine fishes and in fishes from extensive, lake-and-swamp environments such as the Florida Everglades and easternmost North Carolina. A local television news item, 6/11/96, reported that state officials had issued a warning about eating fish caught in Lake Phelps, a popular destination for fishing and waterfowl hunting in eastern North

Carolina. Chemical analyses indicated that Lake Phelps fish had accumulated high-enough mercury levels to possibly be dangerous to human health. Thus low-grade mercury sources, such as coal-burning power plants, are of continued concern. Most coals contain only trace quantities of mercury; some, such as those from the Donetz basin, Ukraine, are notable for their mercury enrichment. Mercury is volatilized at the high temperatures of combustion; some is retained during the course of pollution abatement, but some does escape to the atmosphere. Even if mercury levels in coal are very small, such large tonnages are burned that over time, significant quantities may eventually accumulate in organic-rich surface soils, wetlands, and lakes.

Health and Fitness; Micronutrients and Dietary Supplements

Health and fitness issues command much attention and visibility in today's print and visual media. However, trace elements in rocks, soils, and foodstuffs are not particularly high-profile headline makers. Occasionally references are made to lithium therapy in conjunction with some mental health conditions, and over the past year, a miniflap erupted in the health foods industry about the validity of advertising claims associated with chromium picolinate dietary supplements. The general argument seems to center on which particular form of dietary chromium, if any, gives superior results. Chromium, as the 3+ ion, is known to be essential in Man. An enzyme called the glucose tolerance factor (GTF) contains one Cr ion per molecular unit and adequate/therapeutic dietary chromium intake has been linked to proper insulin function, reduced insulin dependence in some diabetics, reductions in triglycerides and serum cholesterol, and increased ratios of HDLs (high density lipids) to LDLs (low density lipids). The last two effects are generally accepted as reducing risk of cardiovascular disease.

Tin-Bismuth Shotgun Pellets

A news item in local papers in early October, 1996, timed to coincide with the waterfowl hunting season, announced successful completion of extensive trials with recently developed, tin-bismuth shotgun pellets. Use of the traditional lead pellets in shotgun ammunition had been banned or severely restricted in recent years because wounded birds and birds that ingested spent pellets were being weakened and killed by lead poisoning. Steel pellets eliminated the lead-poisoning problem but caused excessive wear and deterioration to shotgun barrels. Softer, non-toxic pellets were needed, and the tin-bismuth pellets seem acceptable on both accounts. They are not poisonous to waterfowl, and they do not cause excessive wear and tear on shotgun barrels. Thus waterfowl are to some extent being better protected and hunters are happy with the new ammunition.

Kesterson Update

A recent commentary on the Kesterson Wildlife Refuge entitled CMC puts up roadblocks to toxic wastewater superhighway (Sheehan, 1996) appeared in the Marine Conservation News (v. 8), a publication of the Center for Marine Conservation, a leading international environmental organization. Selenium is leached from irrigated soils in parts of the San Joaquin Valley, California (p. 521-524). Agricultural wastewaters transport the dissolved selenium to the Kesterson Reservoir and wetlands, which act as a closed hydrologic basin. Thus dissolved contaminants entering the wildlife refuge slowly accumulate over time, and damaging effects of excessive selenium intake eventually showed up as chronic sickness and birth deformities in waterfowl and other aquatic organisms.

Wastewater superhighway is an unflattering reference to a proposed system of canals and tunnels that would convey untreated agricultural wastewaters from the selenium-bedeviled, central valley irrigation districts directly to the Pacific Ocean at Monterey Bay. In addition to being a very, very expensive project, the untreated wastewaters would pose a serious threat to coastal waters and marine life of the Monterey Bay National Marine Sanctuary. The wastewater superhighway idea was proposed, in part, because of strong opposition to channeling the wastewaters into the San Joaquin River, No one it seems, including residents of the San Francisco Bay region, wants anything to do with the potentially poisonous wastewaters. In contrast to this expensive scheme to avoid a cleanup and export the selenium problem elsewhere, the Center advanced the sensible, less costly alternative of cleaning up the agricultural wastewaters in the valley and returning them to the San Joaquin River. The augmented river discharge would then help to restore riparian habitat and salmon and other fisheries destroyed or seriously damaged by reduced flows due to withdrawals for irrigation.

Fluorine in Drinking Water

An article in local newspapers (K. Kolsa, East Carolina University School of Medicine; 7/24/96), focused on the health effects of fluoride ion, F (1-), and fluoride contents of area drinking waters. One to 2 ppm are known to greatly decrease the incidence of dental caries and to be beneficial for bone strength (p. 355-356). Higher levels result in tooth mottling. Greenville's city water is drawn mainly from the Tar River; fluoride addition maintains the content at about 1.3 ppm. Groundwater pumped from late Cretaceous and Paleocene sedimentary aquifers in the area contains 3-5 ppm fluoride; for these waters to meet state standards, dilution with low-fluoride water is necessary. Federal regulation limit fluoride contents of bottled drinking waters to <= 1.2 ppm. Thus individuals who drink bottled water exclusively may benefit from additional dietary fluoride.

Bikinians Finally Can Go Home

An Associated Press article entitled Bikinians look homeward 50 years after atomic tests, 7/14/96, examined the lingering social and environmental consequences of the two, above-ground, American nuclear weapons tests conducted on Bikini atoll 50 years ago. One was dropped from an aircraft and detonated above ground; the other was detonated underwater. Arrogance and ignorance evidently both contributed to underestimating heath-related and environmental dangers associated with the tests. Radiation exposure among military and civilian personnel was deliberate in many cases, accidental in others. The underwater test was evidently sanctioned because the Navy wanted to find out if an underwater atomic bomb blast could destroy an "enemy fleet".

Radiation levels have finally subsided enough for government officials to decide that the island is safe for habitation, and only local areas of "hot soil" remained to be "cleaned up" in preparation for repatriation of ethnic Bikinians, many of whom have never seen their native islands. Once before, Bikini had been declared safe for habitation in 1969, but coconuts, an important food staple in the islands, were found to contain the fission product Cs-137, and the newly resettled islanders were again forced to leave. The isotope, still present in the soil, was absorbed into the roots of the coconut palm and further concentrated in the fruit. Second- and third-generation offspring of island residents exposed to radiation during and/or after the tests are continuing a 50-year-battle to document the long-term health and genetic effects of the nuclear explosions.

Supplemental References

Multiple Authors, 1996, Environmental geochemistry and health with special reference to developing countries: Appelton, J. D., Fuge, R., and McCall, G. J. H., editors, Geological Society Special Publication 113, Geological Society Publishing House, Bath, U. K; available in the U. S. through the AAPG Bookstore, Tulsa, OK, 272 p.

Multiple Authors, 1992, Biogeochemistry of trace metals: Adriano, D., editor, Lewis Publishers, Boca Raton, FL, 528 p.

Multiple Authors, 1983, Applied environmental geochemistry: Thorton, I., editor, Academic Press, Harcourt Brace Jovanovich Publishers, London, U. K. and New York, NY, 501 p.

See especially the articles by Crounse and others, p. 267-308 and 309-334

Environmental Awareness Inventory
Part 4: Minerals, Energy, and Environment

1. Consider the metals aluminum, copper, and lead. Which **one** statement concerning the amount of recycled metal is not correct? 1 pt
a. The banning of leaded gasoline additives and increased emphasis on recycling lead from vehicle batteries has nearly doubled the amount of recycled lead as a percentage of annual consumption between 1960 and 1992.
b. Vigorous recycling of aluminum cans and other scrap has tripled the amount of recycled aluminum as a percentage of annual consumption from 1960 to 1992.
c. Copper prices have remained at very low levels, 40 to 50 ¢/lb, since the mid 1970s, thus the amount of recycled copper as a percentage of annual consumption has declined almost 50 % between 1960 and 1992.

2. Which element has been most widely dispersed throughout Earth's surficial environments by man's twentieth century activities? 1 pt
a. zinc b. platinum c. lead d. silver

3. What mineral objects, locally concentrated on specific areas of the ocean floor, represent a huge, subeconomic resource? 1 pt
a. manganese nodules; Mn, Cu, Ni, and Co b. halite; common salt, NaCl
c. shark teeth and fish bones; phosphates
d. bauxite; oxide and hydroxide ore of aluminum

4. Which areas exhibit long-lasting environmental damage due to release of untreated, metal smelter emissions over many years? **one answer**, 1 pt
a. Oak Ridge, TN, and Chernobyl, Ukraine b. Cairo, Egypt, and Tripoli, Libya
c. Cherbourg, on the Normandy Coast of France, and Naples, Italy
d. Ducktown, TN, and Sudbury, Ontario, Canada

5. Ash from incineration of a sewage/wastewater treatment sludge is found to contain 30 ppm gold, 660 ppm silver, and 8000 ppm copper. What would be a reasonable location and local source for these metals? 1 pt
a. Oklahoma City, OK; large-scale animal feed lot
b. Palo Alto, CA; electronics and photographic facilities
c. Dodge City, KS; very large cattle slaughterhouse d. Golden, CO; Coors brewery

6. Which **one** statement is not correct? 1 pt
a. Space heating represents the largest proportion of low-temperature energy, 50° to 300° C, utilized in the U. S.
b. Petroleum has been the leading energy fuel in the U. S. during the second half of the twentieth century.
c. Lignite coals occur in folded late Paleozoic strata in eastern and central Pennsylvania; anthracite, a coal with relatively low carbon content and high moisture content, is mined in east Texas, North Dakota, and eastern Montana.
d. Acidic groundwater and surface water associated with coal mining derive mainly from oxidation of pyrite, FeS_2, in mine workings and disturbed waste rock.

7. Which two central African nations are large producers of copper and cobalt? 1 pt
a. Chile and Peru b. Zaire and Zambia c. Sudan and Chad d. Angola and Mali

8. What is the maximum efficiency (electrical energy produced/thermal energy released) expected from a new, "state-of-the-art", coal-fired, electrical generating plant, including cogeneration facilities? 1 pt   a. 10 %   b. 75%   c. 95 %   d. 33 %

9. About what percentage of daily or annual petroleum consumption in the U. S. is imported from other countries? 1 pt   a. 0 %   b. 90 %   c. 30 %   d. 55 %

10. Which governmental policy would encourage progress along the so-called "soft, sustainable path" for future energy use in the U. S.? **one answer, 1 pt**
a. Allow tax deductions or credits for energy saving programs and equipment and for improved matching of low-grade energy sources with low-grade end-uses.
b. Eliminate all gasoline taxes; raise federal and state governmental research spending for alternative energy sources such as solar, geothermal, and wind.
c. Legislate that all home heating be by electricity, thus eliminating air pollution and fire dangers due to home oil and gas furnaces.
d. Repeal the Public Utility Regulatory Policy Act of 1978 and other burdensome regulations that stifle free enterprise in the energy industry.

11. Which **one** describes a breeder nuclear reactor? 1 pt
a. uranium-238 transforms to uranium-235; U-235 produces heat by nuclear fission
b. plutonium-239 is formed from uranium-238; Pu-239 fission produces heat
c. tritium and deuterium fuse to helium and vast amounts of thermal energy

12. Which alternative energy resource and its geographic area is incorrect? 1 pt
a. geopressurized systems; hot, pressurized water containing dissolved methane is found in deep sandstone aquifers; Northern Plains, North and South Dakota
b. solar power plant; rotating mirrors follow the Sun, focusing the sunlight on a heat exchange fluid reservoir; the heated fluid is used to generate water vapor to drive turbines and generate electricity; Mojave Desert region, California
c. hot, igneous, geothermal systems; fracture connections through hot rock, 600° to 1000° C, are established between neighboring deep drilled wells; cool water injected into one well is heated during its migration through the hot, broken rock and withdrawn from the second well; Los Alamos area, NM

Answers

| 1. c | 2. c | 3. a | 4. d | 5. b | 6. c |
| 7. b | 8. b | 9. d | 10. a | 11. b | 12. a |

Performance Evaluation; Scoring

Each correct answer is worth one point; add up your total. Now be honest, and subtract one point for each lucky guess!

Score	**Awareness Ranking; advice**	INTERNET Address
9-12	**nerd**; keep up the good work, don't let up	http://envinerd.com
6-9	**blue collar achiever**; earn an A with hard work	http://envblcol.com
3-6	**slowpoke**; don't look back, clueless are right behind	http://envslpoke.com
0-3	**clueless**; work extra hard to make up for a slow start	http://envcluless.com

Mineral Resources and Environment • Chapter 14
Home Study Assignment

1. Which **two** statements are reasonable? The others contain factual errors or serious conceptual flaws.
a. A speculative resource is one form of a reserve.
b. Silver disseminated as a trace element in shale bedrock constitutes a very large silver reserve that can be exploited when higher-grade deposits are exhausted.
c. Reserves include those resources that can be successfully extracted within the present-day context of economic, social, political, and environmental constraints.
d. A subeconomic resource could become a reserve if demand for the specific commodity increases and extraction costs decrease significantly.

2. Which **two** are important constituents of stainless steels?
a. cobalt   b. tin   c. arsenic   d. chromium

3. Which statement concerning copper is incorrect and seriously misleading?
a. A substantial portion of copper used in the U. S. is mined domestically.
b. Glass optical fiber cables are being used in place of copper wire in many modern communications systems.
c. Large-scale, open-pit mines need ore grades of 1 % Cu or higher in order to operate profitably in North America.
d. In the summer of 1996, the Japanese commodities-trading giant, Sumitomo Corp., announced losses exceeding 1.8 billion dollars in unauthorized trading of copper delivery contracts.

4. Which two metallic elements are both present in amounts exceeding 1 weight percent in many common rocks and soils? **one answer**
a. iron, copper   b. silver, carbon   c. iron, aluminum   d. oxygen, potassium

5. Of the following, which **one** has the highest content of that metal in its ore reserves currently being mined?   a. iron   b. lead   c. copper   d. chromium

6. What mineral commodity originates at great depths in the mantle and is carried to the surface as sparse, disseminated crystals in rare, intrusive igneous rocks such as kimberlite?   a. vermiculite   b. graphite   c. gypsum   d. diamond

7. Ores of which **two** metals are mainly formed by intense weathering and leaching of bedrock of appropriate composition under moist, subtropical to tropical conditions? **two answers**   a. copper   b. silicon   c. iron   d. aluminum

8. Which **two** responses describe pegmatites and their associated mineral deposits?
a. small bodies of coarse grained igneous rock formed by crystallization of late-stage, water-rich magma
b. veins and spherical masses of metallic minerals deposited in wall rock around an intrusive igneous rock body
c. deposits of sodium and potassium salts precipitated through evaporation
d. often contain minerals of relatively rare elements such as spodumene (Li), beryl (Be), and tourmaline (B)

9. Which **two** mercury deposits in northern California were named for older, well-known mercury deposits in Europe? **two answers**
a. New Red River   b. New Almaden   c. Nuevo Cerro de Pasco   d. New Idria

10. Massive deposits of copper and zinc sulfide minerals bounded by seafloor basaltic lava flows most likely formed in which environment?
a. regional metamorphic terrane exhumed by subsequent deep erosion
b. wide fault zone representing a former transform plate boundary
c. divergent oceanic plate boundary; ore minerals deposited around submarine hot spring vents
d. in a large gabbro intrusion; layers of ore minerals accumulated by crystal settling

11. Which **two** are appropriate for chromium?
a. The main ore mineral is chromite; major producers include South Africa and Zimbabwe.
b. Most chromium deposits formed as ancient placers in geologically ancient conglomerates.
c. The Stillwater complex, MT, is a major, domestic resource; this complex also contains the only domestic economic reserves of palladium and platinum.
d. Mineable concentrations of chromite grains typically occur in limestone wall rock surrounding bodies of intrusive igneous rock.

12. Which **two** are produced in the U. S. from evaporites in nonmarine sedimentary basins?   a. borates   b. salt; halite   c. potassium salts   d. sodium carbonate

13. In addition to limestone, anhydrite, and halite, what is the fourth mineral commodity to occur in association with some Gulf Coast salt domes?
a. sulfur   b. bauxite   c. ceramic clays   d. phosphate rock

14. Why are identified domestic resources of magnesium listed as unlimited?
a. Mg is a common element in rocks such as dolomite, basalt, and ultramafic rocks.
b. Magnesium salts are produced from seawater and naturally-occurring brines.
c. Secondary enrichment raises the Mg content of soils and partly weathered bedrock to the status of a subeconomic resource.
d. Magnesium is a low-cost byproduct of phosphate mining and phosphate fertilizer manufacture.

15. Which **two** are appropriate for the process of secondary enrichment?
a. Is an important part of the formation and growth of seafloor manganese nodules.
b. Is effected by dense populations of invertebrate organisms clustered around seafloor hot spring vents.
c. Oxidation of pyrite is an important part of the process; much of the iron remains behind in the surface and near-surface bedrock as highly insoluble iron oxide minerals.
d. More soluble metal ions, such as copper, are dissolved in the vadose zone and reprecipitated in the saturated zone just below the water table.

16. In addition to manganese, which other metals are typically enriched to subeconomic resource grade in seafloor manganese nodules? **one answer**
a. Cu, Co, and Ni   b. Ti, Fe, and Pt   c. Au, Ag, and Pb   d. B, Be, and Br

17. Which two statements are appropriate for the metal cobalt?
a. Its chemical symbol is Cb; the metal occurs widely as sulfide-rich mineral veins associated with granitic intrusive and volcanic rocks.
b. Zaire and Zambia, two countries in central Africa, are major suppliers.
c. It is concentrated in seafloor manganese nodules and in surface crusts on submarine rock outcrops in parts of the Pacific Ocean basin.
d. It is a major component of bronze, brass, and most aluminum alloys.

18. Which solvent is commonly used in heap leaching to extract gold from low-grade, gold-bearing rock?   a. sulfuric acid solution   b. sodium cyanide solution
c. sodium bicarbonate solution   d. ammonium nitrate solution

19. Which two responses correctly refer to the Bingham Canyon Mine?
a. one of the world's largest, open-pit copper mines
b. employs cyanide leaching to extract copper from crushed rock
c. much of the previously mined over land has been restored to its original topography and vegetation
d. located near Salt Lake City, Utah; mining and ore processing annually generate an enormous quantity of waste overburden rock and finely-ground, mill tailings

20. Which two statements are true and or reasonably correct? Briefly note why the others are incorrect?
a. Acidic drainage is seldom associated with old, abandoned, underground mine workings because the acidic components in the water are usually neutralized by the bedrock.
b. Underground workings in carbonate-hosted, low-pyrite, lead-zinc sulfide deposits are less likely to generate strongly acidic groundwaters than those opened to mine pyrite-rich copper sulfide ores in veins cutting through granite.
c. Prior to modern environmental regulations, some metal smelters in the U. S. emitted large quantities of SO2 and dangerous quantities of toxic elements such as arsenic, cadmium, and mercury.
d. Metal-rich acidic waters from abandoned open-pit, copper mines of the Tri-State District, Kansas, Colorado, and Nebraska, are a serious environmental threat to surface waters in the region.

21. Which two mineral commodities are most likely to be concentrated in young as well as in geologically ancient placer deposits?
a. gold   b. pyrite   c. native sulfur   d. diamonds

22. For which commodity would transportation costs make up the highest percentage of the purchase price?
a. copper ore concentrates from the U. S. transported by rail and ship to a smelter in Japan
b. native sulfur from the Gulf of Mexico region delivered by ship to an East Coast phosphate fertilizer manufacturing complex
c. diamonds from South Africa delivered by a bonded, well-insured courier to an Amsterdam, Netherlands, gem-cutting facility
d. crushed stone delivered by truck to a construction site near the quarry

23. In a typical marine evaporite basin such as the Michigan Basin, how are the evaporite minerals spatially distributed with respect to the basin's geometry?
a. center, potassium salts; intermediate, halite; peripheral areas, gypsum/anhydrite
b. center, halite; intermediate, gypsum/anhydrite; peripheral areas, potassium salts
c. peripheral areas, gypsum/anhydrite; intermediate, potassium salts; center, halite
d. peripheral areas, halite; intermediate, gypsum/anhydrite; center, potassium salts

24. Which **one** statement concerning recycling of scrap metals in the U. S. is not correct? The other statements are true or represent reasonable inferences.
a. Substantially more aluminum and lead are recycled today than in 1960 and earlier years.
b. Since leaded gasoline has been replaced by unleaded gasoline, the percentage of lead being recycled has increased significantly.
c. Recycled copper has, for many years, made up slightly over one-half of the annual domestic consumption.
d. Iron and steel comprise the bulk of recycled scrap metals.

25. Sewage sludge from an American urban wastewater treatment plant shows anomalously high concentrations of copper, gold, and silver. What is a logical source of origin for the metals?
a. untreated or partly treated wastewater from electronics and photographic industries
b. airborne fallout from a large, modernized copper smelter
c. partly treated wastewater from petroleum refineries
d. partly treated wastewater from breweries and whiskey distilleries

26. Which material is most likely to have little or no value as a metal resource or reserve today?
a. slag from a custom smelter, 1910 to 1925
b. gold-bearing gravels worked only by sluice box and panning in the late 1800s
c. low-grade ore and waste rock used to backfill underground mine workings from 1910 to 1920
d. slag from an American steel mill, 1935 to 1945

27. Which task has not yet been known to have been accomplished by bacteria?
a. convert aqueous cyanide ion to ammonia and nitrate
b. utilize H2S and ferrous iron as energy sources for chemosynthesis
c. oxidize sulfide ions in pyrite to sulfate ions
d. convert dissolved magnesium in chloride brines to magnesium metal

28. Which **two** situations could change a subeconomic gold-bearing, mineral deposit into an economically mineable resource for gold? **two answers**
a. small tonnages of higher grade veins are discovered in the central part of the deposit
b. the world economy collapses, paper money becomes worthless, and gold is the only, trusted, monetary media
c. mining costs increase slightly and the price of gold declines
d. new low-cost mines begin producing in Brazil, Russia, and west Africa

29. Which **two** are nonrenewable resources?
a. petroleum   b. water   c. copper   d. wood and other forest products

30. Ore deposits of which common metal form by prolonged, intense, tropical weathering of specific kinds of bedrock?
a. aluminum   b. magnesium   c. silver   d. mercury

31. Chemical decomposition of what mineral produces acidic soil waters which can cause secondary enrichment in copper and other metallic ore deposits?
a. pyrite   b. quartz   c. bauxite   d. calcite

32. Bauxite is an ore of what metal?
a. gold   b. iron   c. tin   d. aluminum

33. Vein- or fissure-filling deposits of lead and zinc minerals would be more common where?
a. in limestones of a contact-metamorphic zone around a shallow, granite pluton
b. in gneiss and schist formed in a deep, high-grade, regional-metamorphic zone

34. Oxidation of metal sulfide minerals and formation of strongly acidic, downward-percolating waters are important aspects of which ore-forming process?
a. laterite formation and bauxite precipitation
b. secondary enrichment of primary, low-grade, copper deposits
c. accumulation of residual nickel and zinc minerals in very dry, hot, desert areas
d. precipitation of native gold in the A horizons of permafrost soils

35. Which calcium-bearing minerals are important components of plasters and Portland cement? **one answer**   a. quartz and plagioclase   b. calcite and gypsum   c. bauxite and kaolin   d. halite and sylvite

36. Which are important fertilizer minerals with their correct elemental chemical symbols? **one answer**   a. apatite (P), sylvite (K)   b. cassiterite (Sn), albite (Na)   c. halite (Na), calcite (N)   d. sylvite (C), gypsum (P)

37. Which mineral is an essential component of plaster and plaster board?
a. quartz   b. beryl   c. feldspar   d. gypsum

38. Which **two** processes utilize large quantities of limestone?
a. making industrial grade carbonic acid to cure upset stomachs of exam takers
b. producing Portland cement   c. making agricultural lime
d. refining aluminum metal from borax

39. Which **two** are closely associated geologically with "black smokers"?
a. some copper and zinc sulfide mineral deposits
b. titanium and boron deposits in pegmatites
c. underground, coal mine fires   d. submarine, hot spring vents

40. Copper, zinc, and silver sulfides deposited around ancient, seafloor, hot spring vents would be most likely associated with which type of bedrock?
a. basaltic lava flows   b. intrusive granite
c. fragmental kimberlite   d. anhydrite caprock

41. Which mineral and element are used in the manufacture of fiber-optic glass and computer microprocessor chips? **one answer**
a. quartz; silicon   b. halite; sodium   c. galena; lead   d. chalcopyrite; copper

42. Which **one** is not an expected mineral constituent of pegmatites?
a. quartz and minerals of fairly rare elements such as lithium and beryllium
b. potassium and sodium feldspar
c. biotite and muscovite micas   d. gypsum and anhydrite

43. Which **two** minerals are used in the manufacture of sulfuric acid?
a. pyrite   b. calcite   c. chromite   d. native sulfur

44. On the U. S. map provided, plot and label the following areas or locations.
a. Homestake Mine, Black Hills, SD
b. New Almaden and New Idria, CA
c. Tri-State lead-zinc district
d. prominent gold deposits, east of the Mississippi River
e. Bingham Canyon Mine, UT
f. Palo Alto, CA
g. Bone Valley phosphate deposit, FL
h. site of rare, domestic bauxite deposit
i. Stillwater complex, MT
j. Paradox Evaporite basin, UT and CO

45. On the world map provided, accurately plot and label the following locations.
a. Sudbury, Ontario, Canada   b. Zaire and Zambia   c. Malaysia
d. Cyprus   e. Iran   f. Bolivia   g. Jamaica   h. Chile

Environmental Geology Quiz • Chapter 14

1. Which **one** statement contains factual errors or serious conceptual flaws?
a. Silver disseminated as a trace element in shale bedrock constitutes a very large silver reserve that can be exploited when higher-grade deposits are exhausted.
b. Reserves include those resources that can be successfully extracted within the present-day context of economic, social, political, and environmental constraints.
c. A subeconomic resource could become a reserve if demand for the specific commodity increases and extraction costs decrease significantly.

2. Which **two** are important constituents of stainless steels?
a. cobalt   b. tin   c. arsenic   d. chromium

3. Which statement concerning copper is incorrect and seriously misleading?
a. Glass optical fiber cables are being used in place of copper wire in many modern communications systems.
b. A substantial portion of copper used in the U. S. is mined domestically.
c. Large-scale, open-pit mines need ore grades of 1 % Cu or higher in order to operate profitably in North America.

4. What mineral commodity originates at great depths in the mantle and is carried to the surface as sparse, disseminated crystals in rare, intrusive igneous rocks such as kimberlite?   a. gypsum   b. diamond   c. vermiculite   d. bauxite

5. Ores of which **two** metals are mainly formed by intense weathering and leaching of bedrock of appropriate composition under moist, subtropical to tropical conditions? **two answers**   a. aluminum   b. silicon   c. iron   d. copper

6. Which response correctly describes pegmatites and associated mineral deposits?
a. deposits of sodium and potassium salts precipitated through evaporation
b. veins and spherical masses of metallic minerals deposited in wall rock around an intrusive igneous rock body
c. small bodies of coarse grained igneous rock formed by crystallization of late-stage, water-rich magma

7. Which **two** mercury deposits in northern California were named for older, well-known mercury deposits in Europe?
a. New Idria   b. New Almaden   c. Nuevo Cerro de Pasco   d. New Rio Tinto

8. Massive deposits of copper and zinc sulfide minerals bounded by seafloor basaltic lava flows most likely formed in which environment?
a. divergent oceanic plate boundary; ores deposited at submarine hot spring vents
b. wide fault zone representing a former transform plate boundary
c. regional metamorphic terrane exhumed by subsequent deep erosion

9. Which **one** is not appropriate or correct for chromium?
a. The Stillwater complex, MT, is a major, domestic resource; this complex also contains the only domestic economic reserves of palladium and platinum.
b. The main ore mineral is chromite; major producers include South Africa and Zimbabwe.
c. Most deposits formed as placers in geologically ancient conglomerates.

10. Which **two** are produced in the U. S. from evaporites in nonmarine sedimentary basins?   a. sodium carbonate   b. borax, borates   c. salt; halite   d. potassium salts

11. In addition to limestone, anhydrite, and halite, what is the fourth mineral commodity to occur in association with some Gulf Coast salt domes?
a. phosphate rock   b. bauxite   c. ceramic clays   d. sulfur

12. Why are identified domestic resources of magnesium listed as unlimited?
a. Mg salts are produced from seawater and naturally-occurring brines.
b. Mg is a common element in rocks such as dolomite, basalt, and ultramafic rocks.

13. Which **one** best describes secondary enrichment?
a. Seafloor manganese nodules undergo renewed growth after burial as dissolved metals from circulating pore waters are precipitated on the nodules.
b. Biomagnification is effected by dense populations of invertebrate organisms clustered around seafloor hot spring vents.
c. More soluble metal ions, such as copper, are dissolved in the vadose zone and reprecipitated in the saturated zone just below the water table.

14. In addition to manganese, which other metals are typically enriched to subeconomic resource grade in seafloor manganese nodules? **one answer**
a. Cu, Co, and Ni   b. Ti, Fe, and Pt   c. Au, Ag, and Pb   d. B, Be, and Br

15. Which **one** statement concerning cobalt is incorrect?
a. Zaire and Zambia, two countries in central Africa, are major suppliers.
b. It is a major component of bronze and brass.
c. It is concentrated in seafloor manganese nodules and in surface crusts on submarine rock outcrops in parts of the Pacific Ocean basin.

16. Which chemical solution is commonly used in heap leaching to extract gold from low-grade, gold-bearing rock?
a. ammonium nitrate   b. sodium cyanide   c. sodium bicarbonate   d. sulfuric acid

17. Which **one** response correctly describes the Bingham Canyon Mine?
a. employs bicarbonate leaching to extract copper from crushed rock
b. much of the previously mined over land has been restored to its original topography and vegetation
c. located near Salt Lake City, Utah; mining and ore processing annually generate an enormous quantity of waste overburden rock and finely-ground, mill tailings

18. For which finished product/commodity would transportation costs make up the highest percentage of its retail/wholesale purchase price?
a. diamonds from South Africa delivered by a bonded, well-insured courier to an Amsterdam, Netherlands, gem-cutting facility
b. crushed stone delivered by truck to a construction site near the quarry
c. copper concentrates from the U. S. carried by rail and ship to smelters in Japan

19. Which **two** mineral commodities are most likely to be concentrated in placer deposits?   a. pyrite   b. gold   c. native sulfur   d. diamonds

20. Bauxite is an ore of what metal?   a. aluminum   b. gold   c. tin   d. iron

21. In a typical marine evaporite basin such as the Michigan Basin, how are the evaporite minerals spatially distributed with respect to the basin's geometry?
a. center, halite; intermediate, gypsum/anhydrite; peripheral areas, potassium salts
b. center, potassium salts; intermediate, halite; peripheral areas, gypsum/anhydrite

22. Which **one** statement concerning scrap metals in the U. S. is not correct?
a. Since leaded gasoline has been replaced by unleaded gasoline, the percentage of lead being recycled has increased significantly.
b. Iron and steel comprise the bulk of recycled scrap metals.
c. Since about 1960, recycled scrap copper has comprised well over two-thirds of the annual domestic consumption of copper.

23. What is a logical source for anomalously high concentrations of copper, gold, and silver in sewage sludge from an urban wastewater treatment plant?
a. partly treated wastewater from breweries and whiskey distilleries
b. airborne fallout from a large, modernized copper smelter
c. untreated wastewater from electronics and photographic industries

24. Which task is not yet been known to have been accomplished by bacteria?
a. convert dissolved magnesium in chloride brines to magnesium metal
b. convert aqueous cyanide ion to ammonia and nitrate
c. utilize H2S and ferrous iron as energy sources for chemosynthesis

25. Which **two** situations could change a subeconomic gold bearing, mineral deposit into an economically mineable resource for gold? **two answers**
a. mining costs increase slightly and the price of gold declines
b. new low-cost mines begin producing in Brazil, Russia, and the U. S.
c. small tonnages of higher grade ore are discovered in the central part of the deposit
d. the world economy collapses, paper money becomes worthless, and gold is the only, trusted, monetary media

26. Ore deposits of which metal form by prolonged, intense, tropical weathering of specific kinds of bedrock?   a. aluminum   b. mercury   c. silver   d. magnesium

27. Acidic waters formed in copper, lead, and zinc mining areas result mainly from chemical decomposition of what mineral?   a. calcite   b. quartz   c. bauxite   d. pyrite

28. Vein- or fissure-filling deposits of lead and zinc minerals would be more common in which geologic setting?
a. in limestones; a contact-metamorphic zone around a shallow, granite pluton
b. in gneiss and schist; a deep, high-grade, regional-metamorphic zone

29. Oxidation of metal sulfide minerals and formation of strongly acidic, downward-percolating waters are important aspects of which ore-forming process?
a. secondary enrichment of primary, low-grade, copper deposits
b. accumulation of residual nickel and zinc minerals in very dry, hot, desert areas
c. precipitation of native gold in the A horizons of permafrost soils

30. Which are important fertilizer minerals with correct chemical symbols for the nutrient elements? **one answer**   a. halite (Na), calcite (N)
b. cassiterite (Sn), albite (Na)   c. apatite (P), sylvite (K)   d. sylvite (C), gypsum (P)

31. Which mineral is an essential component of plaster and plaster board?
a. quartz b. beryl c. feldspar d. gypsum

32. Which one process does not utilize large quantities of limestone?
a. producing Portland cement   b. refining aluminum metal from borax
c. making agricultural lime

33. Which mineral and element are used in the manufacture of fiber-optic glass and computer microprocessor chips? one answer
a. chalcopyrite; copper b. halite; sodium c. quartz; silicon d. galena; lead

Questions 34-37; Which statements are true/correct (T) or false/incorrect (F)? Briefly note what is wrong/incorrect in the false statements?

34. ( ) Acidic drainage is seldom associated with old, abandoned, underground mine workings because the acidic components in the water are usually neutralized by minerals in the bedrock.

35. ( ) Underground workings in carbonate-hosted, low-pyrite, lead-zinc sulfide deposits are less likely to generate strongly acidic groundwaters than those opened to mine pyrite-rich copper sulfide ores in veins cutting through granite.

36. ( ) Prior to present-day environmental regulations, some metal smelters in the U. S. emitted large quantities of SO2 and dangerous quantities of toxic elements such as arsenic, cadmium, and mercury.

37. ( ) Metal-rich acidic waters from abandoned open-pit, copper mines of the Tri-State District, Kansas, Colorado, and Nebraska, are a serious environmental threat to surface waters in the region.

38. On the map/maps provided, plot and label the following areas or locations.
a. Homestake Mine, Black Hills, SD   b. New Almaden and New Idria, CA
c. Tri-State lead-zinc district   d. Stillwater complex, MT
e. Bingham Canyon Mine, UT   f. Sudbury, Ontario, Canada
g. Cyprus   h. Chile   i. Jamaica   j. Zaire and Zambia

Answers; Home Study Assignment
Mineral Resources and Environment • Chapter 14

1. c, d	2. a, d	3. c	4. c	5. a
6. d	7. c, d	8. a, d	9. b, d	10. c
11. a, c	12. a, d	13. a	14. b	15. c, d
16. a	17. b, c	18. b	19. a, d	20. b, c*
21. a, d	22. d	23. a	24. b, d	25. a
26. d	27. d	28. a, b	29. a, c	30. a
31. a	32. d	33. a	34. b	35. b
36. a	37. d	38. b, c	39. a, d	40. a
41. a	42. d	43. a, d		

*

20. response a; Most silicate rocks have limited capacities to neutralize and buffer acidic waters produced as pyrite and other sulfide minerals undergo weathering and oxidation. Carbonate rocks do have the neutralization capacity, but even in this situation, acidic water percolating along cracks, fissures, and solution channels can be effectively isolated from carbonate wall rock by impervious or only slightly permeable deposits of non-carbonate minerals such as quartz, hematite, feldspars, and clays that line the walls of the openings and channels.

response d; This statement contains lots of substantive errors and would not inspire confidence if offered as a "knowledgeable statement" on the Tri-State District. First of all, the three states are Kansas, Missouri, and Oklahoma; Colorado and Nebraska are not involved. Second, the mines are/were underground workings, not open-pit operations, and lead and zinc, not copper, were the main ore metals produced. Copper was produced locally in some of the deposits. The part about the metal-rich drainage waters threatening regional surface water resources is basically correct. However, in this case, the wall rocks are typically limestones and dolomites, so the acidic drainage problem should be readily manageable.

Answers: Environmental Geology Quiz • Chapter 14

1. a	2. a, d	3. c	4. b	5. a, c
6. c	7. a, b	8. a	9. c	10. a, b
11. d	12. a	13. c	14. a	15. b
16. b	17. c	18. b	19. b, d	20. a
21. b	22. c	23. c	24. a	25. c, d
26. a	27. d	28. a	29. a	30. c
31. d	32. b	33. c	34. F*	35. T
36. T	37. F*			

*

34. Most silicate rocks have limited capacities to neutralize and buffer acidic waters produced as pyrite and other sulfide minerals undergo weathering and oxidation. Carbonate rocks do have the neutralization capacity, but even in this situation, acidic water percolating along cracks, fissures, and solution channels

can be effectively isolated from carbonate wall rock by impervious or only slightly permeable deposits of non-carbonate minerals such as quartz, hematite, feldspars, and clays that line the walls of the openings and channels.

37. This statement contains lots of substantive errors and would not inspire confidence if offered as a "knowledgeable statement" on the Tri-State District. First of all, the three states are Kansas, Missouri, and Oklahoma; Colorado and Nebraska are not involved. Second, the mines are/were underground workings, not open-pit operations, and lead and zinc, not copper, were the main ore metals produced. Copper was produced locally in some of the deposits. The part about the metal-rich drainage waters threatening regional surface water resources is basically correct. However, in this case, the wall rocks are typically limestones and dolomites, so the acidic drainage problem should be readily manageable.

Commentary/Media Watch

Corporate Environmental Image

Mines, smelters, quarries, rock crushers, and tailings ponds are not exactly media darlings. Environmental and economic issues related to the mining industry are most commonly reported. In fairly rare television and print media advertisements, mining companies often feature their environmental and reclamation efforts combined with a soft message noting the societal and economic values of their products. For example, a giant phosphate mining and fertilizer production facility at Aurora, in eastern North Carolina, was developed and operated by Texasgulf Corporation since the mid 1960s. This facility was sold to Potash Corporation of Saskatchewan (PCS) in 1995. Prior to the sale, Texasgulf television adds featured mined-over lands reclaimed for cattle grazing, manmade wetlands habitat, and fish happily swimming in cleaned, treated, wastewater ponds. Over the years, the company's adds had shifted from portraying a mighty industrial and economic giant to showing a good neighbor and corporate citizen proud of its commercial and environmental accomplishments. Without question, public relations were an important motivation for the adds, but the changing images projected by the company track the general public's growing interest in and concern for environmental issues. The current owners are evidently not interested in "wooing" the public; this writer hasn't seen a single media add since they acquired the facility.

Freeport-McMoRan's Grasberg Mine; Gold in Indonesia

From geological, environmental, and social viewpoints, one of the most interesting mining properties in the world today is the Grasberg Mine in Irian Jaya (Van Nort and others, 1991; O'Neill, T., 1996), the Indonesian west half of New Guinea. The Grasberg deposit is one of the world's largest gold reserves; it was discovered and developed by the New York based company Freeport-McMoRan. The deposit is perched atop the central mountain range of New Guinea, 13,000 feet above sea level. To reach it by overland travel, one has to cross some of the wettest, most precipitous, and most inhospitable terrain on Earth. Thus exploration, mine development, transportation, mining (open pit), and tailings disposal presented unusually severe environmental problems. In addition, the project involves complex social impacts swirling around property rights, living habitats, and traditional life styles of relatively primitive native peoples in the hills and canyons downstream from the mine. A class project or some class time devoted to the Grasberg Mine would have something of interest to most students and would dramatically illustrate the complex interplay between "economic development" and "the environment".

  A small Canadian exploration company, Bre-X Minerals Inc., had discovered another world-class gold deposit in Indonesia. In December, 1996, the company was pressured into a government-brokered agreement to join with Barrick Gold Corporation in developing its Indonesian gold deposits. This deal has the potential to turn Barrick into the world's leading gold producer, and Indonesia, with its ample petroleum and huge gold reserves, may become one of southeast Asia's richest nations.

The Wolf Rock-Stone Mountain State Park, NC, Land Deal

  In late June (6/23/96), an interesting agreement between the State of North Carolina and the North Carolina Granite Company received minimal coverage in statewide media. The company, based in Mt. Airy, produces crushed stone and dimension stone. Twenty-five years ago, the company had sold a parcel of land known as Wolf Rock to the State, but had retained mineral rights to the property. Whereas mineral rights on most lands in western states can be claimed or leased, most land owners in the eastern states enjoy exclusive ownership to mineral rights.

  Almost any tough crystalline rock, no matter how highly fractured, can be quarried and processed into crushed stone products. Building and dimension stone quarries require massive, relatively unfractured crystalline rock. The bedrock underlying the property in question is suitable for dimension stone, and of course, that is why the company retained the mineral rights. Lagging economic fortunes in the dimension stone industry and growing public concern for land preservation and parklands set the stage for the deal. The State agreed to pay $500,000 for the mineral rights and the property will become part of Stone Mountain State Park. Thus both parties gained a victory of sorts; the company made some return on its investment and the public gains additional, relatively unspoiled, Southern Appalachian parkland featuring scenic, massive, rounded rock outcrops.

The New World Mining District; The Final Deal?

  The New World Mining District near Cooke City, Montana, lies just a few miles beyond the northern boundary of Yellowstone National Park in the headwaters region of Clark's Fork of the Yellowstone, the Stillwater River, and Soda Butte Creek, a tributary to the Lamar River in the park. The area is thus extraordinarily "sensitive" environmentally. Historically, the mining district was closely tied to the park and to the national parks movement in America. For starters, Cooke City is named for Jay Cooke, the rich and powerful, Great Northern Railway magnate from a prominent Philadelphia family. In anticipation of increased railroad tourist traffic, he and his railroad provided strong support for legislative efforts to establish the park (Yellowstone Park Act, passed by Congress and signed by President Ulysses S. Grant, March 1872).

Cooke was also deeply involved in promoting the newly-discovered (late 1870s) New World Mining District and in building a rail line to support it. Despite heavy financial investments and vigorous promotion, the deposits, mainly gold and copper, were too remote, too small, and too scattered to support a long-term, large-scale mining enterprise; some small, profitable gold mining operations survived but eventually the whole district became inactive. However, mining was extensive enough to leave behind some serious environmental problems such as disturbed land, waste-rock dumps, acidic mine drainage, and mill tailings from which cyanide was leaching into Soda Butte Creek. This writer was fortunate to work on one of the many, small properties in the New World District in 1969. To this day, he remains impressed with the property's potential values in gold and copper.

As technologies and economics change, mining companies take "new looks" at old districts such as New World. Sometimes modern technologies and economic considerations make renewed mining feasible and profitable. Lower mining costs and more efficient milling and recovery processes may change a subeconomic resource into an economic reserve (p. 386-387). This is what happened at New World.

In recent years, encouraging exploration results led the Canadian mining giant Noranda to acquire mineral rights in the district and to set up Crown Butte Mining Company, as a subsidiary of Battle Mountain Gold, to continue with exploration and development work. The specter of a modern mining development within earshot of the park and in the headwaters region of three major rivers touched raw nerves all through "conservation circles". Each time Crown Butte made announcements concerning its planned mine, opposition spread and grew in intensity. For example the organization American Rivers made the "mining threat" a centerpiece of its 1996 fund-raising campaign, and articles concerning the proposed mine appeared in the Sierra Club Legal Defense Fund Quarterly Newsletter on Environmental Law, Winter, 1995-96, Yellowstone Today, Summer/Autumn, 1996 (published by the staff of Yellowstone National Park), the EARTH ALMANAC section of National Geographic Magazine (v. 189, no. 1, January, 1996), and in the news media.

After more than one hundred years, this threat to Yellowstone National Park was finally resolved. Crown Butte will give its claims in the New World District to the federal government in exchange for $65 million worth of, as yet, unspecified federal lands elsewhere. And before pulling out, the company has agreed to complete the task of cleaning up old waste dumps and acid mine drainage in the district.

The final agreement, 8/12/96, was a newsworthy event thanks to presidential election-year politics and its importance to conservation and environmental protection in America. President Clinton's announcement from Yellowstone National Park, timed to coincide with the Republican National Convention in San Diego, was shown widely on CNN and major network newscasts. A small, but enthusiastic crowd attended the outdoor announcement at Buffalo Ranch in the Lamar River Valley. The Eocene-age Absaroka Volcanics of

Specimen Ridge and the Fossil Forests area formed a fitting, geologic backdrop for television coverage of the event.

Gold in Ghana

On October 5, 1996, CNN featured a brief report on Asante Gold, a company engaged in gold mining and exploration in Ghana and neighboring countries in West Africa. The report included interviews with company officials and images of surface mining, underground mining, smelting, and casting of gold bars.

Asbestos Litigation

In late June, business sections of many newspapers carried a brief announcement that Owens Corning was putting aside $1.1 billion as a reserve for settling future asbestos suits and claims against the corporation. As noted in the *special feature* Asbestos, p. 364, the well-documented, carcinogenic and lung-damaging effects of amphibole asbestos fibers, particularly the crocidolite variety, have been confirmed by recent research. However, no such serious health risks have been shown to arise from repeated, heavy exposures to the more widely used, softer, chrysotile/serpentine asbestos fibers. Thus projects involving removal of chrysotile asbestos do little or nothing to protect public health, and the funds would have been better invested elsewhere. However, spending to reduce or eliminate public exposure to amphibole asbestos fibers is money well spent.

Unauthorized Copper Trading at Sumitomo Corporation

The economics and business side of the mineral commodities and metals industry occasionally gets some media coverage, especially if a political shakeup, policy change, or unusual market conditions threaten future supplies and price stability. One such event jolted the world copper markets in mid-June and continued to receive media updates for at least the next four months. On June 15, 1996, Sumitomo Corporation, a Japanese, world-wide commodities trading giant, announced losses of at least $1.8 billion U. S. from "unauthorized" trading in copper futures. The losses, which eventually would rise to $2.6 billion U. S., were amassed over a ten-year period by the heretofore, highly respected copper trader Yasuo Hamanaka, who had somehow managed to conceal his mounting losses from corporate officials. World copper prices plunged on the announcement, the scandal badly shook the Japanese business community, and speculation mounted that major copper exporting countries such as Chile, Peru, Zambia, and Zaire could be adversely affected if copper prices continue to decline.

The scandal also highlighted the relatively lax external oversight policies in force in Japan as compared with those in the United States. Mr. Hamanka was finally indicted for fraud, illegal trades, and forgery in connection with the scandal and arrested on October 22. His indictment and arrest were widely reported in the print media and on television business news programs. The scandal has triggered strong volatility in world copper markets out of fears that Sumitomo might either have to dump massive copper holdings and drive prices lower or have to buy large supplies to fulfill its delivery contracts, thus driving prices up. This story is ongoing; the next episode will focus on how one person could conceal such massive financial losses for so long a time while maintaining such a highly respected position among his corporate and professional associates.

Supplemental References

Craig, J. R., Vaughan, D. J., and Skinner, B., 1996, Resources of the Earth: Origin, use, and environmental impact (second edition): Prentice Hall, Upper Saddle River, NJ, 472 p.

Eliot, J. L., 1996, Gold mine camped on Yellowstone's doorstep: short note, EARTH ALMANAC section, National Geographic Magazine, v. 189, no. 1, January

Kesler, S., 1994, Mineral resources, economics, and the environment: Prentice Hall, Upper Saddle River, NJ, 392 p.

Martinez, J. D., 1991, Salt domes: American Scientist, v. 79, no. 5 ,Sept.-Oct., p. 420-431

O'Neill, T., 1996, Irian Jaya: National Geographic Magazine, v. 189, no. 2, February, p. 2-33

Van Nort, S. D., Atwood, G. W., Collinson, T. B., Flint, D. C., and Potter, D. R., 1991, Geology and mineralization of the Grasberg porphyry copper-gold deposit, Irian Jaya, Indonesia: Mining Engineering, v. 43, no. 3, March, p. 300-303

Energy and Environment • Chapter 15; Home Study Assignment

1. Which **one** statement concerning energy for space heating is not true?
a. accounts for a small fraction of the low-grade energy used in the U. S.
b. is characterized as having a low utilization temperature
c. direct solar space heating is gradually becoming more economically competitive with conventional sources
d. can be significantly reduced by sound building designs and conservation practices

2. Which **two** short paragraphs concerning coal reserves and coal consumption in the U. S. are correct? The others contain basic factual errors.
a. Anthracite comprises a small portion of U. S. coal reserves; the deposit are concentrated in eastern and south-central Pennsylvania. Prior to 1945, anthracite coal was still widely used as a home-heating fuel in New York City, Philadelphia, and eastern Pennsylvania.
b. Large lignite reserves are located in the northern plains and in a belt extending from the Mexican border across Texas to western Tennessee and southern Alabama. Compared to other coals, lignite has relatively high moisture and low carbon contents.
c. High-sulfur coals have sulfur contents exceeding 3 %; bituminous coals from the Rocky Mountain states are generally high-sulfur coals; those from the Appalachian Basin generally have sulfur contents of 1 % or less.
d. Since 1949, coal production in the U. S. has gradually declined by about 50 % as other fuels, such as petroleum and natural gas, replaced coal as the dominant fuel for home heating.

3. Comparatively speaking, how does the United States use energy fuels and minerals?
a. Consumption per person is below that in less industrialized countries like Brazil and Mexico.
b. Consumption per person is above that in other technologically advanced countries such as Japan and Germany.

4. Which use accounts for the majority of the coal burned annually in the U. S.?
a. coke manufactured for the steel industry
b. production of carbon black, synthetic diamonds, and graphite
c. home and industrial space heating   d. generation of electricity

5. Which coal is typically found only in association with tightly folded strata?
a. anthracite   b. peat   c. bituminous   d. lignite

6. The United States, with about 5 % of the world's population, uses about what percentage of the world's total, annual, energy production?
a. 50 %   b. 25 %   c. 15 %   d. 10 %

7. What source supplies the largest percentage of energy consumed annually in the United States?   a. coal   b. uranium   c. hydroelectric   d. petroleum

8. Which fossil fuel is most likely to have the lowest sulfur content when burned?
a. bituminous coal   b. heavy, asphaltic, crude oil   c. natural gas   d. lignite

9. Why are coal and petroleum considered to be fossil fuels?
a. their energy content was derived from ancient sunlight
b. coal beds and petroleum reservoir rocks contain abundant fossils
c. their oxygen and nitrogen content were derived from the ancient atmosphere
d. carbon dioxide, released when they burn, contributes to the greenhouse effect

10. What are two, major, environmental problems associated with coal mining and burning?   a. sulfur dioxide is emitted to the atmosphere
b. Earth's ozone layer is reduced in thickness   c. radioactive carbon-14 is released
d. land and water resources are degraded by underground and surface mining

11. Which one statement is true? The others are flawed in some way.
a. North and South America have about equal reserves of good-grade coals.
b. After the year 2000, U. S. imports of petroleum will probably account for over 50 % of domestic consumption.
c. Coal imports into the U. S. have been rising steadily since about 1950.
d. By the year 2000, Russia and the republics of the former Soviet Union will probably be importing coal and petroleum from African countries.

12. Which ancient environment and associated rocks would pertain to coal and coal formation?
a. freshwater coastal swamps; shale and sandstone
b. shallow, coastal marine basins; halite, gypsum, and anhydrite
c. mid-ocean ridges; submarine basalt lava flows and hot-spring deposits
d. offshore marine basins; black shales and coral reef limestone

13. The terms cap rock and reservoir strata refer to natural accumulations of which nonmetallic resource?
a. oil shale   b. bedded rock salt   c. sedimentary phosphate deposits   d. petroleum

14. Which two explanations are correct? The others are flawed.
a. As contour mining progresses, the depth of overburden generally increases.
b. The area strip mining method is better suited for areas with high, local topographic relief than is the contour strip mining method.
c. Federal regulations now require that strip mined lands in the U. S. should be restored to their premining uses.
d. Lands reclaimed after surface coal mining in the eastern U. S. are relatively free of problems related to acidic groundwaters.

Questions 15-18; Matching; See *case histories* A Tale of Two Mines, p. 420-421.
a. Star Fire Mine   b. Trapper Mine

15. ( ) produces from bituminous coals of the Appalachian Basin

16. ( ) produces from Cretaceous-age bituminous coals of the Colorado Plateau region

17. ( ) operates under a variance from existing law to substantially modify the post-mining relief of reclaimed lands

18. ( ) topographic and climatic conditions are very favorable for restoration of mined lands to their premining natural conditions and for avoiding problems related to acidic drainage

Questions 19-21; Matching: Oil and Gas Maturation

a. below 6 km depth   b. above 1 km depth   c. between depths of 3 to 6 km

19. ( ) zone of thermogenic methane and crude oil generation
20. ( ) zone of biogenic methane generation
21. ( ) zone of thermogenic methane generation

Questions 22-24; Basic Kinds of Crude Oil Traps

a. fault trap   b. anticlinal trap   c. unconformity trap

22. ( ) common oil and gas accumulation site in folded strata
23. ( ) offset, inclined, porous, permeable reservoir stratum is juxtaposed against an impermeable cap rock or source rock
24. ( ) younger, horizontal cap rock overlies older, inclined source and reservoir strata

25. What is the energy source for The Geysers electrical generation facility?
a. sunlight   b. a vapor-dominated geothermal system
c. the wind   d. coal mined in northeastern Arizona

26. Which area has the largest, current production and reserves of crude oil?
a. the North Sea field, northern Europe   b. central and eastern Saudi Arabia
c. offshore Japan   d. United States including Alaska

27. Why is sandstone a much more common reservoir rock for petroleum than is shale? **two answers**
a. Shale is more porous so the oil tends to leak out over time.
b. Sandstone is more porous so the oil can accumulate in open spaces in the rock.
c. Shales, especially black shales, are much richer in primary organic matter than are sandstones.
d. Sandstones are more permeable than shales so subsurface flows of fluids tend to be directed through sandstone strata rather than through shales.

28. Where is the world's largest, commercially developed, tar sand deposit?
a. northern Brunei, Island of Borneo   b. near Lake Athabasca, central Canada
c. northeastern Utah and northwestern Colorado   d. southern Iraq, near Basra

29. Which geologic formation, rich in oil shale strata, was deposited in early Tertiary lakes that once covered parts of Colorado, Utah, and Wyoming?
a. Blue Mountain   b. White Prairie   c. Green River   d. Red Lake

30. Where is the world's largest plant for generating electricity from tides?
a. along the mouth of the Amazon River in Brazil
b. along the coast of California just south of Los Angeles
c. in the Rance River estuary, Brittany, along the Atlantic coast of France
d. in the Netherlands along the Rhine River delta

31. Which element and isotope is the main fuel used in nuclear fission reactors to produce electricity?
a. uranium-239   b. thorium-232   c. thorium-238   d. uranium-235

32. What is the main purpose for building pumped, water-storage systems?
a. to use water saved from wet periods to generate electricity during dry periods
b. to generate electricity from hot, saline, underground water delivered to elevated storage tanks on the surface
c. to convert energy, saved when electrical usage is low, to electricity at times when electrical usage is high
d. to keep the hydroelectric pool filled when the tide is out

33. Where and when was the first, large-scale, commercial, geothermal electrical power-generation facility in the United States brought into production?
a. Yellowstone National Park, 1925   b. Hot Springs, Arkansas, 1970
c. the Geysers area, northern California, 1960   d. Oak Ridge, Tennessee, 1945

34. Which **two** apply to graphite?
a. used as neutron flux moderator rods in Chernobyl-style atomic reactors
b. mineral form of carbon; found in schists and other high-grade metamorphic rocks
c. mostly formed in the biogenic zone of methane generation
d. major organic component of oil shales and tar sands

35. What is the major difference between pressurized water and boiling water nuclear reactors?
a. Boiling water reactors don't require heavy containment structures because the water used for heat exchange is not under pressure.
b. Pressurized water reactors include a secondary heat exchange loop to generate steam for power generation.
c. Boiling water reactors are breeders; most pressurized water reactors are burners.
d. Boiling water reactors are employed for hydrogen fusion: pressurized water reactors employ fission of U-235 as the energy source

36. Which **two** statements concerning nuclear fusion are correct or reasonable?
a. The hydrogen isotopes deuterium and tritium are the major fuel.
b. Nuclear fusion generates very little in the way of dangerous, radioactive by-products.
c. Fusion is expected to soon displace fission as the dominant energy source in commercial reactors.
d. The Three Mile Island, Pennsylvania accident involved an explosion in the molten lithium heat exchange loop of a deuterium fusion reactor.

37. Where are most of the world's natural gas reserves? **one answer**
a. South America and Africa   b. North and South America
c. Russia, former republics of the Soviet Union, and the Persian Gulf countries
d. western Europe and Australia

38. On average, what percentage of the total petroleum in an oil field can be recovered utilizing primary, secondary, and tertiary recovery methods?
a. 20 %   b. 80 %   c. 55 %   d. 90 %

39. Which statement concerning shale oil and oil shale deposits in the U. S. is reasonable and not excessively optimistic nor pessimistic?
a. Each ton of high-grade oil shale will yield about 40 liters of useful petroleum products. Thus land reclamation and safe disposal of the processed, spent shale will be minor factors affecting the cost of the refined products.
b. At world oil prices in the range of $18 to $24 per barrel, large-scale development of U. S. oil shale deposits would have to be subsidized by the federal government since production costs would exceed the market value of the refined products.

40. What is the major function of the control rods in a nuclear reactor?
a. regulate the neutron flux and number of fission events per unit time
b. contain the fissionable uranium isotopes that fuel the chain reaction
c. transport heat from the reactor to vaporize water into steam
d. trap radioactive fission products to prevent them from contaminating the fuel pellets

Questions 41-44; Matching: Radioactive Fission Products of Uranium-235 and their Half-Lives

a. Kr-85; 10 yrs   b. I-131; 8.1 days   c. Sr-90; 28 yrs   d. Cs-137; 30 yrs
41. ( ) behaves biochemically like calcium; accumulates in the bones
42. ( ) very dangerous because it's highly concentrated in the thyroid
43. ( ) inert gas released into the atmosphere from nuclear power plants
44. ( ) accumulates in Arctic tundra mosses; ingested by caribou and reindeer

45. Which region of the U. S. on average has the highest geothermal gradients?
a. New York and the New England states
b. North Dakota, South Dakota, and Nebraska
c. Colorado, Wyoming, and Montana   d. Nevada and southern Idaho

46. Which **one** is inappropriate for geopressurized systems?
a. are vapor-dominant systems   b. contain dissolved methane
c. are formed in interbedded permeable and impermeable sedimentary strata
d. are common along the Gulf of Mexico coastal region, Texas

47. What **one** was a major provision of the 1978 Public Utility Regulatory Policy Act?
a. Public utilities were ordered to buy any excess electrical energy produced by their customers.
b. Public utilities were required to diversify into solar, wind, and other alternate energy sources for producing electricity.
c. Public utilities were forbidden from raising rates to begin recouping capital costs of new generating facilities before a facility actually produced electrical power.
d. Public utilities were ordered to reduce electrical rates for major stock holders.

48. Consider a coal-fired electrical generating plant with no provisions for cogeneration. What percentage of the thermal energy released by coal combustion is converted into electricity?   a. 10 %   b. 75 %   c. 35 %   d. 60 %

49. Which **two** are readily adapted as fuels for relatively nonpolluting fuel cells?
a. methane   b. hydrogen   c. helium   d. bromine

Questions 50-53; Matching: Energy Sources and Appropriate Locations

a. Yellowstone National Park
b. North Dakota
c. Coachella Valley and Salton Sea area, CA
d. Mojave Desert region, CA

50. (   ) possesses substantial lignite reserves and high potential for wind energy
51. (   ) large-scale commercial, facility for producing electricity from sunlight
52. (   ) substantial potential for wind, geothermal, and solar power
53. (   ) world's largest geothermal resource; off limits to commercial development

54. Which **one** is an example of cogeneration?
a. burning coal and using solar cells to make electricity at the same generating plant
b. utilizing energy efficient machinery and electrical devices to reduce the overall consumption of energy in a community or industrial facility
c. production of fissionable plutonium-239 in a breeder reactor supplying heat for steam generation of electricity
d. a process that gainfully utilizes low-grade heat, otherwise wasted, from a high-temperature heat source such as a coal-fired, electrical generation plant

55. With well-designed provisions for cogeneration, the efficiency of a new thermoelectric generation facility can be raised to about what value?
a. 10 %   b. 40 %   c. 90 %   d. 70 %

56. In regard to new, thermoelectric power plants, which setting would be more conducive to gainfully utilizing cogeneration?
a. a new, near-mine power plant in northeastern Arizona delivering electricity to Phoenix, AZ, Los Angeles, CA, and Las Vegas, NV
b. a new plant in a highly industrialized area such as northern New Jersey or the lower Delaware Valley, Pennsylvania

57. Which **two** statements are correct? The other two contain factual errors.
a. The LUZ solar power plant in the Mojave Desert region of California has a backup natural gas burner to generate steam during cloudy days. Costs per kilowatt hour are competitive with conventional sources.
b. The ocean thermal energy conversion process is most efficient in shallow waters with a small temperature difference between surface and bottom waters.
c. High bird mortality is an environmental consequence of some wind farms.
d. Burning municipal waste to produce electricity is less common in western Europe than in the U. S. Western European countries, particularly France, are less dependent on electricity generated by nuclear power plants than is the U. S.

Questions 58-61; Do you agree (A) or Disagree (D) with the following statements? State the basis and reasons for your disagreements.

58. (   ) In general, the so-called "hard" pathway, or business as usual energy policy is less likely to reduce future energy demands in the U. S. than a "softer" policy incorporating more diversified, less centralized, renewable energy sources such as solar, wind, and biomass.

59. ( ) In place of electricity drawn from large-scale regional grids, U. S. energy conservation policy should encourage utilization of local, low-grade energy sources, such as solar, for low-grade, end requirements such as home hot-water heating and space heating.

60. ( ) Groundwater is a very useful heat exchange medium for space heating and cooling in tropical and subtropical regions where seasonal variations in air temperature are relatively small.

61. ( ) Shrill media attacks by environmentalists and the irrational public hysteria over nuclear plant safety and disposal of spent reactor fuel are largely responsible for lowered growth rates and lowered projected rates of growth of energy usage in the U. S. and for lower profit margins and stagnant stock prices of America's investor-owned public utilities.

62. On the U. S. map provided, plot and label the following areas or locations.
a. Three Mile Island plant site, PA
b. The Geysers power plant site, CA
c. Yellowstone National Park, WY
d. Mojave Desert region, CA
e. Jemez Mountains, NM
f. Salton Sea area, CA
g. Black Mesa Basin coal field, AZ
h. Kentucky coal fields
i. Battle Mountain geothermal high
j. Powder River Basin coal fields, WY

63. On the world map provided, accurately plot and label the following locations.
a. Iraq
b. Saudi Arabia
c. Athabasca tar sands, Canada
d. Persian Gulf
e. Chernobyl, Ukraine
f. Rance River estuary, France
g. Qatar
h. Bahrain
i. United Arab Emirates

Environmental Geology Quiz • Chapter 15

1. Which **one** statement concerning energy used for space heating is not true?
a. can be significantly reduced by sound building designs and conservation practices
b. is characterized as having a low utilization temperature
c. accounts for only a small fraction of the low-grade energy used in the U. S.

2. Which **one** response concerning coal reserves and coal consumption in the U. S. is correct? The others contain basic factual errors.
a. Since 1949, coal production in the U. S. has gradually declined by about 50 % as petroleum and natural gas replaced coal as the dominant fuel for home heating.
b. High-sulfur coals have sulfur contents exceeding 3 %; bituminous coals from the Rocky Mountain states are generally high-sulfur coals; those from the Appalachian Basin generally have sulfur contents of 1 % or less.
c. Anthracite comprises a small portion of U. S. coal reserves; the deposit are concentrated in eastern and south-central Pennsylvania. Prior to 1945, anthracite coal was the prime home-heating fuel in major cities of the east and northeast.

3. How does the United States per/person consumption rate of energy fuels compare to that in other countries?
a. U. S. rate is lower than that in less industrialized countries like Brazil and Mexico.
b. U. S. rate is above that of other technologically advanced countries such as Japan.

4. Which use accounts for the majority of coal burned annually in the U. S.?
a. coke manufacture   b. generation of electricity   c. space heating

5. Which coal is typically found only in association with tightly folded strata?
a. lignite   b. bituminous   c. peat   d. anthracite

6. What source supplies the largest percentage of energy consumed annually in the United States?   a. coal   b. uranium   c. petroleum   d. hydroelectric

7. Which fossil fuel is most likely to have the lowest sulfur content when burned?
a. natural gas   b. heavy, asphaltic, crude oil   c. bituminous coal   d. lignite

8. What are **two**, major, environmental problems associated with coal mining and burning?   a. carbon-14 is released to the atmosphere
b. Earth's ozone layer is depleted   c. sulfur dioxide is emitted to the atmosphere
d. land and water resources are degraded by underground and surface mining

9. Which **one** statement is probably correct? The others are flawed in some way.
a. Coal imports into the U. S. have been rising steadily since about 1950.
b. By the year 2000, Russia and the republics of the former Soviet Union will probably be importing large amounts of petroleum from South Africa.
c. North and South America have about equal reserves of good-grade coals.
d. By 2010, imported petroleum will account for over 50 % of U. S. consumption.

10. Which ancient environment and associated rocks pertain to coal formation?
a. freshwater coastal swamps; shale and sandstone
b. shallow, coastal marine basins; halite, gypsum, and anhydrite
c. offshore marine basins; black shales and coral reef limestone

11. Cap rock and reservoir strata refer to accumulations of which resource?
a. sedimentary phosphates   b. petroleum   c. oil shale   d. bedded rock salt

12. Which **one** statement is incorrect and/or basically flawed?
a. Lands reclaimed after surface coal mining in the western U. S. are relatively free of problems related to acidic groundwaters.
b. The area strip mining method is better suited for areas with high, local topographic relief than is the contour strip mining method.
c. Federal regulations now require that strip mined lands in the U. S. should be restored to their premining uses.

Questions 13-14; Matching; Two Coal Strip Mines
a. Star Fire Mine   b. Trapper Mine

13. ( ) mines bituminous coals of the Appalachian Basin; operates under a variance to substantially modify the post-mining relief of reclaimed lands

14. ( ) produces from Cretaceous-age bituminous coals; topographic and climatic conditions are very favorable for successful restoration of mined lands

15. At what burial depths is all primary, sedimentary organic matter subjected to thermogenic methane generation?
a. below 6 km   b. between 3 to 6 km   c. above 1 km

16. Which **one** refers to an anticlinal crude oil trap?
a. younger, horizontal cap rock overlies older, inclined source and reservoir strata
b. offset, inclined, porous, permeable reservoir stratum is juxtaposed against an impermeable cap rock or source rock
c. common oil and gas accumulation site in folded sedimentary rock strata

17. What is the energy source for The Geysers electrical generation facility?
a. wind   b. coal from Arizona   c. vapor-dominated geothermal system   d. sunlight

18. Why are sandstones more common as petroleum reservoir rocks than shales?
a. Shales, especially black shales, are much richer in organic matter than sandstones.
b. Sandstones are more permeable than shales so subsurface fluids tend to flow through sandstone strata rather than through shales.

19. Where is the world's largest, commercially developed, tar sand deposit?
a. northeastern Utah and northwestern Colorado   b. southern Iraq, near Basra
c. northern Brunei, Island of Borneo   d. near Lake Athabasca, central Canada

20. Which geologic formation, rich in oil shale strata, was deposited in early Tertiary lakes that once covered parts of Colorado, Utah, and Wyoming?
a. Red Lake   b. Blue Mountain   c. White Prairie   d. Green River

21. Which **one** is not correct for graphite?
a. mostly formed in the biogenic zone of methane generation
b. mineral form of carbon; found in schists and other high-grade metamorphic rocks
c. used as neutron flux moderator rods in Chernobyl-style atomic reactors

22. Which isotope is the main fuel used in nuclear fission reactors to produce electricity?   a. thorium-232   b. uranium-239   c. uranium-235   d. thorium-238

23. What is the main purpose for building pumped, water-storage systems?
a. to use water saved from wet periods to generate electricity during dry periods
b. to use electricity produced at times of low demand to supplement electricity production when electrical usage is high
c. to keep the hydroelectric pool filled when the tide is out so electricity can be produced during the whole tidal cycle

24. Which **one** statement about nuclear electrical power generation is correct?
a. Boiling water reactors don't require heavy containment structures because the water used for heat exchange is not under pressure.
b. Pressurized water reactors include a secondary heat exchange loop to generate steam for power generation.
c. The Three Mile Island, Pennsylvania accident involved an explosion in the molten lithium heat exchange loop of a deuterium fusion reactor.

25. Where are most of the world's natural gas reserves? **one answer**
a. North America   b. South America and Africa
c. Russia, former republics of the Soviet Union, and the Persian Gulf countries

26. On average, what percentage of the total petroleum in an oil field can be recovered utilizing primary, secondary, and tertiary recovery methods?
a. 20 %   b. 80 %   c. 55 %   d. 90 %

27. Which statement concerning shale oil and oil shale deposits in the U. S. is reasonable and not excessively optimistic nor pessimistic?
a. At world oil prices in the range of $18 to $24 per barrel, large-scale development of U. S. oil shale deposits would have to be subsidized by the federal government since production costs would exceed the market value of the refined products.
b. Each ton of high-grade oil shale will yield about 40 liters of useful petroleum products. Thus land reclamation and safe disposal of the processed, spent shale will be minor factors affecting the cost of the refined products.

Questions 28-29; Matching: Fission Products of Uranium-235 and their Half-Lives
    a. Sr-90; 28 yrs   b. Cs-137; 30 yrs
28. (   ) accumulates in Arctic tundra mosses; ingested by caribou and reindeer
29. (   ) behaves biochemically like calcium; accumulates in the bones

30. Which region of the U. S. has the highest geothermal gradients?
a. New York, Pennsylvania   b. Colorado, Wyoming   c. Nevada, southern Idaho

31. What was a major provision of the 1978 Public Utility Regulatory Policy Act?
a. Public utilities were forbidden from raising rates to begin recouping capital costs of new generating facilities before the facility produced any electrical power.
b. Public utilities were ordered to buy excess electrical energy produced by customers.

32. Which **one** is inappropriate for geopressurized systems?
a. are vapor-dominant systems   b. contain dissolved methane
c. develop in interbedded, coastal, permeable and impermeable sedimentary strata

33. In a modern, coal-fired electrical generating plant without cogeneration, what percentage of the combustion thermal energy is actually converted into electricity?   a. 60 %   b. 75 %   c. 35 %   d. 10 %

34. Which **two** are readily adapted as fuels for relatively nonpolluting fuel cells?
a. methane   b. bromine   c. helium   d. hydrogen

35. Which **one** statement is incorrect?
a. The LUZ solar power plant in the Mojave Desert region of California has a backup natural gas burner to generate steam during cloudy days. Costs per kilowatt hour are competitive with conventional sources.
b. High bird mortality can be an environmental consequence of wind farms.
c. Western European countries, particularly France, are less dependent on electricity generated by nuclear power plants than is the U. S.

36. Which **one** is an example of cogeneration?
a. gainfully utilizing low-grade heat, otherwise wasted, from a high-temperature heat source such as a coal-fired, electrical generation plant
b. utilizing energy efficient machinery and electrical devices to reduce the overall consumption of energy in a community or industrial facility
c. generating electricity from coal combustion and solar cells at the same facility

37. Which area possesses substantial lignite reserves, high potential for wind energy, modest potential for solar energy, and little potential for geothermal resources?
a. Mojave Desert area, CA   b. Yellowstone National Park, WY
c. North Dakota   d. Coachella Valley/Salton Sea area, CA

Questions 38-40; Do you agree (A) or Disagree (D) with the following statements? State the basis and reasons for any disagreements.

38. ( ) In general, the so-called "hard" pathway, or business as usual energy policy is less likely to reduce future energy demands in the U. S. than a "softer" policy incorporating more diversified, less centralized, renewable energy sources.

39. ( ) U. S. energy policy should encourage the use of electricity from large-scale, regional grids for low-grade energy requirements, such as home hot-water heating and space heating, because local, low-grade energy sources, such as solar and wind, are notoriously inefficient.

40. (   ) Groundwater is a very useful heat exchange medium for space heating and cooling in tropical and subtropical regions where seasonal temperature variations are relatively small.

41. On the map/maps provided, plot and label the following areas or locations.
a. Three Mile Island plant site, PA
b. Battle Mountain geothermal high
c. Yellowstone National Park, WY
d. Mojave Desert region, CA
e. Jemez Mountains, NM
f. Kentucky coal fields
g. Persian Gulf
h. Athabasca tar sands, Canada
i. Chernobyl, Ukraine
j. United Arab Emirates

Answers; Home Study Assignment
Energy and Environment • Chapter 15

1. a	2. a, b	3. b	4. d	5. a
6. b	7. d	8. c	9. a	10. a, d
11. b	12. a	13. d	14. a, c	15. a
16. b	17. a	18. b	19. c	20. b
21. a	22. b	23. a	24. c	25. b
26. b	27. b, d	28. b	29. c	30. c
31. d	32. c	33. c	34. a, b	35. b
36. a, b	37. c	38. c	39. b	40. a
41. c	42. b	43. a	44. d	45. d
46. a	47. a	48. c	49. a, b	50. b
51. d	52. c	53. a	54. d	55. d
56. b	57. b, d	58. A	59. A	60. D*
61. D*				

*

60. Groundwater is useful as a heat exchanger in areas characterized by strong, seasonal, variations in air temperature, not in tropical or subtropical areas as stated. Groundwater stays at a constant temperature year-round, equal to the mean annual temperature. Thus on cold winter days, the groundwater is warmer than the air and the "extra heat" can be utilized for space heating. On hot summer days, the groundwater is substantially cooler than the air, so the water can be efficiently utilized as a heat exchange fluid in air conditioning systems.

61. There is no question that the nuclear reactor industry in the U. S. is "dead in the water" and public fears over safety issues are a major factor. It however, can be vigorously debated as to whether these fears are justified and realistic or unjustified and hysterical. The public's disenchantment with nuclear energy and technology is clearly at the forefront of today's contentious, highly emotional

debates over radioactive waste storage and disposal. Most communities are even hostile to relatively safe, well-designed landfills for low-level, radioactive wastes. The Yucca Mountain site is almost sure to be the final resting repository for high-level waste generated in power plant reactors and weapons manufacture. Transporting the waste to the Nevada site is fraught with dangers and it will take a well-oiled selling job to convince the public that the processes of transportation and final storage are safe. Good will and favorable publicity will instantly vanish with one serious accident.

The trends toward higher efficiencies, better conservation practices, and alternative energy sources have all helped to reduce growth in the demand for electricity in the U. S in the last decade or so. Previous projections of the electricity growth rate, in retrospect, were highly inflated. Thus utility companies that vastly over built generating capacity 20 years ago have had plenty of time to reflect on their "low growth rates" and "stagnant stock prices".

As for stock prices of public utilities, many analysts "don't like" companies with an aging nuclear generation capacity, given the substantial expenditures needed for maintenance and safety-related improvements. Liabilities arising from a serious reactor accident could easily bankrupt a company. So stick with fossil-fuel fired utilities for the time-being. France and Germany, with far more extensive commitments to nuclear electricity than the U. S., will tell the tale over the next few decades. Safe operation and disposal practices may promote another look at nuclear energy in this country. On the other hand, a Chernobyl-scale accident in western Europe will probably seal the fate of America's nuclear power industry.

Answers: Environmental Geology Quiz • Chapter 15

1. c	2. c	3. b	4. b	5. d
6. c	7. a	8. c, d	9. d	10. a
11. b	12. b	13. a	14. b	15. a
16. c	17. c	18. b	19. d	20. d
21. a	22. c	23. b	24. b	25. c
26. c	27. a	28. b	29. a	30. c
31. b	32. a	33. c	34. a, d	35. c
36. a	37. c	38. A	39. D*	40. D*

*

39. On the contrary, domestic energy policy should encourage local, alternative energy sources for low-grade uses such as home heating, hot-water heating, and air conditioning. Thus high-grade energy, mainly in the form of electricity, would be reserved for uses that demand this high-quality energy. The capital cost savings and environmental gains realized by not building and not fueling additional, unneeded, new, generating capacity would be a significant, positive result of a so-called "soft energy policy" emphasizing diversified, decentralized energy sources. Region-wide blackouts and brownouts would also be much less likely and less damaging.

Alternate energy sources are not necessarily more or less efficient than electricity from a regional grid. We can be sure that as more effort is devoted to alternative energy sources, efficiencies will improve and costs will come down.

40. Groundwater is useful as a heat exchanger in areas characterized by strong, seasonal, variations in air temperature, not in tropical or subtropical areas as stated. Groundwater stays at a constant temperature year-round, equal to the mean annual temperature. Thus on cold winter days, the groundwater is warmer than the air and the "extra heat" can be utilized for space heating. On hot summer days, the groundwater is substantially cooler than the air, so the water can be efficiently utilized as a heat exchange fluid in air conditioning systems.

Commentary/Media Watch

The 70s Oil Crisis; Is an Encore Looming in the Near Future?

The 1970s oil supply crisis was a much-needed wakeup call to oil importing and exporting nations alike. World demand was growing faster than world production; OPEC production limits made the discrepancy worse, and world oil prices rose. Petroleum exporting countries, such as Mexico, went on lavish spending sprees in anticipation of $38/barrel oil and growing demand. In the United States, autos lined up at gasoline stations when supplies were available or stayed parked for days on end. The crisis ended as suddenly as it began. Evidently the law of supply and demand was still in effect; world production rose, world-wide economic recession and cutbacks induced by high energy prices reduced consumption, big OPEC producers switched focus to raising "market share", and oil prices collapsed. Peso devaluation and rocketing consumer prices in Mexico were so devastating that pinto beans were cynically renamed "Mexican caviar".

The "energy crisis" sparked renewed interest in energy conservation, support for alternative, domestic energy sources, reduced dependency on foreign sources, and increased efficiency in energy use. The 55 mi/hr national speed limit was instigated under the banner of energy independence, reduced fuel consumption, and saving lives. The federal government invested heavily in "synfuels" such as gasoline from oil shale, and fuels from biomass such as ethanol from corn and sugar and methanol from forest products. Instant subdivisions, complete with all the suburban amenities, sprang up in Rifle and Battlement Mesa, CO, near the, prime, shale-oil deposits in western Colorado (p. 426-428; Figure 15.17). The inevitable oil-shale bust left home owners with no jobs and no buyers; the new communities became ghost towns, bringing to mind the boom/bust dynamics of a 19th century, western, silver-mining camp.

This positive energy policy momentum has since largely evaporated. Growing world-wide demand and competition for supplies have prompted many petroleum industry experts and analysts to recommend buying oil company and oil industry service company stocks as prudent investments for the twenty-first century. Imported oil comprises well over 50 % of domestic consumption and makes a significant negative contribution to America's trade deficit. Fuel consumption and saved lives didn't help the 55 mi/hr national speed limit survive the 104th Congress; and, nary a whisper on energy was heard from the three major candidates in the 1996 presidential election. Even Ralph Nader and his Green Party have focused on environmental issues other than a national energy policy (450-451).

World oil prices respond more to Iraq's internal politics and American cruise missile attacks, political intrigue in Nigeria, Saudi Arabia, and Indonesia, and internal strife in Chechnya than to a rational, national energy policy in Washington, DC. High prices and taxes act to hold down gasoline consumption in Europe and Japan; in the United States a temporary 10-cent per gallon increase in the gasoline price, such as happened early in 1996, was grounds for congressional outcries and calls for repeal of the inconsequential 4.3 cent/gallon tax added in 1993. Hot-selling new vehicles are not small, fuel efficient models, but relatively low-gasoline-mileage minivans, sport utility vehicles, and light trucks. This trend threatens to derail legislative mandates for gradually increasing average fleet-wide gasoline mileage ratings of all new vehicles sold by individual companies, but well-financed attempts in the 104th Congress to roll back these goals were eventually defeated.

Significant advances however, have been made in energy use efficiency and conservation. Progressive public utility companies are investing in diverse, alternative energy sources and shying away from building new, massive, generating facilities that present huge upfront capital requirements, less-than-glittering long-term financial returns, and dubious environmental consequences. New appliances that use electricity more efficiently than older models and high-tech management of energy usage in factories and office buildings have helped slow down projected growth rates in electricity usage. Air quality considerations in cities subjected to seasonal or year-round problems with photochemical smog (p. 494-496), such as Denver and Los Angeles, have led to the mandated introduction of oxygen-enriched gasoline to reduce unburned hydrocarbon emissions, and new vehicles utilizing non-gasoline energy sources, such as electricity from batteries and propane, are beginning to appear in most urban areas. As of June 1, 1996, reformulated, highly refined gasolines, designed to reduce air pollution and improve air quality, were required statewide in California, and with "sedate fanfare", General Motors rolled out its new electric car in early November. Furthermore, rest assured that American ingenuity is "out there" and hard at work. A local, television news report, 6/11/96, on the Big Rock Blue Marlin Tournament off North Carolina's coast, featured a fishing boat operating entirely on a soybean oil-based diesel fuel.

Methane: Fossil Fuel of Choice and Global Climate Modifier

Methane (natural gas) is emerging as a fossil fuel of choice. Its combustion adds only one "pollutant", carbon dioxide, to the atmosphere. Enormous quantities are locked up in shallow marine sediments as gas hydrates, crystalline, icelike, clathrate compounds stabilized by sea bottom hydrostatic pressures and cold temperatures; gas hydrates containing methane also occur in terrestrial, permafrost areas. This methane represents an enormous, subeconomic energy resource.

Methane is also generating new interest as an important atmospheric gas in the global warming scene (Chapter 16). Methane molecules are about ten times more efficient absorbers of infrared radiation than carbon dioxide; thus methane is attracting increasing attention as an important atmospheric greenhouse gas. The methane gas hydrates are easily decomposed by lowered pressures and/or higher temperatures, releasing methane to the surroundings and eventually to the atmosphere. Thus lowering of sea level, such as would happen during an interval of continental glaciation, would reduce sea bottom hydrostatic pressures, resulting in methane release from gas hydrates. In this case the additional methane would produce "negative feedback", reducing the percentage of energy lost from Earth by infrared radiation to space (Figure 16.4, p. 464) and causing the atmosphere to cool more slowly or to warm. Global and ocean warming are of greater concern. Increased bottom-water temperatures would trigger releases of enormous quantities of gas-hydrate methane now locked up in marine sediments. In this scenario, the additional methane would produce a "positive feedback" and act to warm the atmosphere even more.

Environmental Issues and the Nuclear Power Industry

Environmental issues and policies concerning public utilities attract plenty of media coverage, especially when nuclear electrical power generation is involved. On a visit to southern Maine, mid August, 1996, this writer was impressed with the extent of media coverage devoted to Yankee Power's Brunswick Number Two plant in Brunswick, Maine. This is one of the oldest, still-operating, nuclear generating plants in the nation. Escalating maintenance problems and mounting safety concerns form the core of widespread, public opposition to continued operation of the plant. Also, the potentially contentious and very costly issue of what to do with aging nuclear plants and their dangerously radioactive components has been yanked from under the rug and into the public arena.

Duke Power Company, one of the largest public utility companies in the Carolinas, has a major investment in nuclear generating plants. Thus the company is especially interested in issues and policies that impact on the nuclear power industry. An innovative suggestion concerning nuclear reactor fuel made by Duke Power officials was widely reported, 6/23/96, in statewide media. They proposed that large existing stocks of weapons-grade plutonium (Pu-239) be utilized in MOX, a mixed uranium and plutonium oxide reactor fuel for nuclear generating plants. The pro arguments are simple and straightforward; this would help dispose of excess, unneeded plutonium, reduce fuel costs to utilities, and by inference reduce electrical power costs to consumers.

Environmental groups criticized the proposal as self-serving, motivated only by visions of higher profits for utilities with heavy investments in nuclear generating plants. Opposition also focused on the notably dangerous characteristics of plutonium. In addition to its radioactivity, plutonium is highly toxic and burns readily in air; a "plutonium fire" spreading toxic, highly radioactive smoke and dust over nearby areas is, next to a runaway fission explosion as happened at Chernobyl, a nuclear engineer's worst nightmare. Inhaling plutonium-laced smoke is an express ticket to lung cancer. It is true that plutonium metal readily oxidizes and burns in air; it is equally obvious that plutonium in plutonium oxide has already been oxidized so the fissionable metal oxides in MOX do not represent fire hazards. Opponents also misled the public about differences between a chemical explosion, an extremely fast, exothermic, chemical oxidation in air, and a runaway, nuclear fission explosion as is evidenced by the following statement quoted from the newspaper article; "plutonium is extremely explosive and must be kept in quantities smaller than a critical mass".

The specter of burning plutonium metal is bad enough, but it palls beside the horror envisioned in an accidental nuclear explosion. Strangely enough, long-term storage and disposal problems posed by spent reactor fuel and dangerous radioactive fission products were not raised in opposition to the MOX proposal. These and other safety-related issues (p. 337-342) have effectively halted new investments in nuclear generating plants in the United States, and the State of Washington's massive default on municipal bonds sold to finance overly ambitious nuclear generating capacity hardly helped inspire public confidence in the industry.

The industry was involved in other media coverage this past summer, 1996. Congressional opponents of "pork barrel" funds designated for the Advanced Light Water Reactor project, led by Representative Mark Foley (R., Florida), argued against additional spending on the project before the House Appropriations Committee. Since 1992, the project has cost the taxpayers $378 million. A request for continued funding was part of the Energy and Water Appropriations Bill under consideration by the Committee. This request was made despite the facts that no new nuclear generating facilities had been started in the United States since the 1970s and most of the money went directly to corporate giants such as General Electric, Westinghouse, and Asea Brown Boveri Combustion Engineering.

On 7/24/96, the print and other media also covered a court decision of special interest to the nuclear power industry. The U. S. Circuit Court of Appeals for the District of Columbia ruled that the Department of Energy, DOE, must begin accepting spent nuclear fuel rods by 1998, despite the fact that no viable storage facility yet exists for this extremely hazardous material. The best candidate for such a facility is still the proposed Yucca Mountain, Nevada site (p. 340-342). Evidently, the decision hinged on the court's interpretation of the word "dispose". Utility companies could, in effect, "dispose" of their reactor waste by turning over to the DOE. The decision seemed to sidestep and ignore the substantive and highly complex social, technological, and environmental issues confronting the whole issue of high-level radioactive waste disposal.

Energy Conservation

CNN's program Science and Technology Week often include features on energy conservation and other environmental issues. A program in mid-October, 1996, featured a home in Georgia designed to be as energy efficient as possible, given today's knowledge and technology. The roof was covered with photo-voltaic shingles; on sunny days, these generated enough electricity to "run the home", charge large batteries for nighttime and cloudy days, and at times sell a small amount of electricity to the local utility. Water is initially heated by solar energy and stored in insulated underground tanks. The "pre-warmed" water can be supplied to a conventional hot-water heater and used for supplemental space heating. Thus stored solar energy is used to reduce electricity consumption on cloudy days and at night. The home was designed and positioned to efficiently exploit seasonal variations in solar radiation; year-round, solar electrical production and wintertime solar absorption are maximized and summertime solar energy absorption is minimized. The home is heavily insulated and carefully designed and constructed to minimize heat losses in cold weather and heat gains in hot weather. Finally, wells on the property provide constant-temperature groundwater for space heating and air conditioning. All groundwater utilized in the home is recharged back into the local groundwater system. In the winter months, groundwater is usually warmer than the air and is used for space heating. In the summer, the groundwater is normally cooler than the air and is employed as a heat sink for the air conditioning system.

A portion of the November 10 program focused on the looming deregulation of the utility industry, bringing intense competitive pressures for increased efficiency, reduced costs, and diverse, alternative sources of electricity. Individual homeowners and businesses will reap big savings in energy costs by investing in onsite generation of electricity, such as with solar cells, and in energy conservation. Low-wattage fluorescent lighting can reduce electricity use and lower summer air-conditioning costs because less inside heat is generated. Also mentioned was the strategy of "ice cooling". During times of low electricity demand, such as between midnight and 6:00 A. M., electricity is used to make ice. During the day when electrical usage and air-conditioning demands are highest, the melting ice contributes to interior cooling, thus lowering air-conditioning demands during peak electricity usage periods. Local power generation, conservation, and innovative methods of load management can reduce peak demand and thus reduce the need for new power plants and the risks for grid-wide brownouts or power failures.

Deregulation of Utilities

Given the strong pro-deregulatory mood in Congress, legislation to deregulate utilities is soon expected. Environmentalists are concerned that deregulation will revive a "sell as much as you can" mentality among utility industry officials, and scuttle, hard-won, largely successful programs to save energy and lower the rate of increase in energy usage. Most utility sponsored conservation programs are subsidized by slightly higher prices to some consumers, often large-volume corporate users. These programs could disappear if utilities negotiate special, lower-rate, sweetheart deals with large-volume, industrial consumers. General Motors has already announced its opposition to small rate surcharges it now pays in Michigan to fund conservation programs. In addition, some utilities are seen to be gearing up for massive advertising campaigns to promote use of their products and increase market share. On the brighter side, deregulation will increase diversity in the industry and offer more opportunities for innovative companies promoting energy savings and alternative energy sources. In addition, a viable solution to the severe problem of the massive capital expenditures required to construct new generating capacity seems nowhere in sight.

Supplemental References

Ahearne, J., 1993, The future of nuclear power: American Scientist, v. 81, no. 1 (Jan.-Feb.), p. 24-35

Asmus, P., 1993, Saving energy becomes company policy: Utilities – and the environment – profit from efficiency: The Amicus Journal, v. 14, no. 4, p. 38-42

Carlowicz, M., 1996, Gas hydrates fuel diverse interests: EOS, v. 77, no. 23, p. 219

Carlowicz, M., 1996, Warming by methane? EOS, v. 77, no. 33, p. 322

Edwards, M., 1994, Chornobyl: National Geographic Magazine, v. 186, no. 2, p. 100-115; note the different spelling from that usually shown

Kartha, S. and Grimes, P., 1994, Fuel cells; Energy conversion for the next century: Physics Today, v. 47, p. 54-61

Kvenvolden, K. A., 1993, A primer of gas hydrates: in The Future of Energy Gases, Howell, D. M., editor, U. S. Geological Survey Professional Paper 1570, p. 279-291

Kvenvolden, K. A., 1993, Gas hydrates as a potential resource – A review of their methane content: in The Future of Energy Gases, Howell, D. M., editor, U. S. Geological Survey Professional Paper 1570, p. 555-561

Environmental Awareness Inventory
Part 5: Global Change, Land Use, and Decision Making

1. Which statement concerning Earth's stratospheric ozone layer is correct? 1 pt
a. The ozone molecules selectively absorb infrared radiation from the Sun.
b. Chlorine photodissociated from manmade CFCs are its main threat.
c. In recent years during the Northern Hemisphere winter months, a hole in the ozone layer has developed over the North Pole.
d. Continued manufacture and world-wide use of ozone-destroying compounds has virtually insured that the ozone layer will disappear by the year 2050.

2. In about 1950, annual direct measurements of atmospheric carbon dioxide content were initiated at a site in Hawaii. How was the carbon dioxide content of Earth's atmosphere estimated for each of the previous 2000 years? 1 pt
a. widths of exactly-dated, annual growth rings in trees were compared from different parts of the world
b. air bubbles trapped in well-dated, annual ice layers in the Antarctic Ice Cap were analyzed for their proportion of carbon dioxide
c. supercomputers were used to calculate past values based on the Perfect Gas Law and computed average temperatures and pressures for each year in question
d. were computed from chemical equilibrium concentrations of bicarbonate ion in seawater determined by counting and measuring widths of annual or seasonal growth patterns in calcium carbonate shells of marine invertebrates such as clams, oysters, and gastropods

3. When did the last remnants of the large, Pleistocene, continental ice sheets and glaciers melt away in North America and what year marked one of the coldest years of the so-called Little Ice Age? **one answer**, 1 pt
a. about 12,000 years ago; 1400 A. D.   b. about 120,000 years ago; 3000 B. C.
c. about 12,000 years ago; 5000 B. C.   d. about 150,000 years ago; 1800 A. D.

4. Which are the two major gases that contribute to anthropogenic acidic rain, fog, and snow? **one answer**, 1 pt
a. sulfur dioxide, SO_2, and ozone, O_3 b. carbon dioxide, CO_2, and methane, CH_4
c. nitrogen dioxide, NO_2, and sulfur dioxide, SO_2
d. methane, CH_4, and hydrogen chloride, HCl

5. Which meteorological scenario is responsible for coastal atmospheric inversions, such as those that affect the Los Angeles basin area, that lead to dangerous buildups of ground-level air pollutants? 1 pt
a. Cool, ground-level sea breezes blow into the basin from the ocean while hot, dry winds aloft are blowing into the basin in a seaward direction after having crossed the mountains surrounding the basin.
b. Hot, dry, ground-level winds are blowing into the basin from the ocean and cool, moist air aloft is blowing into the basin in a seaward direction after having crossed the mountains surrounding the basin.

6. Which **one** is not a significant greenhouse gas in Earth's atmosphere? 1 pt
a. methane, CH_4 b. carbon dioxide, CO_2 c. gaseous CFCs   d. nitrogen, N_2

7. Which **one** of the following legislative goals included in the 1990 amendments to the Clean Air Act will probably be the most difficult to achieve, given our current levels of scientific understanding and technology? 1 pt
a. Establish a system of air pollution allowances/credits involving sulfur dioxide emissions that can be traded (bought and sold) under open market conditions.
b. Reduce overall sulfur dioxide emissions by 50 % by the year 2000.
c. Attain specified reductions in overall emissions of nitrogen oxides.

8. Which **one** response characterizes the pitiful, environmental story of aquatic organisms and water fowl in the Kesterson Wildlife Refuge, California? 1 pt
a. gross birth defects, poisonings, and reproductive failures due to excessive levels of organic mercury compounds accumulated from seed fungicides used in agriculture and contamination from the New Almaden mercury mine
b. birth deformities, reproductive failures, and deaths attributed to an influx of sodium cyanide-laced wastewater accidentally released from a gold-leaching facility at a gold mining operation in the foothills of the Sierra Nevada
c. numerous birth deformities and deaths due to excessive selenium dissolved in agricultural wastewaters from irrigated fields in the San Joaquin Valley

9. Which response correctly translates acronyms in the following sentence? 1 pt
Important characteristics of ERUs are included in a GIS database for inclusion in an EIS to be submitted to the EPA.
a. ERU, Environmental Risk Unit; GIS, Geologic Information Service; EIS, Environmental Issue Study; EPA, Environmental Progress Agenda
b. ERU, Environmental Resource Unit; GIS, Graphic Insight System; EIS, Environmental Information Statute; EPA, Environmental Protection Association
c. ERU, Environmental Resource Unit; GIS, Geographic Information System; EIS, Environmental Impact Statement; EPA, Environmental Protection Agency

10. Which method of site selection incorporates an analysis of ecological and aesthetic factors in addition to costs and expected benefits? 1 pt
a. physiographic determinism   b. biotic terrain analysis
c. comprehensive siting technology   d. cultural and ecological resource utilization

Answers

1. b	2. b	3. a	4. c	5. a
6. d	7. c	8. c	9. c	10. a

Performance Evaluation; Scoring

Each correct answer is worth one point; add up your total. Now be honest, and subtract one point for each lucky guess!

Score	**Awareness Ranking; advice**	INTERNET Address
8-10	**nerd**; keep up the good work, don't let up	http://envinerd.com
5-8	**blue collar achiever**; earn an A with hard work	http://envblcol.com
2-5	**slowpoke**; don't look back, clueless are right behind	http://envslpoke.com
0-2	**clueless**; work extra hard to make up for a slow start	http://envcluless.com

Global Change and Earth System Science • Chapter 16
Home Study Assignment

1. Which **two** are best suited for providing information on second-order, global-system changes?
a. chemical analyses of Amazon River water
b. studies of Native American shellfish middens, coastal areas, Pacific Northwest
c. studies of Sumerian agricultural practices
d. studies of coal, sandstone, and shale depositional patterns in the Appalachian Basin

Questions 2-7; Matching: Paleoclimatic Studies

a. studies of pollen in lake, swamp, and river floodplain sediments
b. bubbles in ice of the Greenland Ice Sheet   c. carbon-14   d. CFCs
e. dendrochronology; analyses of tree growth rings
f. microfossils in ocean floor sediments

2. ( ) might be used to estimate groundwater recharge rates in shallow, unconfined aquifers

3. ( ) can be used to estimate the past CO2 content of the atmosphere

4. ( ) provide detailed hydrologic and climatic data back to 12,000 years B. P. in some areas

5. ( ) provide information on past vegetation patterns, from which past climatic conditions and changes may be inferred

6. ( ) can be studied to estimate marine temperature conditions in the geologic past

7. ( ) useful for estimating ages of charcoal and other once-living organic materials for the past 50,000 years

8. Which response is not logically appropriate for global circulation models?
a. temperature, pressure, and wind speeds are some of the variables computed for Earth's surface and lower atmosphere; can go forward or backward in time
b. based on estimated initial values of variables and accepted equations of energy and momentum conservation
c. numerical output based on repeated computations of all variables in three-dimensional, computational cells
d. appear in fashion shows in Milan, Paris, New York, Hong Kong, and Tel Aviv, all in the same year

9. Which **two** gases comprise 99 % of Earth's atmosphere?
a. carbon monoxide   b. oxygen   c. nitrogen   d. methane

10. Which response shows the correct configuration of Earth's atmosphere?
a. tontosphere, lowermost 10 km; stratosphere; 10 to 50 km up
b. stratosphere, lowermost 10 km; topmostsphere, 10 to 50 km up
c. troposphere, lower 10 km; stratosphere, 10 to 50 km up
d. stratosphere, lower 10 km; mesosphere, above 10 km

11. Which **one** is not a contributor to the so called greenhouse warming of Earth's atmosphere?
a. N2　　　　　　　b. CH4　　　　　　c. CFCs　　　　　　d. CO2

12. Which **two** responses correctly and/or rationally describe known or projected trends in atmospheric carbon dioxide content starting with 1500 A. D.?
a. remained steady at between 270 and 280 ppm until 1820; has been rising at an accelerating rate since then
b. decreased from 300 ppm in 1500 to 155 ppm in 1850; has been steadily rising since then, reaching 350 ppm in 1990
c. projected to reach 450 ppm by 2050
d. projected to decline from current levels to 200 ppm by the year 2100

13. Which **two** trace components of today's atmosphere may have been fairly abundant in Earth's primitive, oxygen-poor atmosphere?
a. methane   b. ammonia   c. ozone   d. argon

14. Which **two** processes are driven mainly by Earth's internal heat energy?
a. plate tectonic movements   b. convective circulation of the oceans
c. volcanism   d. atmospheric wind circulation

15. Which **two** responses regarding the speed of electromagnetic radiation are correct?
a. is equal to the speed of light in empty space
b. is equal to 300,000 km/sec in transparent glass
c. yellow light travels more slowly through a transparent mineral, such as diamond, than it travels in a vacuum or empty space
d. in a given liquid or solid material, all wavelengths of electromagnetic energy travel at the same speed; that speed is slower than the speed in empty space

16. Which **one** statement concerning electromagnetic radiation is incorrect?
a. Wavelengths of weather radar are in the range from 1 to 10 cm; Doppler radar exploits the relative movements of air masses, containing liquid water mists and rain droplets, toward or away from the radar station.
b. Gamma radiation is the most energetic in the electromagnetic spectrum; gamma rays are only produced during nuclear processes, such as radioactive decay.
c. Among the wavelengths of visible light, red (0.7 µm) has a shorter wavelength than blue.
d. Electromagnetic radiation emitted from the Sun is mainly in the form of visible and ultraviolet wavelengths; Earth radiates energy in the long infrared wavelengths.

17. Assume that radiant energy is being emitted by two objects with different temperatures. Which statement is correct?
a. Radiant energy from the higher temperature object has a shorter wavelength energy spectrum than that emitted by the cooler object.
b. Radiant energy from the higher temperature object has a longer wavelength energy spectrum than that emitted by the cooler object.

18. A mass, heated only by absorbing radiant energy of specific wavelengths, emits radiant energy of its own. Which explanation is correct?
a. The emitted radiant energy has shorter wavelengths than the absorbed energy.
b. The emitted radiant energy has longer wavelengths than the absorbed energy.

19. Consider Earth's energy budget. Which **two** explanations are correct? The others are factually incorrect and/or clearly based on ignorance.
a. Approximately 35 % of incoming solar radiation is reflected back into space; an increase in cloud cover would lower that value because more heat would be trapped in the lower atmosphere.
b. Earth's internal heat arriving at the surface each year totals about 100 exojoules; this energy has little or no effect on processes in the atmosphere and hydrosphere.
c. The so-called greenhouse gases strongly absorb ultraviolet radiation; with more such gases in the atmosphere, more of Earth's nighttime radiation is absorbed in the atmosphere and prevented from being lost to outer space.
d. The ozone, O_3, molecule strongly absorbs ultraviolet light. Thus a decrease in stratospheric ozone would raise the quantity of ultraviolet radiation arriving at Earth's surface.

20. Which physical law describes the quantity of radiant energy, Q, emitted per unit area from an object with a surface temperature represented by T?
a. $Q = aT$; a is a constant b. $Q = c/T$; c is a constant
c. $Q = e^{-aT}$; a is a constant d. $Q = cT^4$; c is a constant

21. Which term refers to the percentage of light reflected from an object's surface?
a. libido b. albedo c. alfredo d. liprido

22. Which **two** statements are correct? The other two contain serious factual mistakes and/or erroneous conclusions.
a. An increase in the area covered by highly reflective clouds would increase Earth's overall albedo.
b. The formerly extensive, Pleistocene, continental ice sheets in the Northern Hemisphere were associated with global high sea levels and unusually low, average albedo in land and water-covered areas between 45° and 60° N. latitude.
c. Higher contents of the greenhouse gases will result in gradual warming of the atmosphere; this is because a higher fraction of Earth's radiant energy will be absorbed in the atmosphere.
d. The significant characteristic of greenhouse gases is that they strongly absorb radiant energy in the ultraviolet portion of the electromagnetic spectrum. Thus the flux of ultraviolet energy arriving at Earth's surface will decrease as the atmospheric content of greenhouse gases increases.

23. Which compounds are most implicated in deterioration of Earth's stratospheric ozone layer? **one answer**
a. methane and other hydrocarbon gases and vapors b. nitrogen oxides
c. carbon dioxide and sulfur dioxide d. chlorofluorocarbons

24. What is a Dobson unit?
a. 1 ppb O_3 in air b. 1 ppm O_2 in air c. 1 ppm CFC in air d. 1 ppb SO_2 in air

25. Which **two** statements concerning climate and climatic variations are supported by scientific, paleoclimatological, and/or historical studies?
a. The little Ice Age extended from about 1300 to 1500 A. D.; the coldest temperatures represented about a 1.5° C drop from the norm for the previous 400 years.
b. Barley and other crops, not capable of living in the cold climates of today, were routinely cultivated in Viking settlements in Iceland, Greenland, and North America between 800 and 1200 A. D.
c. Since 1900, average annual temperatures have increased by 2° C for North America and 1° C for land areas of the Southern Hemisphere.
d. Continued melting and reduction in the volume of the Antarctic ice sheet is a predicted result of future global warming. This melting will raise sea level and cool the surface waters of the southern hemisphere oceans.

26. Which is the **one** incorrect statement concerning ozone?
a. The chemical formula is O_3; ozone is a powerful oxidizing agent.
b. Ozone in the troposphere is produced by photochemical reactions involving nitrogen dioxide and hydrocarbon molecules; much of this tropospheric ozone is thus anthropogenic.
c. Production of stratospheric ozone is controlled by photodissociation of O_2 molecules and recombination of the two, single O molecules with O_2 molecules.
d. Ozone molecules strongly absorb electromagnetic energy in the red part of the visible spectrum; that is why a clear sky appears to be blue during daylight hours.

27. Which **two** statements concerning ozone are based on scientific studies? The others contain elements of hearsay and gross ignorance.
a. CFCs are very unstable chemical compounds; for that reason, they only persist for short times in the troposphere and stratosphere; thus increased, surface UV radiation due to ozone depletion will only last for a few years at most.
b. The southern hemisphere spring hole in Earth's protective ozone layer over Antarctica was an unexpected consequence of complex climatic and atmospheric processes; the general threat to the ozone layer had been predicted in the 1970s.
c. American support for international efforts to reduce and phase out manufacture, use, and release of CFCs represent just another surrender of sovereign American rights to an ultraliberal, international conspiracy led by the United Nations. Each American has a constitutional right to use freons without imposed limits.
d. Chlorine, derived from CFCs and other sources, is the major ozone-destroying constituent in the stratosphere. However, bromine, Br, compounds and nitrous oxide, N_2O, can also participate in ozone-destroying reactions.

28. Which **one** is probably not a serious possible consequence of depleted stratospheric ozone and increased ultraviolet radiation at Earth's surface?
a. Cyanobacteria (the photosynthetic blue-green algae) would be highly stressed and drastically reduced in abundance, leading to collapse of the marine food chain.
b. Human skin cancers and other sun-caused skin disorders would increase.
c. Important food-crop plants, such as corn, soybeans, and wheat, could be damaged, resulting in lowered per plant yields.
d. Organisms higher up the oceanic food pyramid could be adversely affected by lowered productivity of photosynthetic marine plankton.

29. Which two statements concerning ozone depletion over Antarctica are based on scientific studies and observations? The others contain one or more factual errors or serious flaws in scientific interpretation.
a. During the dark, southern hemisphere winter, compounds such as chlorine nitrate and hydrochloric acid react to form chlorine molecules and nitric acid.
b. During the southern hemisphere winter, wind speeds at heights from 10 to 30 km are very strong over the South Pole and fairly weak and variable in direction in an annular zone roughly above the circumference of the Antarctic continent.
c. Ozone depletion rates are highest during the middle of the southern hemisphere summer when sunlight is most intense and chlorine nitrate and hydrochloric acid concentrations are at a maximum in the lower part of the troposphere.
d. With the first sunlight of the southern hemisphere spring, Cl2 molecules photo dissociate into Cl atoms; these in turn react with O2 and/or O3 molecules, destroying existing ozone molecules and severely retarding ozone formation.

Questions 30-35; Matching; Molecules and Chemical Formulas
a. O3 b. ClONO2 c. HCl d. HNO3 e. CL2 f. ClO
30. ( ) chlorine   31. ( ) ozone   32. ( ) hydrochloric acid
33. ( ) nitric acid 34. ( ) chlorine oxide 35. ( ) chlorine nitrate

36. Which one denotes the acid-base scale applicable to aqueous solutions?
a. Ha   b. pH   c. Ac   d. pA

Questions 37-41; Natural Aqueous Solutions/Acidity Scale Terms and Values
a. moderately to strongly basic solution   b. weakly basic solution
c. distilled water   d. weakly acidic solution   e. strongly acidic solution
37. ( ) neutral; pH = 7   38. ( ) acid solution in a car battery; pH = 1
39. ( ) ammonia solution in water; pH = 12
40. ( ) dilute, aqueous, sodium bicarbonate solution; pH = 8.2
41. ( ) vinegar and some acidic precipitation; pH = 3 to 4

42. Which are the correct names and chemical formulas for the two major acidic gases and components of strongly acidic precipitation? **two answers**
a. sulfur dioxide, SO2; sulfuric acid, H2SO4
b. hydrogen chloride, HCl; hydrochloric acid, HCl
c. carbon dioxide, CO2; carbonic acid, H2CO3
d. nitrogen dioxide, NO2; nitric acid, HNO3

43. Which two processes now account for most of the sulfur dioxide and nitrogen oxides in the atmosphere?
a. smelting and refining of aluminum and titanium metal   b. coal burning
c. ionization from lightning strikes and around ultrahigh voltage power lines
d. combustion of gasoline and fuel oil

44. Which **one** is the international agreement to limit future emissions and manufacture of CFCs?   a. London Protocol, 1985   b. Paris Accord, 1975
c. Montreal Protocol, 1987   d. Helsinki Accord, 1982

45. Which **two** statements about acidic precipitation are supported by sound investigations? The others are incorrect and/or lacking in merit.
a. Red fir and red spruce trees have been dying from New England to North Carolina; acidic rain, acidic fog, and ozone are implicated as major suspects.
b. Government-owned utilities, such as the Tennessee Valley Authority, TVA, in general, have far better records of reducing SO2 emissions than private sector, investor-owned utilities.
c. Since 1990, emissions of SO2 from metal smelters and coal-burning electric power plants in the Unites States have increased significantly, worsening the problem of acidic precipitation.
d. Ammonia, released into the atmosphere from large, cattle and hog raising facilities, eventually oxidizes to NO2 and contributes to acidic precipitation.

46. Which **two** responses correctly state how mineralogical and chemical properties of soil or bedrock impact on the soil moisture and shallow groundwater?
a. Carbonate-rich bedrock and soils have very high buffering capacities; however, the carbonates will slowly dissolve away as acidic waters are neutralized.
b. Toxic metals, such as cadmium and lead, are effectively immobilized in well-buffered, low-pH soil water and groundwater.
c. Granitic bedrock and naturally-acidic forest soils in New England and the Scandinavian countries provide a strong buffering capacity to maintain soil moisture and shallow groundwater with a high pH.
d. Areas in the western plains and mountainous west owe their low, acid precipitation sensitivity to well-buffered, calcite-bearing, B soil horizons. Calcium carbonate does not usually accumulate in soils formed under the more humid conditions of the eastern United States.

47. Which **two** statements concerning excessively acidified, surface, aquatic environments are supported by scientific studies and observations. The others are hearsay and/or without substantive, supporting evidence.
a. Birth defects and reproductive failures in fish and amphibians usually occur before a lake or river is acidified enough for the water to be lethal to the adult organisms.
b. Most spring-fed lakes and ponds occupying Florida sinkholes are weakly buffered against acidification and have experienced severe ecological deterioration due to acidic precipitation.
c. Most phosphates are more soluble in acidified waters than in neutral or slightly basic water; thus in acidified lakes, an important nutrient for supporting growth of algae and aquatic plants is lost in lake overflow, instead of remaining behind to naturally fertilize the lake aquatic system.
d. Aluminum and other metals, such as copper, zinc, and iron, are much less soluble in acidic, low-pH waters than in neutral or slightly alkaline waters.

48. Which **two** effects were observed during the next few years after the powerful volcanic eruptions of Tambora, 1815, El Chichón, 1982, and Pinatubo, 1991?
a. abnormal global warming b. brilliantly colored sunsets and sunrises
c. some global cooling
d. unusually rapid deterioration of stone buildings in Europe and North Africa

49. Which one statement concerning airborne particulates is decidedly contrary to known observations and scientific findings?

a. In general, fine size particulates, such as those in tobacco and oil-fire smokes, are more dangerous to human lungs than coarser particulates such as pollen.

b. The volcanic eruptions of Pinatubo, Philippine Islands, 1991, and El Chichón, Mexico, 1982, propelled large quantities of NO_2 aerosol mists into the lower stratosphere, resulting in worldwide acidic clouds and fog for the next few years following the eruptions.

c. The oil field fires in Kuwait, 1991, produced low-level, dark regional plumes of unburned, particulate matter; these smoke clouds had little effect on worldwide weather or regional weather patterns.

d. Atmospheric particulates act as condensation nuclei for raindrops and fog droplets.

50. On the U. S. map provided, plot and label the following areas or locations.

a. Adirondack Mountains
b. Ohio River Valley
c. Wheeling, WV
d. northeastern Iowa
e. Delaware River Basin

51. On the world map provided, accurately plot and label the following locations.

a. Sweden
b. Norway
c. Kuwait
d. Finland
e. Nova Scotia, Canada
f. Greece
g. northern Persian Gulf

Environmental Geology Quiz • Chapter 16

1. Which **one** will not provide insight on second-order, global-system changes?
a. chemical analyses of Amazon River water   b. studies of agriculture in Sumeria
c. studies of Native American shellfish middens, coastal areas, Pacific Northwest

Questions 2-5; Matching: Paleoclimatic Studies
a. bubbles in the Greenland Ice Sheet   b. carbon-14   c. CFCs   d. dendrochronology

2. (   ) might be used to estimate recharge rates in shallow, unconfined aquifers

3. (   ) can be used to estimate the CO2 content of the atmosphere in the past

4. (   ) may provide detailed, hydrologic and climatic data for the last 12,000 years

5. (   ) useful for estimating ages of once-living materials up to 50,000 years old

6. Which **one** response is not logically appropriate for global circulation models?
a. numerical output based on repeated computations of variables, such as temperature and pressure, in three-dimensional, computational cells
b. based on estimated initial values of variables and accepted equations of energy and momentum conservation
c. appear annually in fashion shows in Milan, Paris, New York, and Tokyo

7. Which response shows the correct configuration of Earth's atmosphere?
a. tontosphere, lowermost 10 km; stratosphere; 10 to 50 km up
b. stratosphere, lowermost 10 km; topmostsphere, 10 to 50 km up
c. troposphere, lower 10 km; stratosphere, 10 to 50 km up
d. stratosphere, lower 10 km; mesosphere, above 10 km

8. Which **two** gases comprise 99 % of Earth's atmosphere?
a. methane   b. oxygen   c. carbon monoxide   d. nitrogen

9. Which **one** is not a contributor to the so called greenhouse warming of Earth's atmosphere?   a. CO2   b. CH4   c. CFCs   d. N2

10. Which **one** response correctly describes known or projected trends in atmospheric carbon dioxide content?
a. remained fairly steady at about 275 ppm between 1500 and 1820; has been rising at an accelerating rate since then
b. decreased from 300 ppm in 1500 to 155 ppm in 1850; has been steadily rising since then, reaching 350 ppm in 1990; projected to reach 450 ppm by 2050
c. projected to rise from current levels to 200 ppm by the year 2100

11. Which **two** gases may have been fairly abundant in Earth's primitive, oxygen-poor atmosphere?   a. ammonia   b. ozone   c. methane   d. argon

12. Which **two** responses regarding the speed of electromagnetic radiation are correct?   a. is equal to 300,000 km/sec in transparent glass
b. is equal to the speed of light in empty space
c. in a given liquid or solid material, all wavelengths of electromagnetic energy travel at the same speed; that speed is slower than the speed in empty space
d. yellow light travels more slowly through a transparent mineral, such as diamond, than it travels in a vacuum or empty space

13. Which **one** statement concerning electromagnetic radiation is incorrect?
a. Wavelengths of weather radar are in the range from 1 to 10 cm; Doppler radar exploits the relative movements of air masses, containing liquid water mists and rain droplets, toward or away from the radar station.
b. Gamma radiation is the least energetic in the electromagnetic spectrum; gamma rays are only produced during nuclear fission processes.
c. Radiation emitted from the Sun is mainly in the form of visible and ultraviolet wavelengths; Earth radiates energy in the long infrared wavelengths.

14. A mass, heated only by absorbing radiant energy of specific wavelengths, emits radiant energy of its own. Which explanation is correct?
a. The emitted radiant energy has longer wavelengths than the absorbed energy.
b. The emitted radiant energy has shorter wavelengths than the absorbed energy.

15. Which **one** explanation concerning Earth's energy budget is correct?
a. Earth's internal heat arriving at the surface each year totals about 100 exojoules; this energy has little or no effect on processes in the atmosphere and hydrosphere.
b. Approximately 35 % of incoming solar radiation is reflected back into space; an increase in cloud cover would lower that value because more heat would be trapped in the lower atmosphere.
c. The so-called greenhouse gases strongly absorb ultraviolet radiation; with more such gases in the atmosphere, more of Earth's nighttime radiation is absorbed in the atmosphere and prevented from being lost to outer space.

16. Which physical law describes the quantity of radiant energy, Q, emitted per unit area from an object with a surface temperature represented by T?
a. $Q = aT$; a is a constant   b. $Q = c/T$; c is a constant
c. $Q = e^{-aT}$; a is a constant   d. $Q = cT^4$; c is a constant

17. Which term refers to the percentage of light reflected from an object's surface?
a. liprido   b. alfredo   c. albedo   d. libido

18. Which **one** statement is correct?
a. The Earth's albedo will decrease as the atmospheric content of CO_2 increases.
b. The formerly extensive, Pleistocene, continental ice sheets in the Northern Hemisphere were associated with global high sea levels and unusually low, average albedo in land and water-covered areas between 45° and 60° N. latitude.
c. An increase in the area of highly reflective clouds would increase Earth's albedo.

19. Which gases are most implicated in deterioration of Earth's stratospheric ozone layer? **one answer**   a. nitrogen oxides   b. methane and other hydrocarbon gases
c. carbon dioxide and sulfur dioxide   d. chlorofluorocarbons

20. Which is the **one** incorrect statement concerning ozone?
a. Ozone molecules strongly absorb electromagnetic energy in the red part of the visible spectrum; that is why a clear sky appears to be blue during daylight hours.
b. Ozone in the troposphere is produced by photochemical reactions involving nitrogen dioxide and hydrocarbons; tropospheric ozone is thus anthropogenic.
c. Production of stratospheric ozone is controlled by photodissociation of O_2 molecules and recombination of the two, single O atoms with O_2 molecules.

21. Which **two** processes now account for most of the sulfur dioxide and nitrogen oxides in the atmosphere?   a. combustion of gasoline and fuel oil
b. smelting and refining of aluminum metal   c. coal burning
d. ionization from lightning strikes and ultrahigh voltage power lines

22. Which **one** statement is not supported by scientific and/or historical studies?
a. The little Ice Age extended from about 1300 to 1500 A. D.; the coldest temperatures represented about a 1.5° C drop from the norm for the previous 400 years.
b. Barley and other crops, not capable of surviving in the cold climates of today, were routinely cultivated in Viking settlements in Iceland, Greenland, and North America between 800 and 1200 A. D.
c. Continued melting and reduction in the volume of the Antarctic ice sheet is a predicted result of future global warming. This melting will raise sea level and cool the surface waters of the southern hemisphere oceans.

23. What is a Dobson unit?   a. 1 ppb O3 in air   b. 1 ppm O2 in air   c. 1 ppb SO2 in air

24. Which **one** statement concerning ozone is not supported by scientific studies?
a. CFCs are very unstable chemical compounds; for that reason, they only persist for short times in the troposphere and stratosphere; thus increased, surface UV radiation due to ozone depletion will only last for a few years at most.
b. The southern hemisphere spring hole in Earth's protective ozone layer over Antarctica was an unexpected consequence of complex climatic and atmospheric processes; the general threat to the ozone layer had been predicted in the 1970s.
c. Chlorine, derived from CFCs and other sources, is the major ozone-destroying constituent in the stratosphere. However, bromine, Br, compounds and nitrous oxide, N2O, can also participate in ozone-destroying reactions.

25. Which **one** is probably not a serious possible consequence of depleted stratospheric ozone and increased ultraviolet radiation at Earth's surface?
a. Cyanobacteria (photosynthetic blue-green algae) would be highly stressed and drastically reduced in abundance, leading to collapse of the aquatic food chain.
b. Human skin cancers and other sun-caused skin disorders would increase.
c. Important food-crop plants, such as corn, soybeans, and wheat, could be damaged, resulting in lowered world food production.

26. Which **one** statement concerning ozone depletion over Antarctica is not supported by scientific studies and observations?
a. During the dark, southern hemisphere winter, compounds such as chlorine nitrate and hydrochloric acid react to form chlorine molecules and nitric acid.
b. With the first sunlight of the southern hemisphere spring, Cl2 molecules photo dissociate into Cl atoms; these in turn react with O2 and/or O3 molecules, destroying existing ozone molecules and severely retarding ozone formation.
c. Ozone depletion rates are highest during the middle of the southern hemisphere summer when sunlight is most intense and chlorine nitrate and hydrochloric acid concentrations are at a maximum in the lower part of the troposphere.

27. Which international agreement limited emissions and manufacture of CFCs?
a. London Treaty, 1985   b. Helsinki Accord, 1982   c. Montreal Protocol, 1987

Questions 28-30; Matching; Molecules and Chemical Formulas

a. O3   b. ClONO2   c. ClO

28. ( ) chlorine oxide   29. ( ) ozone   30. ( ) chlorine nitrate

Questions 31-35; Natural Aqueous Solutions/Acidity Scale Terms and Values

a. weakly basic solution   b. neutral solution   c. weakly acidic solution

31. ( ) distilled water   32. ( ) dilute, sodium bicarbonate solution

33. ( ) pH = 4 to 5   34. ( ) vinegar and some acidic precipitation   35. ( ) pH = 7

36. Which are the **two** major gases and acids of strongly acidic precipitation?
a. SO2, H2SO4   b. NO2, HNO3   c. HCl, HCl   d. CO2, H2CO3

37. Which **one** statement about acidic precipitation is incorrect and lacking in merit?
a. Ammonia, released into the atmosphere from large, cattle and hog raising facilities, eventually oxidizes to NO2 and contributes to acidic precipitation.
b. Since 1990, SO2 emissions from metal smelters and coal-burning power plants in the U. S. have increased significantly, worsening the acid precipitation problem.
c. Red fir and red spruce trees have been dying from New England to North Carolina; acidic rain, acidic fog, and ozone are implicated as major suspects.

38. Which **two** responses are supported by scientific studies and observations?
a. Birth defects and reproductive failures in fish and amphibians usually occur before a lake is acidified enough for the water to be lethal to the adult organisms.
b. Toxic metals, such as cadmium and lead, are effectively immobilized in well-buffered, low-pH soil water and groundwater.
c. Humus-rich, moist forest soils and granitic bedrock provide strong buffering to maintain slightly alkaline pH conditions in the soil and shallow groundwater.
d. Most phosphates are more soluble in acidified waters than in neutral or slightly alkaline waters; thus acidified lakes lose this important nutrient in stream outflow much more readily than do normal lakes.

39. Which **two** effects were observed during the next few years after the powerful volcanic eruptions of Tambora, 1815, El Chichón, 1982, and Pinatubo, 1991?
a. weakly acidic rains in the tropics   b. brilliantly colored sunsets and sunrises
c. slight global cooling   d. unusually rapid deterioration of stone buildings in Asia

40. Which **one** statement concerning airborne particulates is not correct?
a. In general, coarser particulates such as pollen are more dangerous to human lungs than finer size particulates, such as those in tobacco and oil-fire smoke.
b. The oil field fires in Kuwait, 1991, produced low-level, regional, smoke plumes of unburned, particulate matter; these had little effect on regional weather patterns.

41. Which **one** denotes the acid-base scale applicable to aqueous solutions?
a. pA   b. Ha   c. pH   d. Ac

42. On the map/maps provided, plot and label the following areas or locations.
a. Adirondack Mountains   b. northeastern Iowa   c. Delaware River Basin
d. Greece   e. Norway   f. Kuwait   g. Finland

Answers; Home Study Assignment
Global Change and Earth System Science • Chapter 16

1. b, c	2. d	3. b	4. e	5. a
6. f	7. c	8. d	9. b, c	10. c
11. a	12. a, c	13. a, b	14. a, c	15. a, c
16. c	17. a	18. b	19. b, d	20. d
21. b	22. a, c	23. d	24. a	25. a, b
26. d	27. b, d	28. a	29. a, d	30. e
31. a	32. c	33. d	34. f	35. b
36. b	37. c	38. e	39. b	40. d
41. a	42. a, d	43. b, d	44. c	45. a, d
46. a, d	47. a, c	48. b, c	49. b	

Answers: Environmental Geology Quiz • Chapter 16

1. a	2. c	3. a	4. d	5. b
6. c	7. c	8. b, d	9. d	10. a
11. a, c	12. b, d	13. b	14. a	15. a
16. d	17. c	18. c	19. d	20. a
21. a, c	22. c	23. a	24. a	25. a
26. c	27. c	28. c	29. a	30. b
31. b	32. a	33. c	34. c	35. b
36. a, b	37. b	38. a, d	39. b, c	40. a
41. c				

Commentary/Media Watch

Global Climatic Change; Water Vapor and Methane

Current ideas (Carlowicz, 1996a) related to abrupt, significant shifts in global climates are focused on changes in the water vapor content of the atmosphere. During the Pleistocene, mountain glaciers in tropical areas evidently reached fairly low elevations before melting, suggesting that air temperatures aloft in the tropics were much cooler than today. After the northern hemisphere ice caps had disappeared, much warmer conditions soon prevailed, preventing the glaciers from forming or causing them to melt at much higher elevations.

Wallace Broecker, Columbia University, identifies water vapor as the only logical candidate with the required heat capacity and potential for high-enough abundance variations to drive such recently-documented, rapid changes in global climate. Warmer, tropical, ocean-surface water temperatures would raise evaporation rates, atmospheric water vapor contents, and poleward heat transfer. Higher water vapor contents would also contribute to greenhouse warming of the atmosphere. Cooler temperatures would lower evaporation rates, atmospheric water vapor contents, and air-mass dew points, thus reducing net atmospheric heat transfer and greenhouse warming associated with water vapor. Thus once a temperature-change trend is established, increasing water vapor contents in the atmosphere would provide positive feedback for warming while decreasing abundances would provide positive feedback for cooling.

Current research findings on global climatic change (Carlowicz, 1996b) also focus on methane, a gas with ten time the infrared absorption rate as carbon dioxide. During a warming event, increased ocean-water temperatures would cause decomposition of methane clathrate compounds (gas hydrates) in shallow sediments, resulting in increased atmospheric methane contents and greenhouse warming. Thus methane feeds back positively on global warming. Methane's effect during a cooling trend is more complicated, especially if sea level is lowered by ice cap growth. In this case, cooler seawater temperatures stabilize gas hydrates, so biogenic methane is retained in marine sediments. However, as sea level drops, lowered hydrostatic pressures on shallow-water sediments destabilize existing gas hydrates and allow the methane to escape to the atmosphere. Thus enough methane might be released from gas hydrates in shallow-marine environments to slow down or even reverse a global cooling trend associated with glaciation.

The Ozone Hole and Stratospheric Chlorine Contents

The world-wide ban on freon manufacture and significant declines in freon use are finally showing dividends as stratospheric chlorine contents are leveling off and beginning to declined. In early June, newspapers and CNN both reported that ozone-destroying molecules in the stratosphere were beginning to decline for the first time since the Antarctic ozone hole was observed. Both sources also predicted significant improvement in the ozone depletion problem within the next ten years. On November 7, 1996, Dr. Rumen Bojkov, of the World Meteorological Organization, stated that the southern hemisphere spring, 1996, ozone hole was slightly smaller than the record-size hole observed in 1995, but the size difference was insignificant. However, he predicted that the ozone depletion rate should peak sometime between 2001 and 2005, assuming continued strict compliance with international agreements. One ominous development in the ozone-depletion situation is that illicit drug manufacturers are shifting to freon (a chlorofluorocarbon) as the solvent of choice in the chemical processing of methamphetamines and other illegal drugs. The chlorofluorocarbons are being smuggled into the U. S., raising concerns that they are still being manufactured, illegally or under government protection, in one or more foreign countries. Needless to say, freon used in criminal drug laboratories is not being recycled. Authorities hope that such usage is on a small-enough scale that it will not derail the positive progress made to date.

Worst Drought in Years; Yet No Dust Bowl on the Southern Plains

The severe drought conditions of fall 1995 through spring 1996 in the southwest and southern plains temporarily drove wheat futures prices to record levels in the early summer of 1996 and merited an article in Time Magazine (1st week June), comparing the then current drought threat to the infamous Dust Bowl days of the 1930s. A week or so later, an Associated Press article on the same topic was featured in the Sunday Business Section of Greenville's local newspaper. The article chronicled the severe economic and personal losses on the southern plains brought on by the extremely dry winter and spring, specifically in the Oklahoma Panhandle, an area in the heart of the original Dust Bowl. The writer suggested that despite the record-low rainfall totals, sound conservation practices had prevented a second Dust Bowl. However, only passing mention was made of the unusually long-lasting period of strong winds during the original Dust Bowl drought. It remains to be seen how effective these conservation practices would have been should Dust Bowl wind conditions have accompanied the recent drought.

Drought and Famine in the Sahel

On the international scene, brief media coverage in late July reported that drought-induced famine conditions were threatening the west African country of Burkina Faso. Reports from the capital city, Ouagadougou, complained that legally and otherwise, greedy merchants were acquiring food distributed by the government to avert starvation and selling it for big profits in neighboring Mali, also suffering from drought conditions and a scarcity of food. Both countries lie in the Sahel, a region on the southern edge of the Sahara that is well-known for periodic, extended drought conditions and ongoing problems with desertification (see p. 78-79 and Figure 3.18).

Global Warming

Global warming attracts subdued, but steady media attention. Earth Matters, shown each Sunday on CNN, devoted its September 1, 1996, program entirely to global warming. Earth Matters features well-documented, educational programming focused on the earth and environmental sciences. Information on past and future programming can be assessed through CNN's home page at http://CNN.com (click Earth).

Local newspapers, 7/8/96, covered a follow-up meeting to the 1992 Rio de Janeiro conference of the United Nations Intergovernmental Panel on Climatic Change. Meeting panelists predicted a 3.6 °C rise in temperature over the next 100 years and noted the many dangers posed by global warming. The potential spread of diseases and parasites now restricted to the tropics was specifically cited as being of grave concern. Local newspapers, 7/14/96, also found space for an Associated Press article based on a scientific paper by NASA scientist, Jim Hansen (Geophysical Research Letters, 1996). On the basis of worldwide temperature data, he concluded that average temperatures during 1995 exceeded those of 1990, making 1995 the warmest year yet on record. He further noted that by 1995, global temperatures had completely recovered from a temporary downward jog caused by atmospheric particulates from the 1991 Pinatubo eruption.

Supplemental References

Alley, R. and others, 1996, Twin ice cores from Greenland reveal history of climatic change, more: EOS, v. 77, no. 22, p. 209-210

Ausubel, J. H., 1996, Can technology spare the Earth? American Scientist, v. 84, no. 2, March/April, p. 166-178

Bates, R. and Wolff, E. W., 1995, Interpreting natural climate signals in ice cores: EOS, v. 76, no. 47, November, p. 477 & 482-483

Berner, E. K. and Berner, R. A., 1996, Global environment: Water, air, and geochemical cycles: Prentice Hall, Upper Saddle River, NJ, 376 p.

Broad, W. J., 1995, Hot vents in the sea floor may drive El Niño: in Themes of the Times; The Changing Earth, Fall 1996 edition; distributed by Prentice Hall, Upper Saddle River, NJ, p. 3

Carlowicz, M., 1996a, Did water vapor drive climatic cooling? EOS, v. 77, no. 33, p. 321-322

Carlowicz, M., 1996b, Warming by methane? EOS, v. 77, no. 33, p. 322

Gates, W. L., 1992, AMIP: The atmospheric model intercomparison project: Bulletin American Meteorological Society, v. 73, p. 1962-1970; http://www.agu.org/eos_elecas96069e.html

Ingram, B. L., Ingle, J. C., and Conrad, M. E., 1996, A 2000 yr record of Sacramento-San Joaquin river inflow into San Francisco Bay estuary, California: Geology, v. 24, no. 4, p. 331-334

Issar, A. S., 1995, Climatic change and the history of the Middle East: American Scientist, v. 83, no. 4 (July-August), p. 350-355

Kinzig, A. P. and Socolow, R. H., 1994, Human impacts on the nitrogen cycle: Physics Today, v. 47, p. 24-31

Kiehl, J. T., 1994, Clouds and their effects on the climate system: Physics Today, v. 47, p. 36-42

Lynch, C. F., 1996, Global warming; Warm up to the idea; Global warming is here: The Amicus Journal, v. 18, no. 1, p. 20-25

MacKenzie, F. T. and MacKenzie, J. A., 1995, Our changing planet Earth: Earth system science and global environmental change: Prentice Hall, Upper Saddle River, NJ, 387 p.

Stevens, W. K., 1995, In rain and temperature data, new signs of global warming: in Themes of the Times; The Changing Earth, Fall 1996 edition; distributed by Prentice Hall, Upper Saddle River, NJ, p. 5

Sullivan, W., 1995, New theory on ice sheet catastrophe is the direst one yet: in Themes of the Times; The Changing Earth, Fall 1996 edition; distributed by Prentice Hall, Upper Saddle River, NJ, p. 6

Toggweiler, J. R., 1994, The ocean's overturning circulation: Physics Today, v. 47, p. 45-50

Turekian, K. K., 1996, Global environmental change: Prentice Hall, Upper Saddle River, NJ, 200 p.

Vitousek, P., D'Antonio, C., Loope, L., and Westbrook, R., 1996, Biological invasions as global environmental change: American Scientist, v. 84, p. 468-478

Air Pollution • Chapter 17; Home Study Assignment

Questions 1-4; Matching; Dangerous Smog Events

a. hometown of sports greats Tony Dorsett, Stan Musial, and the killer smog of 48
b. site of dangerous air pollution event in 1930
c. methyl isocyanate leaks, some small and inconsequential, one large and disastrous
d. site of sulfuric acid-rich, killer smog of 1952

1. ( ) Meuse River Valley, Belgium   2. ( ) Bhopal, India and West Virginia
3. ( ) Donora, Pennsylvania; Monongahela River Valley   4. ( ) London, U. K.

5. Which two factors are most responsible for greatly reducing the likelihood of a repetition of the London killer smog event and its numerous deaths attributed to acute respiratory stress?

a. Since that event, heating oil and natural gas have largely replaced coal in home heating and many industrial uses.
b. Over the next few years after the event, private industry in the U. K. voluntarily invested heavily in air pollution abatement.
c. Londoners have since been equipped with particulate filters and breathing apparatus, protecting them from the dangerous effects of strongly acidified smog.
d. The dense fogs still roll in, but government-imposed air quality standards now help limit quantities of particulates and sulfur dioxide emitted to the atmosphere.

6. Which two statements are correct and/or valid interpretations concerning the 1952 London smog event?

a. Sulfur dioxide levels peaked at about 0.75 parts per million on the fourth day of the smog event.
b. Airborne particulate concentrations of about 1.75 mg/cubic meter lasted for over 24 hours. According to TSP ratings used in U. S. cities today, this level of airborne particulates would be regarded as highly injurious to health.
c. Highest daily death rates occurred in the first two days of the event.
d. An eight-day-long, atmospheric inversion event, featuring warm moist surface air overlain by cold, dry air aloft, provided weather conditions conducive to trapping atmospheric pollutants near the surface.

7. Which air pollutant is mostly anthropogenic, being formed during combustion of fossil fuels?   a. CO   b. SO2   c. hydrocarbons   d. NO2

8. Which air pollutant in the lower atmosphere is naturally produced during lightening strikes and by photochemical reactions involving hydrocarbons, oxygen, and nitrogen?   a. SO2   b. O3   c. particulates   d. CO

9. Which one is a primary gaseous air pollutant emitted from a stationary, point source?

a. sulfur dioxide; stack of a coal-burning electrical power plant
b. ozone; ground-hugging, Los Angeles smog
c. nitrogen oxides; autos on I-95 between Miami, FL, and Maine
d. particulate-laden smoke from a poorly tuned diesel locomotive

Questions 10-17; Review of Air Pollutants and Their Chemical Formulas

a. C8H18   b. NH3   c. SO2   d. NO2
e. CO   f. O3   g. H2S   h. HF

10. ( ) produced when hydrocarbons and other fossil fuels are burned at high temperature in air; nitrogen dioxide

11. ( ) produced by combustion of pyrite and organically-bound sulfur in coals; sulfur dioxide

12. ( ) released in some high temperature industrial processes and combustion; ingestion of contaminated leaves and grasses can seriously damage teeth and bones of deer and other wildlife; hydrogen fluoride

13. ( ) a powerful chemical oxidizer; a reaction product of photochemical smog that irritates eyes and lungs and is very detrimental to healthy plant growth; ozone

14. ( ) odoriferous gas produced naturally in oxygen-deficient environments such as bottom waters and sediments in swamps and bogs; hydrogen sulfide

15. ( ) a hydrocarbon; a major component of gasoline; octane

16. ( ) a silent killer because of its strong affinity for hemoglobin; produced during inefficient, less-than-complete combustion of hydrocarbons and other fossil fuels; carbon monoxide

17. ( ) produced naturally by bacterial action in outdoor latrines; basic ingredient used in fertilizer manufacture; leaking railcars and tank trucks can be big trouble; ammonia

18. Which two statements concerning the likelihood of inversions favorable for smog buildup are correct and based on sound meteorological principles?
a. Along the East Coast from Florida to Maine, inversion conditions occur, on average, less than 10 days each year, usually in the fall and winter months.
b. Summer, intermontane valley inversions are very common in the heavily urbanized Salt Lake Valley, Utah.
c. Spring and summer month inversions occur, on average, four or five days each year in Texas and the high plains of eastern New Mexico.
d. The Los Angeles basin has its most frequent and severe inversion events during the late summer and early fall when cool air from onshore sea breezes is overridden by hot, dry air flowing seaward from interior desert areas.

19. Which two characterize the brown air type smog?
a. photochemical reactions involving auto-produced hydrocarbons and nitrogen oxides are of central importance
b. includes ozone as a nasty air pollutant
c. mainly related to burning sulfur-bearing fuels such as coal and heavy fuel oil
d. smoke particulates are a major component

20. As of 1996, which large urbanized area on average has the better air quality and can look forward to continued gradual improvements of air quality?
a. Mexico City, Valley of Mexico   b. Los Angeles, Los Angeles Basin

21. Which **two** statements concerning urban microclimatological effects are fairly well-documented and not just based purely on a lack of knowledge and/or personal beliefs?
a. Heavily built, urban areas emit enough local, non-solar heat to slightly warm the ground level air compared to air in contact with more open, vegetated lands in the surrounding rural and suburban residential areas.
b. Highly urbanized areas generally have higher, average, near-surface wind speeds than outlying rural areas because there are far fewer trees and shrubs to obstruct the moving air.
c. The urban dust dome effect refers to cool, dust-laden, low-level breezes blowing into a highly urbanized area to replace warmer, relatively clear air rising over the central city area.
d. Particulate-laden air moving from a major urban area through a nearby rural area can increase precipitation from storm events and increase the incidence of fog and low-level cloudiness.

22. Why should mid-latitude, highly urbanized areas receive slightly less incoming ultraviolet energy than nearby rural areas at the same elevation? **two answers**
a. Relative to nearby rural areas, cities experience slightly longer periods of cloudiness and fog.
b. Warmer, ground-level city air absorbs more UV radiation per unit volume than cooler, ground level rural air.
c. Manmade building materials such as asphalt, roofing materials, and concrete reflect more UV radiation than grass, shrubs, and trees.
d. In general, concentrations of atmospheric particulate matter are usually higher over cities than over nearby rural areas.

23. Which **one** is not a specific goal or legislative action set forth in the 1990 Federal Clean Air Act?
a. reduce overall SO2 emissions by 50 % by the year 2000
b. reduce emissions of most toxic and carcinogenic air pollutants by 90 % by the year 2000
c. sets up allowances and tradable credits concerning sulfur dioxide emissions
d. calls for national, annual, pollution fees on all vehicles by the year 2000; the fees will depend on a vehicle's fuel consumption rating and the measured effectiveness of its emission control systems

24. Which response is most appropriate concerning fluidized-bed combustion and furnace injection air pollution abatement methods?
a. natural gas is the fuel; silica sand is added to immobilize unburned hydrocarbons as silicon carbide
b. waste paper and combustible refuse are the fuel; calcium sulfate is added to immobilize carbon dioxide and carbon monoxide as calcium carbonate
c. coal is the fuel; SO2 is the air pollutant; calcium carbonate, limestone is added to immobilize the SO2 as calcium sulfate
d. coal is the fuel; nitrogen dioxide is the air pollutant; ammonia is added to immobilize the NO2 as ammonium nitrate

25. What is meant by the acronym NAAQS?
a. national ambient air quality standard   b. national air alert quality statute

26. What is meant by the acronyms PSI and TSP in an air quality context?
a. particulates per square inch; tons of sulfur dioxide particles
b. pollution stipend interchange; tactical suspect population
c. particulates standard issue; technical support personnel
d. pollution standards index; total suspended particulates

27. Which two statements concerning pollution abatement are correct? The others contain factual errors.
a. Ammonium sulfate fertilizer is a possible useful byproduct of SO2 pollution abatement.
b. Precipitators are installed to efficiently remove sulfur dioxide from coal-combustion flue gases.
c. Lime, CaO, and limestone, CaCO3, are used to precipitate SO2 as insoluble calcium sulfates in flue gas scrubbers.
d. In the post 1990s Japan, personal particulate masks and sulfur dioxide absorption filters are still widely used to protect against severe, day-to-day air pollution.

28. Which two correspond to Emergency Air Quality Levels? **two answers**
a. exposure to 800 µg/cubic meter of SO2 for 24 hours   b. PSI value of 400
c. exposure to 1000 µg/cubic meter of ozone for 1 hour
d. exposure to 300 µg/cubic meter TSP over a continuous, 24-hour period

29. Which are the two most reasonable, defensible statements concerning air pollution, pollution abatement, and associated costs?
a. Air pollution abatement is just another tax-and-spend, liberal subterfuge to incur additional governmental spending, an enlarged bureaucracy, raise taxes, and assault private business and the free enterprise system.
b. Significant reductions in acid precipitation will result in lowered maintenance costs for most exterior surfaces, outdoor metal structures such as lamp and sign posts, highway guard rails, and exposed steel in bridges.
c. High-paying jobs are more important than good health and clean air. Thus outdated, obsolete industries and new, entrepreneurial enterprises should be exempted from compliance with heavy-handed air pollution regulations.
d. As SO2 from identified point sources, such as metal smelters and coal-fired generating plants is cleaned up, further reductions from dispersed, mobile and fugitive sources will probably be more difficult and expensive to accomplish.

30. On the U. S. map provided, plot and label the following areas or locations.

a. Chicago, IL	b. West Virginia	c. Four-Corners area
d. Los Angeles, CA	e. Ohio River Valley	f. northwestern Indiana
g. Donora, PA; lower Monongahela River Valley		h. southern Appalachians

31. On the world map provided, accurately plot and label the following locations.

a. Japan	b. Belgium	c. London, U. K.	d. Bhopal, India
	e. Mexico City		f. Germany

Environmental Geology Quiz • Chapter 17

Questions 1-4; Matching; Dangerous Smog Events

a. sulfuric acid-rich, killer smog, 1952   b. dangerous air pollution event, 1930
c. methyl isocyanate leaks, some small and inconsequential, one large and disastrous
d. hometown of sports greats Tony Dorsett, Stan Musial, and the killer smog of 1948

1. ( ) Meuse River Valley, Belgium   2. ( ) Bhopal, India and West Virginia
3. ( ) Donora, Pennsylvania; Monongahela River Valley   4. ( ) London, U. K.

5. Which **one** factor greatly reduces the likelihood of a repetition of the London killer smog event?
a. Since the deadly smog event, lower sulfur fuels, such as fuel oil and natural gas, have largely replaced coal in home heating and many industrial uses.
b. Over the next few years after the event, private industry in the U. K. voluntarily invested heavily in air pollution abatement.
c. Londoners have since been equipped with particulate filters and breathing apparatus, protecting them from the dangerous effects of strongly acidified smog.

6. Which acidic air pollutant is mostly anthropogenic, being formed during combustion of fossil fuels?   a. CO2   b. SO2   c. hydrocarbons   d. CH4

7. Which gaseous air pollutant is produced by photochemical reactions involving hydrocarbons, oxygen, and nitrogen?   a. SO2   b. CO   c. O3   d. particulates

8. Which **one** is a primary gaseous air pollutant emitted from a stationary, point source?   a. nitrogen oxides; autos on I-95 between Miami, FL, and Maine
b. ozone; ground-hugging, Los Angeles smog
c. sulfur dioxide; stack of a coal-burning electrical power plant
d. particulate-laden smoke from a poorly tuned diesel locomotive

Questions 9-12; Matching; Air Pollutants and Chemical Formulas

a. C8H18   b. NH3   c. CO   d. HF

9. ( ) released in some high temperature industrial processes and combustion; ingestion of contaminated leaves and grasses can seriously damage teeth and bones of deer and other wildlife; hydrogen fluoride

10. ( ) a hydrocarbon; a major component of gasoline; octane

11. ( ) a silent killer because of its strong affinity for hemoglobin; produced during inefficient, combustion of hydrocarbons and other fossil fuels; carbon monoxide

12. ( ) formed naturally by bacterial action in latrines; basic ingredient used in fertilizer manufacture; leaking railcars and tanks can be big trouble; ammonia

13. Which **two** characterize the brown air type smog?
a. mainly related to burning sulfur-bearing fuels such as coal and heavy fuel oil
b. includes ozone as a nasty pollutant   c. smoke particulates are a major component
d. photochemical reactions of auto-produced hydrocarbons and nitrogen oxides

14. What is meant by the acronym NAAQS?
a. national air alert quality statute   b. national ambient air quality standard

15. Which **one** statement concerning urban microclimatological effects is fairly well-documented and not just based on hearsay and/or personal beliefs?
a. Highly urbanized areas generally have higher, average, near-surface wind speeds than outlying rural areas because fewer trees and shrubs obstruct the moving air.
b. The urban dust dome effect refers to cool, dust-laden, low-level breezes blowing into a highly urbanized area to replace warmer, relatively clear, rising air.
c. Particulate-laden air from a major urban center moving through a nearby rural area can increase precipitation from storm events and increase the incidence of fog and low-level cloudiness.

16. Why should mid-latitude, highly urbanized areas receive slightly less incoming ultraviolet energy than nearby rural areas at the same elevation? **one answer**
a. Warmer, ground-level city air absorbs more UV radiation per unit volume than cooler, ground level rural air.
b. Manmade building materials such as asphalt, roofing materials, and concrete reflect more UV radiation than grass, shrubs, and trees.
c. In general, concentrations of atmospheric particulate matter are usually higher over cities than over nearby rural areas.

17. Which **one** is not a specific goal or action in the 1990 Federal Clean Air Act?
a. reduce overall SO2 emissions by 50 % by the year 2000
b. sets up allowances and tradable credits concerning sulfur dioxide emissions
c. requires national, annual, pollution fees on all vehicles by the year 2000; the fees will depend on a vehicle's fuel consumption rating and the measured effectiveness of its emission control systems

18. Which **one** response concerning fluidized-bed combustion and air pollution abatement is correct?
a. natural gas is the fuel; silica sand is added to immobilize unburned hydrocarbons as silicon carbide
b. waste paper and combustible refuse are the fuel; calcium sulfate is added to immobilize carbon dioxide and carbon monoxide as calcium carbonate
c. coal is the fuel; SO2 is the air pollutant; calcium carbonate, limestone is added to immobilize the SO2 as calcium sulfate

19. What do the acronyms PSI and TSP mean in an air quality context?
a. particulates per square inch; tons of sulfur dioxide particles
b. particulates standard issue; technical support personnel
c. pollution standards index; total suspended particulates

20. Which **two** statements concerning pollution abatement are correct?
a. Lime, CaO, and limestone, CaCO3, are used to precipitate SO2 as insoluble calcium sulfates in flue gas scrubbers.
b. Ammonium sulfate fertilizer is a possible useful byproduct of SO2 pollution abatement.
c. Precipitators are installed to efficiently remove sulfur dioxide from coal-combustion flue gases.
d. In the 1990s, personal particulate masks and sulfur dioxide absorption filters are still widely used in Japan to protect against severe, day-to-day air pollution.

21. Which are the two most reasonable, defensible statements concerning air pollution, pollution abatement, and associated costs?
a. Significant reductions in acid precipitation will result in lowered maintenance costs for most exterior surfaces and outdoor metal fixtures such as lamp and sign posts, highway guard rails, and many bridges.
b. High-paying jobs are more important than good health and clean air. Thus outdated, obsolete industries and new, entrepreneurial enterprises should be exempted from compliance with heavy-handed air pollution regulations.
c. As SO2 from point sources such as metal smelters and coal-fired generating plants is cleaned up, further reductions from dispersed, mobile and fugitive sources will probably be more difficult and expensive to accomplish.
d. Air pollution abatement is just another tax-and-spend, liberal subterfuge to incur additional governmental spending, an enlarged bureaucracy, raise taxes, and assault private business and the free enterprise system.

Questions 22-24; True (T) or False (F). What is incorrect about the false statements?

22. ( ) The Los Angeles basin has its most frequent and severe inversion events during the late summer and early fall when cool air from onshore sea breezes is overridden by hot, dry air flowing seaward from interior desert areas.

23. ( ) Maximum daily death rates and sulfur dioxide concentrations of about 0.75 parts per million coincided on the fourth day of the 1952 London smog event.

24. ( ) Warm, moist air at the surface overlain by cold, dry air provides weather conditions conducive for atmospheric inversions and trapping air pollutants.

25. On the map/maps provided, plot and label the following areas or locations.
a. Chicago, IL   b. Donora, PA   c. Four-Corners area
d. Los Angeles, CA   e. southern Appalachians   f. northwestern Indiana
g. Japan   h. Belgium   i. London, U. K.   j. Bhopal, India

Answers; Home Study Assignment
Air Pollution • Chapter 17

1. b	2. c	3. a	4. d	5. a, d
6. a, b	7. d	8. b	9. a	10. d
11. c	12. h	13. f	14. g	15. a
16. e	17. b	18. a, d	19. c, d	20. b
21. a, d	22. a, d	23. d	24. c	25. a
26. d	27. a, c	28. b, c	29. b, d	

Answers: Environmental Geology Quiz • Chapter 17

1. b	2. c	3. d	4. a	5. a
6. b	7. c	8. c	9. d	10. a
11. c	12. b	13. a, c	14. b	15. c
16. c	17. c	18. c	19. c	20. a, b
21. a, c	22. T	23. T	24. F*	

*

24. Atmospheric inversions arise when denser air near the surface is overridden by less dense air aloft, thus producing a temporarily stable, stratified, near-surface air mass that collects pollutants and inhibits dilution and dispersion of pollutants that normally accompany mixing of different air masses. The typical inversion features cool, moist air near the surface and warmer, drier air aloft. The effect of temperature contrast on density more than compensates for the density-lowering effect of increased relative humidity. Near-surface fog that often accompanies these inversions reflects sunlight, keeping the air cool, and gradually evolve into smog as the trapped air pollutants are concentrated in the stagnant, near-surface air.

Sooner or later, the inversions are disrupted by vigorous weather systems that break down the density stratification and renew normal atmospheric circulation patterns. Under these conditions, convective overturn of cooler, denser air aloft and warmer, less dense air near the land surface results in brisk winds and strong breezes that promote mixing and dispersion of air pollutants.

Commentary/Media Watch

Reducing Sulfur Dioxide Emissions in the U. S.

The Environmental Defense Fund's role in promoting clean air over the scenic lands of northern Arizona and southern Utah includes specific goals for reducing SO2 emissions by the year 2000 (1996, Restoring habitat and clean air in the Grand Canyon: EDF Letter, v. XXVII, no. 4, p. 1 & 3). These include an absolute cap on emissions enforced by allowing SO2 emission credits to be traded as currency, and encouraging companies to preferentially shut down older, polluting facilities first and, if necessary, replace them with new clean facilities. Both issues are prominently mentioned in Chapter 17 of the text!

As summarized by Carlowicz (1996, In Brief; Cleaning the air: EOS, v. 77, no. 34, p. 330), the dirtiest 445 SO2-emitting facilities in the U. S. have been cleaned up one way or another (no longer in use or fitted with pollution abatement equipment) resulting in a decline in total SO2 emissions from 25.9 million tons in 1980 to 18 million tons in 1995. This decline translated into a 10 to 25 % reduction in acidic precipitation. The improved air quality was estimated to produce over $10 billion in health care savings; other savings resulting from less corrosion and less deterioration of out-of-doors materials from acidic precipitation will add substantially to the total.

Note that pollution abatement costs are born by a small number of governmental and corporate organizations. Savings are dispersed throughout the society and will not show up directly in corporate ledgers or individual checking accounts. This is a strong argument for taxpayer-supported, pollution abatement efforts. The savings go to the society at large, and the individual taxpayer does get a visible return on investment.

Reducing Sulfur Dioxide Emissions at Sudbury

On the international front, Canadians have long argued that acidification of lakes in eastern Canada was largely due to sulfur dioxide from south of the border, particularly from the Ohio Valley and other heavily industrialized parts of the Midwest. Nevertheless in the past, smelters in the Sudbury, Ontario mining district have spewed out enormous quantities of sulfur dioxide and toxic particulates, resulting in strongly acidified lakes and severely degraded lands downwind from the smelters. However, progress in reducing sulfur dioxide emissions in the United States is being matched in Canada. CNN's morning news, September 11, included a short segment on INCO's (International Nickel Company) pollution abatement program in progress at Sudbury. By installing particle collectors and SO2 scrubbers (Figures 17.11 & 17.12), switching to low-sulfur fuels for space and industrial heating, and reducing the proportion of metal-poor, sulfur- and arsenic-rich minerals such as pyrrhotite (FeS) and arsenopyrite (FeAsS) in the smelter feed, sulfur dioxide emissions have been reduced by 90 %. Efforts to capture the remaining 10 % are continuing.

CNN's news segment focused on visible signs of environmental recovery. Normal pHs are returning to local aquatic systems, and fish and other organisms have begun to re-establish themselves in formerly sterile, strongly acidified, lakes and streams. Grasses and small plants are beginning to revegetate soils made barren by acidic soil moisture and trace-metal poisoning. Lime is applied to neutralize acidic soils and to protect plants by reducing the mobility and bioavailability of trace metals.

Air Quality; New, Tougher Standards for Ozone and Tiny Particulates Proposed

An Associated Press article, 6/21/96, based on a study by the American Lung Association, reported that hospital admissions for severe respiratory stress increase by 10 % on hot summer days when atmospheric ozone levels exceed the 0.12 ppm standard set by EPA. Children and the elderly are particularly vulnerable. This article also clearly illustrates how "hidden costs" of pollution are passed from polluting entities to the public at large. Study after study like this show that atmospheric ozone and tiny particulates are detrimental to public health and dangerous and sometime life-threatening to asthmatics and others with respiratory impairments. Results like these are behind EPA's proposal in late November, 1996, for tougher regulatory standards for these two pollutants. Despite huge anticipated savings in public health and improved air quality, General Motors and the other big auto manufacturers are expected to oppose the new standards.

The Breathing Wall; An Innovative Approach to Improved Indoor Air Quality

A news item shown at least twice on CNN in early October, featured an experimental aquatic garden complete with tropical vegetation, pools, and dripping water was built on the lower floors of a new office building in Toronto, Ontario, Canada, with the intent of improving indoor air quality and saving energy. The aquatic garden functions as a" breathing wall" to regulate relative humidity, absorb excess CO_2, add O_2, and hasten the absorption and/or chemical degradation of indoor air contaminants. The idea is to substitute a pleasing, functioning, "natural", indoor ecosystem for fans, ducts, motors, electricity, and air filters, and to reduce the need for introducing outside air, which itself may require cleaning. In addition during the winter months, air introduced from outside has to be heated and humidified at considerable expense. If the concept proves successful, costs for ventilation and good indoor air quality can be substantially lowered.

A mysterious problem with "disappearing carbon dioxide" derailed careful plans to depend entirely on ecological methods to regulate indoor air composition in the Biosphere II complex, southern Arizona. Subsequent study showed that the missing gas was reacting with "new concrete" exposed inside the structure. With time, CO_2 absorption declined considerably and an excess CO_2 problem developed. Thus ecological regulation of indoor air quality may initially be affected by extraneous factors, and a long-term monitoring period is advisable to evaluate such experiments.

Conservation Groups Sue EPA

An Associated Press article, 7/19/96, reported that the National Wildlife Federation, Sierra Club Legal Defense Fund, and the Chesapeake Bay Foundation had filed suit in federal court against the Environmental Protection Agency, claiming that the agency was stalling and failing to develop effective programs to control emissions of toxic substances, such as mercury, lead, and dioxin, to the atmosphere. Such emissions can add significantly to the levels of these toxic substances in aquatic environments. This suit illustrates the complexities and ambiguities inherent in formulating and enforcing environmental policies. At the same time that powerful industrial and congressional interests are trying to slow down or roll back environmental regulations, environmental organizations are using the courts to push the EPA toward adopting tougher regulations and enforcement policies. These circumstance demand sound, informed, workable, science-based decisions on environmental issues.

Supplemental References

Turco, R. P., 1996, Earth under siege; From air pollution to global change: Oxford University Press, New York, NY, 480 p.

Landscape Evaluation and Land Use • Chapter 18
Home Study Assignment

1. Which **two** statements concerning New Mexico are correct? The others contain blatant, factual errors.
a. Large copper mines and smaller lead-zinc mines in Grant County present little or no threat of polluting the groundwater because in most parts of the county, the water table is more than 500 feet below the surface.
b. Irrigation runoff and agricultural wastewater could relatively easily infiltrate into the shallow groundwater system of the Rio Grande Valley in Doña Ana County.
c. Groundwater, in general, flows to the south and east in the area of oil and gas fields and potash deposits in the southeastern corner of the state.
d. Groundwaters in the irrigated agricultural lands in Roosevelt County are relatively safe from pollution from cattle feed lots because the water table in most areas is more than 200 feet below the surface.

2. Look over the block diagram and map (Figures 18.2 and 18.3) of the area near Morrison, CO. Which ERU would you recommend for the site of a proposed tennis, golf, hiking, mountain biking, etc. recreational complex? Why?
a. mountain forest: cool in summer; good view to the east; requires extensive topographic modification; moderate hazards from forest fires and mass wasting; little danger from flash floods
b. Pleistocene grasslands: requires little topographic modification; nice and secluded, nestled between the mountains and hogbacks; little danger from flash floods; minor hazard from grass/brush fires
c. hogback wood and grasslands; central location; requires extensive topographic modification; would lend itself to a very difficult, world-class golf course; open vista, nice views in all directions; no threat from flash floods; dip slopes have some potential for soil slips and other forms of mass wasting
d. floodplain forest; cool and secluded; little topographic modification required; golf course design could incorporate challenging aspects of the creeks, floodplain surface, and vegetation; potential hazards from fire and flash floods

3. Which **one** gives the correct meaning of the acronym ERU?
a. Environmental Response Unit   b. Emergency Resource Unit
c. Environmental Resource Unit   d. Emergency Response Unit

4. Which **one** gives the correct meaning of the acronym GIS?
a. Geographic Impact System   b. Geologic Information System
c. Geologic Impact Statement   d. Geographic Information System

5. Which **one** gives the correct meaning of the acronym EIS?
a. Environmental Impact System   b. Environmental Information Statement
c. Environmental Impact Statement   d. Environmental Information System

6. Which **one** gives the correct meaning of the acronym EPA?
a. Environmental Policy Agency   b. Environmental Policy Act
c. Environmental Protection Agency   d. Environmental Protection Act

7. Which **two** responses concerning cost-benefits analysis are generally widely-accepted and not highly controversial and/or politically incendiary?
a. Proposed projects with estimated costs that greatly exceed derived benefits over the planned lifetime of the project should only be built with public, taxpayer-provided funds because private-sector companies would not contemplate building such a project.
b. Proposed projects with long-term benefits-to-cost ratios of less than one are much more difficult to justify than competing projects for which long-term benefits exceed costs.
c. Extraneous items such as maintenance, beautification, and wildlife habitat should be excluded from costs associated with long-term projects such as dams, highways, and bridges.
d. Intangible values such as aesthetic qualities, social impacts, and long-term environmental impacts are much more difficult to quantify than items such as construction costs and the anticipated revenue stream and enhanced land values resulting from completion and ongoing operation of the project.

8. Which **one** gives the correct meaning of the acronym CEQA?
a. Certified Energy Quantity Agency   b. California Environmental Quality Act
c. Certified Environmental Quality Agency   d. California Energy Quality Act

9. Consider a proposed flood control project designed to protect against a river stage associated with the 25-year flood in a particular area. Which **one** would **not** seriously compromise the original cost-to-benefits analysis of the project?
a. Extensive development of the formerly flood-prone areas substantially raises flood damages associated with a low probability, unusually severe, flood event such as the 100-year flood.
b. As part of the proposed project, the formerly flood-prone areas are zoned for open space, parks, and outdoor recreational lands, thus effectively prohibiting developers and real estate tycoons from generating millions of additional dollars by building high-value assets such as luxury hotels, condominiums, convention centers, and shopping malls.
c. Dikes designed to provide the flood protection are built too close to the river, effectively constricting the floodplain, raising water surface elevations associated with given discharges, and shortening recurrence intervals of floods computed on the basis of pre-project hydrologic data.
d. As the project was nearing completion, an unassuming naturalist discovered a heretofore unknown species of aquatic snail that lived only in the area impacted by the project, forcing expensive, late-stage design changes and ecological modifications to protect the snail's habitat.

10. Which agency has overall administrative jurisdiction over Cape Hatteras National Seashore, North Carolina?
a. National Park Service, Department of Interior
b. Bureau of Reclamation, Department of the Interior
c. Environmental Protection Agency, Executive Branch of U. S. Government
d. National Oceanic and Atmospheric Administration, Department of Commerce

11. What is **physiographic determinism**? How does it differ from cost-benefits analysis?
a. a widely-held, nineteenth century notion that the western frontier lands of America were preordained to be occupied and settled by whites, no matter how high the costs or meager the benefits
b. a philosophical basis for landscape evaluation; in addition to purely economic considerations, the values of different lands are also dependent on bedrock, soil, topography, climatic factors, and location
c. a near paranoid, anarchistic form of neonationalism; a planning process instigated by the Montana Freemen and other American militia groups to artificially devalue their land holdings and avoid paying property taxes
d. a set of principles utilized in site selection; in addition to purely economic considerations, ecological and aesthetic principles and concerns are included in the planning process

Matching; Questions 12-24; Names/Locations; Significant Environmental Issues
one response in used twice

a. Storm King	b. Oregon	c. Franconia	d. California
e. Kansas City	f. San Joaquin	g. Flaming Gorge	h. Ducktown
i. Denver	j. Yosemite	k. San Luis	l. Florissant

12. ( ) focus, in 1904, of one of America's first, highly-visible, court cases concerning the legal rights of property owners subjected to injurious environmental effects of a nearby metal smelter

13. ( ) dam, reservoir, and scenic area where the Green River cuts through the eastern Uinta Mountains, northern Utah; this and Glen Canyon are the only, very large-capacity reservoirs in the Colorado River Basin above Hoover Dam (Lake Mead)

14. ( ) a canal that carried polluted agricultural wastewater directly into Kesterson National Wildlife Refuge, California

15. ( ) national monument in Colorado noted for delicate, Tertiary, plant and insect fossils preserved in fine-grained, volcanic-ash-rich lake beds; focus of an appeals court case won by preservationists based on the Trust Doctrine and the Ninth Amendment

16. ( ) now a well-known national park; first scenic public land in the U. S. set aside for preservation in its natural condition and for the enjoyment of future generations

17. ( ) an inlet spanned by a high-rise, highway bridge in 1975, fueling major ocean-front development and population growth on the Outer Banks, North Carolina

18. ( ) city in Virginia just south of Washington, DC; notable for its pioneering application of Geographic Information System technology to land-use issues arising from explosive growth and urbanization

19. ( ) one of the first states to require comprehensive planning at the local level; a department of land conservation and development at the state level reviews each local plan

20. ( ) major urban area in the midwest; cool, stable, underground space left over from limestone mining was put to good use for warehouses, offices, and manufacturing facilities

21. ( ) rich, irrigated, agricultural valley of central California: notable environmental issues include very shallow, perched water tables necessitating field drainage; pesticides, herbicides, and excessive selenium in agricultural wastewater; occasional severe dust storms; declining water tables, and land subsidence due to aquifer dewatering

22. ( ) cities, counties, and other local jurisdictions are encouraged to embrace comprehensive planning and to link this strategy with compliance with the state's SEQA

23. ( ) highly controversial, proposed, hydropower project on the Hudson River, 1962-1981, pitting striped bass supporters and scenic preservationists against Con Edison, a major east coast utility company

24. ( ) mile-high city; NBA Nuggets play in coliseum complex built over old gravel quarries and sanitary landfill sites

25. Which **two** statements concerning the Kesterson National Wildlife Refuge, California, are supported by the known facts?

a. The Kesterson ponds were an integral part of a much larger project, designed and managed by the Bureau of Reclamation, to protect rich, irrigated, agricultural lands from excessive salt buildup in the soils.

b. Elevated selenium contents of soils are more characteristic of the eastern side of the San Joaquin Valley; soils on the western side of the valley, near to the Sierra Nevada, are not noted for elevated selenium contents.

c. Fatalities and birth deformities documented for Kesterson's water birds and other wildlife are attributed to excessive intake of selenium.

d. Toxic chemicals observed to be concentrated in sediments, aquatic plants, and fish of the Kesterson ponds were mainly derived from poorly treated sewage released into the San Luis Drain from the cities of Merced, Los Banos, and Modesto.

26. Which **one** would be an environmentally sound, viable, "soft" solution, acceptable to all interested parties, to the problems associated with the San Luis Drain and Kesterson situation?

a. Extend the San Luis Drain to the San Joaquin River, and allow the selenium and other dangerous chemicals in the wastewater to be diluted and carried northward into San Francisco Bay.

b. Build an extensive canal, pipeline, and tunnel system to discharge the polluted Central Valley agricultural wastewater directly into the Pacific Ocean at Monterey Bay.

c. Treat the polluted wastewater to comply with standards acceptable to the California Water Resources Control Board and use the clean, recycled water to augment the flow of the San Joaquin River enough to restore salmon and other native fishes to the river system.

d. To increase the flow of freshwater into the San Luis Drain, divert the Kings River to the west side of the valley and prohibit all irrigated farming west of the drain.

27. Which **one** response concerning engineering geology and safety considerations involved in tunnel construction contains obvious factual errors or potentially dangerous, geotechnical misconceptions?
a. Highly fractured bedrock makes for relatively safe and rapid tunnel construction because less blasting is required for each foot of tunnel advance.
b. Driving a tunnel across an anticlinal axis in folded sedimentary rock generally results in more stable roof conditions than in driving an equivalent tunnel across a synclinal axis.
c. Steeply inclined to vertical, open, fracture systems encountered in tunnels being driven deep under mountain ridges may contain large quantities of groundwater under considerable pressure, especially in humid areas.
d. Concrete-lined tunnels excavated from bedrock are generally safe from earthquake-shaking damage; however if at all possible, a tunnel should not be driven across a known, active fault zone.

28. What is meant by the term **grouting**?
a. allowing groundwater to drain from a saturated fracture system encountered in an underground mining operation or tunnel construction project
b. mitigating against evaporative losses from spray irrigation by installing porous, pipes to release the water beneath the surface in the plant root zone
c. pumping or otherwise injecting cement or other fluidized material into open fractures in bedrock; the injected material seals the fractures and greatly reduces the movement of water through the fractures
d. bypassing lengthy, highly technical, detailed studies in EISs in favor of focusing directly on the most obvious and controversial environmental issues

29. Which is the correct description of a **negative declaration** issued by an appropriate governmental agency involved in assessing the environmental impact statement being prepared for a specific, proposed project?
a. The project does not pose significant environmental impacts and, in the environmental impact statement, thorough consideration of various alternative sites, designs, and operating methods is unnecessary.
b. The project presents significant potential for adverse, environmental impacts and thus, the environmental impact statement requires a thorough evaluation of alternatives.

30. Which is the **one** correct example of **mitigation** in an environmental impact/planning context?
a. An equivalent area and quality of waterfowl habitat is created nearby to compensate for that lost in the proposed project.
b. A panel of independent experts, unaffiliated in any way with the plaintiffs and defendants, is convened to decide on an equitable, fair, out-of-court settlement to an unusually contentious environmental issue.
c. Scoping by a grassroots organization of concerned citizens leads to extensive mediation and negotiations based mainly on the legal doctrine of Balancing Trust.
d. A concerned citizens group uses the Public Trust Doctrine as the basis for a lawsuit to prevent a major forest products company from clear-cutting in National Forest areas highly susceptible to gullying, mass wastage, and landslides.

31. Who was a justice of the U. S. Supreme Court that decided the suit by the State of Georgia against the Tennessee Copper Company? The decision, 1915, forced the company to reduce sulfur dioxide emissions from its Ducktown, TN, smelter.
a. Roger B. Taney   b. John Marshall   c. Earl Warren   d. Oliver Wendell Holmes, Jr.

32. Which two responses concerning the Cape Hatteras National Seashore are reasonable and/or essentially correct? The others contain serious factual errors or are based on gross ignorance and/or strongly-held personal feelings.
a. Based on the Balancing Doctrine, the U. S. Government and the NPS have the responsibility and obligation to protect highways, bridges, and private property, new and old, from storm damage and coastal erosion.
b. An emergency evacuation of Hatteras Island, ordered in response to an approaching hurricane, could be stopped or severely impaired by traffic accidents and/or vehicle failures on the Oregon Inlet Bridge.
c. Overwash damage to the Hatteras Island Highway, NC 12, now accompanies most tropical storms and "noreasters", especially in the very narrow portions of the island between the villages of Avon and Salvo.
d. The oceanside beaches and dunes of Hatteras Island were still largely natural and unmodified until 1964 when completion of the Oregon Inlet highway bridge spurred a surge in oceanfront building and development.

33. On the U. S. map provided, plot and label the following areas or locations.
a. State of New Mexico
b. Outer Banks, North Carolina
c. Flaming Gorge Dam, UT
d. Kesterson National Wildlife Refuge, CA
e. Monterey Bay, CA
f. State of Oregon
g. Storm King Mountain, NY
h. Florissant National Monument, CO

Environmental Geology Quiz • Chapter 18

1. Which statement is more logical, reasonable, and defensible?
a. Unconfined groundwaters systems with water tables more than 50 feet below the surface are fairly well-protected against being polluted by infiltrating wastewaters from surface mines, animal feed lots, and irrigated farmlands.
b. Wastewaters from feed lots and fields can relatively easily infiltrate shallow, groundwater systems of alluvium-filled valleys of dominantly influent streams.

2. Which **one** gives the correct meaning of the acronym ERU?
a. Environmental Response Unit
b. Emergency Resource Unit
c. Environmental Resource Unit
d. Emergency Response Unit

3. Which **one** gives the correct meaning of the acronym GIS?
a. Geographic Impact System
b. Geologic Information System
c. Geologic Impact Statement
d. Geographic Information System

4. Which **one** statement exhibits correct usage of the contained acronyms?
a. In an EIS submitted to EPA, ERU data may be incorporated into GIS.
b. In an ERU submitted to EIS, GIS data may be incorporated into EPA.
c. In an EPA submitted to GIS, EIS data may be incorporated into ERU.
d. In a GIS submitted to ERU, EPA data may be incorporated into EIS.

5. Which **two** responses concerning cost-benefits analysis are generally widely-accepted and not highly controversial and/or politically incendiary?
a. Intangible values such as aesthetic qualities, social impacts, and long-term environmental impacts are much more difficult to quantify than items such as construction costs and anticipated revenues and enhanced land values resulting from completion and operation of the project.
b. Proposed projects with estimated costs that greatly exceed derived benefits should only be built with public, taxpayer-provided funds because private-sector companies would not contemplate building such a project.
c. Proposed public projects with long-term benefits-to-cost ratios of less than one are much more difficult to justify financially and politically than competing projects for which long-term benefits exceed costs.
d. Maintenance, beautification, and wildlife habitat should be excluded from costs associated with long-term projects such as dams, highways, and bridges.

6. Which federal agency has overall administrative jurisdiction over Cape Hatteras National Seashore, North Carolina?
a. Environmental Protection Agency, Executive Branch of U. S. Government
b. National Oceanic and Atmospheric Administration, Department of Commerce
c. National Park Service, Department of Interior

7. Which is a city in Virginia, just south of Washington, DC, notable for its pioneering application of Geographic Information System technology to land-use issues arising from explosive growth and urbanization?
a. Quantico   b. Franconia   c. Fairfax   d. Arlington

8. Consider a proposed flood control project designed to protect against a river stage associated with the 25-year flood. Which **one** would **not** eventually compromise assumptions used in the original cost-to-benefits analysis?
a. New development occurred in the now-protected, formerly flood-prone area.
b. The flood-prone areas were zoned for open space, parks, and outdoor recreation.
c. As the project neared completion, a naturalist discovered a heretofore unknown species of aquatic snail that lived only in the area impacted by the project.

9. What is **physiographic determinism**?
a. a philosophical basis for landscape evaluation; in addition to purely economic considerations, the values of different lands are also dependent on bedrock, soil, topography, climatic factors, and location
b. a near-paranoid, anarchistic form of neonationalism; a planning process instigated by the Montana Freemen and other American militia groups to artificially devalue their land holdings and avoid paying property taxes
c. a widely-held, nineteenth century notion that the western frontier lands of America were preordained to be occupied and settled by whites, no matter how high the costs or meager the benefits
d. a set of principles utilized in site selection; in addition to purely economic considerations, ecological and aesthetic principles and concerns are included in the planning and selection process

10. Which was **one** of the first to require comprehensive planning at the local level? A state land conservation and development department reviews each local plan.
a. Oregon   b. California   c. Virginia   d. Maryland

11. Which was the highly controversial, proposed, hydropower project on the Hudson River, 1962-1981, pitting striped bass supporters and scenic preservationists against Con Edison, a major New York utility company?
a. Sleepy Hollow   b. Forest Gorge   c. Catskill Falls   d. Storm King

12. Which canal or drain carried dangerously polluted agricultural wastewaters directly into Kesterson National Wildlife Refuge, California?
a. Dos Rios   b. Aqua Verde   c. San Luis   d. La Florida

13. Which is the rich, irrigated, agricultural valley of central California notable for environmental problems such as very shallow, perched water tables, excessive selenium in agricultural wastewater, occasional dangerous dust storms, declining water tables, and severe land subsidence?
a. San Fernando   b. San Joaquin   c. Owens   d. Napa

14. Which mining area was the focus, in 1904, of one of America's first, highly-visible, court cases concerning the legal rights of property owners subjected to injurious environmental effects of a nearby metal smelter?
a. Ducktown, TN   b. Butte, MT   c. Brigham City, UT   d. Deadwood, SD

15. Under which large, bustling midwestern city were cool, stable, spacious, underground openings left over from limestone mining put to good use for warehouses, offices, and manufacturing?
a. Kansas City, MO   b. Omaha, NE   c. Sioux Falls, SD   d. Madison, WI

16. Which is the dam, reservoir, and scenic area where the Green River cuts through the eastern Uinta Mountains, northern Utah? This and Glen Canyon are the only, very large-capacity reservoirs in the Colorado River Basin above Hoover Dam (Lake Mead).
a. Rainbow Canyon   b. Flaming Gorge   c. Escalante Rim   d. Vernal Falls

17. Which is a national monument in Colorado noted for delicate, Tertiary, plant and insect fossils preserved in fine-grained, volcanic-ash-rich lake beds? The area was the focus of an Appeals Court case won by preservationists based on the Trust Doctrine and the Ninth Amendment.
a. Arapahoe   b. Fossil Butte   c. Maroon Bells   d. Florissant

18. Which is the well-known national park in California that includes the first, scenic public lands in the U. S. set aside for preservation in their natural condition and for the enjoyment of future generations?
a. Mt. Lassen   b. Yellowstone   c. Yosemite   d. Redwood

19. Which **one** statement concerning the Kesterson National Wildlife Refuge, California, is not supported by the known facts?
a. The Kesterson ponds were an integral part of a much larger project, designed and managed by the Bureau of Reclamation, to protect rich, irrigated, agricultural lands from excessive salt buildup in the soils.
b. Toxic chemicals observed to be concentrated in sediments, aquatic plants, and fish of the Kesterson ponds were mainly derived from poorly treated sewage released from the cities of Merced, Los Banos, and Modesto.
c. Fatalities and birth deformities documented for Kesterson's water birds and other wildlife are attributed to excessive intake of selenium.

20. Which **one** would be an environmentally sound, viable, "soft" solution acceptable to all interested parties entangled in the Kesterson situation?
a. Extend the San Luis Drain to the San Joaquin River, allowing the dangerously polluted wastewater to be diluted and carried northward into San Francisco Bay.
b. Build an extensive canal, pipeline, and tunnel system to convey polluted Central Valley agricultural wastewater directly into the Pacific Ocean at Monterey Bay.
c. Treat the polluted wastewater to comply with standards acceptable to the California Water Resources Control Board; use the recycled water to augment flows in the San Joaquin River enough to restore salmon and other native fishes.

21. Which U. S. Supreme Court justice helped decide the suit, 1915, that forced a mining company to reduce sulfur dioxide emissions from its smelter?
a. Roger B. Taney   b. John Marshall   c. Earl Warren   d. Oliver Wendell Holmes, Jr.

22. Which is the correct description of a **negative declaration** issued by an appropriate governmental agency involved in assessing the environmental impact statement being prepared for a specific, proposed project?
a. The project does not pose significant environmental impacts and the EIS does not require thorough consideration of possible alternatives.
b. The project poses potential for significant, adverse, environmental impacts and the EIS requires a thorough evaluation of possible alternatives.

23. Which **one** response concerning engineering geology and safety considerations involved in tunnel construction contains obvious factual errors or potentially dangerous, geotechnical misconceptions?
a. Highly fractured bedrock makes for relatively safe and rapid tunnel construction because less blasting is required for each foot of tunnel advance.
b. In driving a tunnel through folded sedimentary rock, more stable roof conditions are generally encountered across an anticlinal axis than across a synclinal axis.
c. Steeply inclined to vertical, open, fracture systems encountered in tunnels being driven deep under mountain ridges may contain large quantities of groundwater under considerable pressure, especially in humid areas.

24. What is meant by the term **grouting**?
a. allowing groundwater to drain from a saturated fracture system encountered in an underground mining operation or tunnel construction project
b. bypassing lengthy, highly technical, detailed studies in EISs in favor of focusing directly on the most obvious and controversial environmental issues
c. pumping or otherwise injecting cement or other fluidized material into open fractures in bedrock; the injected material seals the fractures and greatly reduces the movement of water through the fractures

25. Which is the **one** correct example of **mitigation** in environmental planning?
a. Scoping by a grassroots organization of concerned citizens leads to extensive changes in an EIS based mainly on the legal doctrine of Balancing Trust.
b. An equivalent area and quality of waterfowl habitat is created nearby to compensate for that lost in the proposed project.

26. Which **one** response concerning the Cape Hatteras National Seashore is correct?
a. Based on the Balancing Doctrine, the U. S. Government and the NPS have the legal responsibility and obligation to protect highways, bridges, and private property, new and old, from storm damage and coastal erosion.
b. An emergency evacuation of Hatteras Island, ordered in response to an approaching hurricane, could be stopped or severely impaired by traffic accidents and/or vehicle failures on the Oregon Inlet Bridge.
c. The oceanside beaches and dunes of Hatteras Island were still largely natural and unmodified until 1964 when completion of the Oregon Inlet highway bridge spurred a surge in oceanfront building and development.

27. On the map/maps provided, plot and label the following areas or features.
a. Monterey Bay, CA   b. Outer Banks, North Carolina
c. Storm King Mountain, NY   d. Kesterson National Wildlife Refuge, CA

Answers; Home Study Assignment
Landscape Evaluation and Land Use • Chapter 18

1. b, c	2. b*	3. c	4. d	5. c
6. c	7. b, d	8. b	9. b	10. a
11. d	12. h	13. g	14. k	15. l
16. j	17. b	18. c	19. b	20. e
21. f	22. d	23. a	24. i	25. a, c
26. c	27. a	28. c	29. a	30. a
31. d	32. b, c			

*
2. The other choices involve much more extensive earth-moving activities and topographic modifications and present real risks from forest fires, floods, and mass wasting. The Pleistocene grasslands ERU is relatively safe from flooding, can easily be modified to protect against grass/brush fires, and has only a small risk of mass wasting events.

Answers: Environmental Geology Quiz • Chapter 18

1. b	2. c	3. b	4. a	5. a, c
6. c	7. b	8. b	9. d	10. a
11. d	12. c	13. b	14. a	15. a
16. b	17. d	18. c	19. b	20. c
21. d	22. a	23. a	24. c	25. b
26. b				

Commentary/Media Watch

Another Kesterson?
On October 8 and 9, 1996, newspapers and CNN both confirmed heretofore widely scattered reports of deformed frogs in Minnesota, Wisconsin, and other neighboring upper midwestern states. The cause or causes are unknown. Possibilities include pesticides, herbicides, toxic chemicals, and increased ultraviolet radiation. The situation is somewhat reminiscent of the Kesterson Wildlife Refuge, where severe deformities in waterfowl were eventually traced to excessive selenium intake (p. 521-524). However, the frog deformities have been observed over a much larger area and are not limited to a any single hydrologic basin.

Environmental Stewardship
A television add sponsored by forest-products giant Georgia-Pacific, a large land owner in North Carolina, was being shown regularly on CNN, fall, 1996. The video featured quiet, placid waters of the lower Roanoke River in eastern North Carolina with fall colors of the floodplain hardwood forests in the background. Close-up images reveal an adult bald eagle, colorful leaves floating on the water, and sparkling dew drops on a spider web. A boy and his dad are shown fishing from the river banks. The verbal message extols the natural beauty, the uniqueness of flora, and the diversity of wildlife habitat in the area. The father tells his son how his employer and owner of these lands, Georgia-Pacific, has joined

hands with the Nature Conservancy to preserve the area for outdoor recreation and wildlife habitat. The dialogue may be slightly over dramatized, but it does carry a progressive message of corporate responsibility and a land ethic. The Nature Conservancy is a leading, national and international conservation organization dedicated to preserving special lands with unusually rich, floral diversity, either by entering into a long-term management agreement with the owners or by buying the land outright.

Another Nature Conservancy project in North Carolina was featured in an Associated Press article, 8/4/96, with the seemingly contradictory byline Logging Company Clear-Cutting Land for Environmentalists. In 1977, the Conservancy purchased a 900 acre tract called the Green Swamp Preserve. The long-term goal is to restore the area to a longleaf pine-turkey oak forest, similar to that which existed prior to European settlement. As a first step, the Conservancy hired Carolina Wood, a timber/tree harvesting company, to clear-cut parts of the tract and replant with seedling longleaf pines that will mature in about 50 years. Small wood lot owners in this part of North Carolina have been aggressively planting faster growing pine species for many years, and the slower growing longleaf pine and hardwoods have declined drastically in some areas and disappeared entirely from others. The impetus behind the restoration comes from efforts to save the red-cockaded woodpecker (see Peters, 1996), a bird that lives only in mature, longleaf forests and that has been seriously threatened by loss of habitat in recent years. In a real sense, the red-cockaded woodpecker is the "spotted owl" of the Southeast, but, it hasn't sparked the intense debate that swirls around the more famous, old-growth forest species. As part of the timber cutting agreement, Carolina Wood is under contract to adhere to strict environmental rules designed to minimize soil erosion, stream siltation, and overall land degradation.

Statewide newspapers, 10/3/96, reported a $1 million grant from the North Carolina Heritage Trust to establish a 9,000-acre state forest in the western part of the state. The tract is astride the Bevard Zone between the Piedmont and Blue Ridge geologic terranes and features diverse landscapes and habitats ranging from low, rounded, granite bedrock domes to bogs and swamp forests. The Trust granted the funds to The Conservation Fund, a Virginia-based, non-profit organization. The Conservation Fund will use the proceeds to purchase the land from its current owners, the DuPont Corporation. Trusts like the Heritage Fund operate in most parts of the country, and the state forest land purchase is a typical example of their activities. A land trust works to amass an ample "war chest"; the money is necessary for the trust to be legitimate financial player when appropriate tracts of land are available for sale or are suddenly put "on the block". Financial power is a necessity. A trust must be able to move quickly, especially if it's in competition with other potential purchasers with different goals. Otherwise, the land may be permanently lost to conservation and preservation.

CNN, October 12, reported on final passage of a $275 million federal appropriation to begin restoring the Kissimmee River-Lake Okeechobee-Everglades hydrologic system to its pre-engineered, natural configuration. The goals of the restoration are to reduced excessive agricultural runoff, reduce loads of dissolved nutrients, increase unchanneled, overland water flows through Everglades

National Park, and restore aquatic, wetland, and riparian habitats in the region. However in the November elections, Florida voters defeated a ballot initiative to levy a one cent per pound tax on cane sugar produced in the region to help finance the restoration project.

Federal money, $750,000, was also approved, 7/19/96, by the Senate Appropriations Committee for land purchases and property easements to stop unsightly development along the Blue Ridge Parkway in western North Carolina and to preserve the Parkway's natural beauty and scenic vistas. This proposed legislation was vigorously supported by North Carolina's two senators, despite their less-than-sterling voting records concerning environmental issues in general. Also in an editorial titled On Firmer Ground, local newspapers, 6/13/96, applauded progressive legislation sponsored by North Carolina's 3rd District congressman concerning land uses, conservation, restoration, and property exchanges involving wetlands.

Recent legislation (Associated Press article, 10/3/96) legalizing land swaps and tradeoffs for wetlands and wetlands mitigation have sparked a thriving, but quiet, real estate market in North Carolina wetlands. Here's how it works. Say group A's planned development infringes on a parcel of state-designated wetlands. Previously, said development was so to speak, dead in the water. Now, as a precondition for a permit to proceed, group A can purchase and preserve or mitigate an equivalent acreage of wetlands elsewhere in the state. The main goal of the new regulations is to stop the net loss of wetlands that has been going on steadily for many years, State officials estimate that only 4.7 million acres remain of the state's original 11 million acres of wetlands. About 1.2 million acres were lost during the 1980s alone. The new laws are designed to foster a "market place" and "currency" for wetlands values in the same way that sulfur dioxide emission credits represent a currency to be used in the marketplace for good air quality.

Public Participation in Land Use Mapping

Recently, this writer participated in a public meeting held for the purpose of allowing citizen participation in his city's land use planning project. To varying degrees, similar efforts are in progress in most municipalities and counties in the United States. Regional or statewide planning efforts are important in some states. In accordance with basic American ideals of grassroots democracy and public participation in policy making, citizen input is a basic tenet of the planning process.

The idea was for randomly-chosen groups of participants to come to some consensus on appropriate uses for undeveloped lands recently incorporated into the city or for lands expected to eventually be incorporated during the next five to ten years. The meeting was held in a city hall conference room; it was widely publicized in advance. Attendance was completely voluntary. Although developers. land owners, and other business interests with significant "stakes" in the planning process were well represented, the majority of attendees were "average citizens" basically concerned about the future well-being of their community.

Nicely-done, table-size, computer-printed base maps, soil maps, and maps depicting floodplains, wetlands, and other lands with planning restrictions were provided by the city. The meeting was organized and directed by the planning consultants hired by the city to complete a comprehensive, city-wide planning study. After about an hour of background presentations and instructions, the groups went about the task of drawing up an appropriate land use map for designated land tracts. Most group members were complete strangers; all races, genders, and probably creeds were represented. Over the next three hours, with the help of an "experienced facilitator", the groups went about their tasks of partitioning the designated lands into tracts for commercial, high-density residential, single-family, low-density residential, and industrial uses. After some prodding and much discussion, the groups arrived at a consensus land use map. The biggest handicap to this process was a lack of map-reading skills. Participants displayed detailed knowledge of the land areas, their geomorphic limitations, accessibility to transportation routes, and locations of probable employment, but many had trouble visualizing where these were on the map and how they all fit together.

Environmental geology instructors can participate in similar exercises in their home areas and easily incorporate a group land use mapping exercise into their course. Most cities or counties, etc. would be happy to provide appropriate base maps for that purpose. Site studies (a field trip!) to the area before and/or after making the land use map would add the important element of field checking and verification. A good geologic or land use map is based on sound field work and office work. Don't neglect field observations. The capabilities of GPS, GIS, and computer cartography are dazzling, but don't let them keep you permanently glued to the keyboard.

Supplemental References

Clarke, K. C., 1997, Getting started with geographic information systems: Prentice Hall, Upper Saddle River, NJ, 350 p.

Jensen, J, R., 1996, Introductory digital image processing; A remote sensing perspective: Prentice Hall, Upper Saddle River, NJ 07458, 315 p.

Multiple Authors, 1994, Geological and landscape conservation: O'Halloran, D., Green, C., Harley, M., Stanley, M. and Knill, J., Geological Society Publishing House, Bath, U. K; available in the U. S. through the AAPG Bookstore, Tulsa, OK, 544 p.

Peters, R. L., 1996, Hope for the red-cockaded? Defenders; Conservation Magazine of Defenders of Wildlife, v. 71, no. 4, p. 27-32

Sabins, F. F., 1996, Remote sensing; Principles and interpretation, third edition: W. H. Freeman and Company, New York, NY, 494 p.

Walker, J. D. and others, 1996, Development of geographic information systems-oriented databases for integrated geological and geophysical applications: GSA Today, v. 6, no. 3, March, p. 1-7

MAP ART; World
Cartesia Software

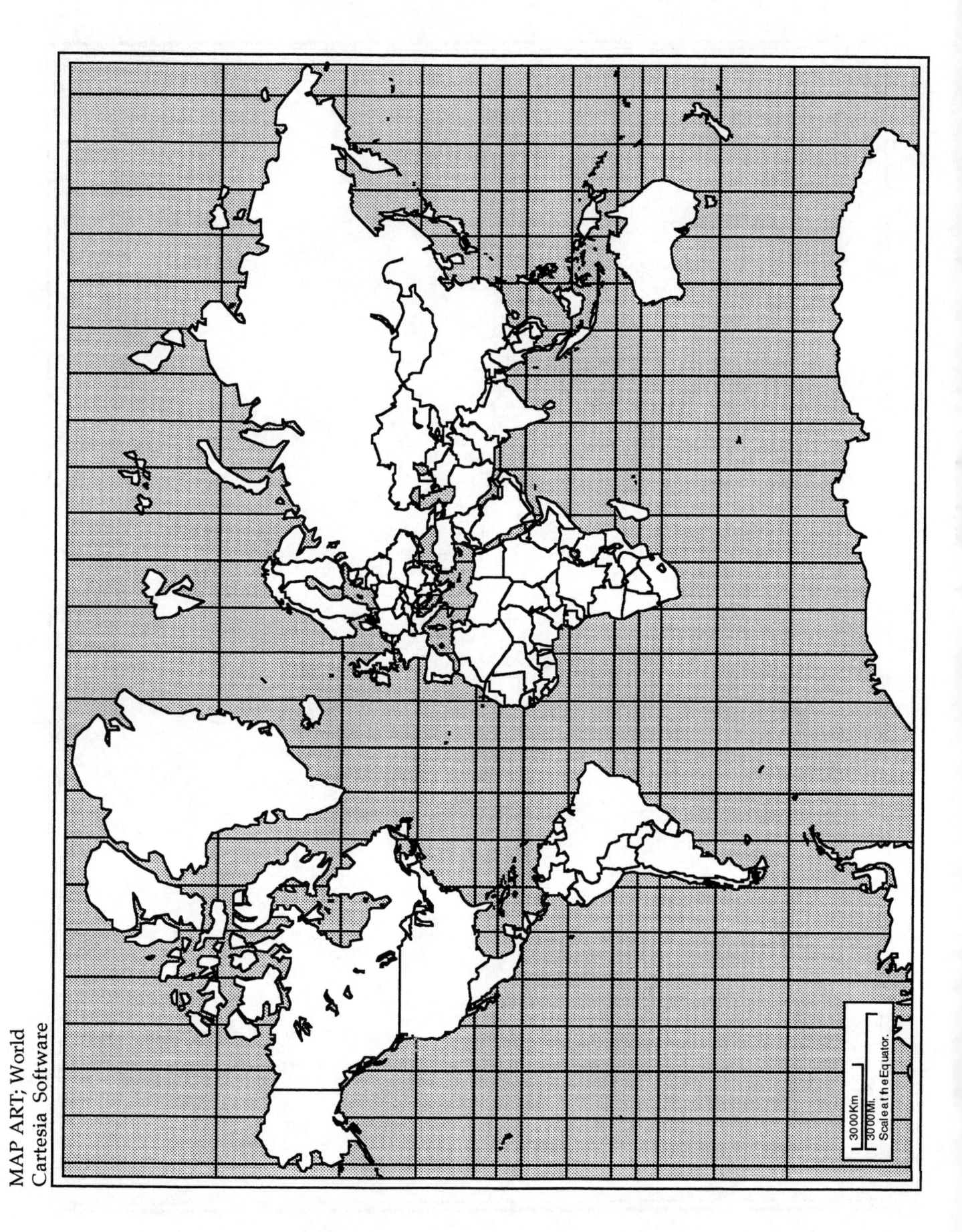

289

USA Map
Cartesia Software